# Concise Color Encyclopedia of
# Science

Left: Interior of the linear atomic particle accelerator at Stanford University, California. Built underground, the linear accelerator is two miles long.
Right: The Red Planet, Mars, as seen from Earth through a telescope. Note the prominent polar 'ice' cap.
Page 8: The tall fractionating towers of an oil refinery at night. Crude oil provides the major part of the world's energy. It is also an essential raw material for the chemical industry.

**Encyclopedia compiled and written by**
Robin Kerrod

**Science and Technology Consultant**
Eric Davies, Senior Lecturer, Polytechnic of South Bank

**Schools Adviser**
David George M.A., B.Sc., Dip. Ed.

**Research Assistants**
Sheila Kerrod, Caroline Bonella

**Editor-in-Chief**
Michael W. Dempsey B.A.

**Executive Editor**
Angela Sheehan B.A.

**Art Editor**
Derek Aslett

**Picture Researchers**
Marion Gain, Penny Warn

First United States publication 1975.

All rights reserved. Except for use in a review, the reproduction or utilization of this work in any form or by any electronic, mechanical, or other means, now known or hereafter invented, including xerography, photocopying, and recording, and in any information storage and retrieval system is forbidden without the written permission of the publisher. Published in Canada by Fitzhenry & Whiteside Limited, Toronto.

Printed in the United States of America

**Library of Congress Cataloging in Publication Data**
Main entry under title:

Concise color encyclopedia of science.

  1. Technology.  2. Science.  I. Kerrod, Robin, ed.
T47.C63 1973    600    73-13688
ISBN 0-690-00683-7

# Concise Color Encyclopedia of
# Science

Robin Kerrod

**Thomas Y. Crowell Company**
New York • Established 1834

# Contents

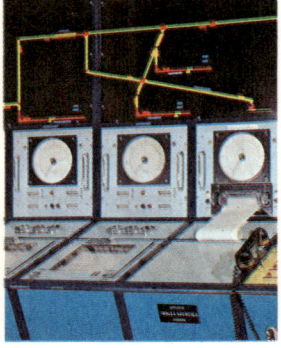

| | |
|---|---|
| Rockets into Space | 9 |
| Satellites into Orbit | 12 |
| Space Probes | 16 |
| Man in Space | 18 |
| Steps into Space | 20 |
| Destination Moon | 24 |
| Atoms and Elements | 26 |
| Radioactivity | 29 |
| Pioneers of the Atomic Age | 32 |
| Atomic Energy | 33 |
| Common Plastics | 36 |
| Plastics Technology | 38 |
| The Electromagnetic Spectrum | 42 |
| Properties of Light | 44 |
| Color | 46 |
| Photography | 48 |
| Railways and Locomotives | 52 |
| Pioneering Days of Rail | 56 |
| Modern Railway Operation | 58 |
| Underground Railways—Subways | 59 |
| Timber | 60 |
| Paper | 64 |
| Rubber | 66 |
| Printing | 68 |
| Heat | 70 |
| Steam Engines and Turbines | 72 |
| Coal and Coal Gas | 74 |
| Petroleum and Natural Gas | 76 |
| Electricity and Magnetism | 80 |
| Power Stations | 85 |
| Pottery | 88 |
| Industrial Ceramics | 92 |
| Glass | 94 |
| Telegraphy | 100 |
| Telephony | 102 |

Copyright © 1973 by Purnell & Sons Ltd.

| | |
|---|---|
| Sound Recording | 104 |
| Electronics | 106 |
| Radio and Television | 108 |
| Computers | 112 |
| Sound | 114 |
| Hydrostatics | 116 |
| Ships | 118 |
| Ship Construction and Propulsion | 120 |
| Submarines | 122 |
| Hydrofoils and Hovercraft | 123 |
| Navigation | 124 |
| Docks | 125 |
| Important Minerals | 126 |
| Mining and Quarrying | 128 |
| Extracting and Refining Metals | 130 |
| Steel and Other Alloys | 134 |
| Shaping Metals | 136 |
| Joining and Cutting Metals | 140 |
| Metals and their Uses | 143 |
| Machining Metals | 146 |
| Civil Engineering Construction | 148 |
| Preparing for Construction | 150 |
| Bridges | 154 |
| Dams | 156 |
| Roads | 158 |
| Skyscrapers | 159 |
| Tunnels | 160 |
| Canals and Harbors | 161 |
| Facts and Feats of Construction | 162 |
| Cement and Concrete | 164 |
| Water Supply and Treatment | 165 |
| Chemical Combination | 166 |
| The Chemical Industry | 168 |
| Soaps and Detergents | 172 |
| Explosives | 173 |
| Paint | 174 |
| Pharmaceuticals | 176 |
| Astronomy | 178 |
| The Earth | 179 |
| The Moon and the Sun | 180 |
| The Planets | 182 |
| The Stars | 186 |
| Textile Fibers | 190 |
| Spinning and Weaving | 194 |
| Dyeing and Finishing | 198 |
| Pioneers of the Textile Industry | 200 |
| Clocks | 201 |
| The Automobile | 204 |
| Parts of the Engine | 206 |
| Fuel and Ignition Systems | 208 |
| Lubrication and Cooling | 210 |
| Transmission | 211 |
| Steering, Braking and Suspension Systems | 212 |
| Diesel and Rotary Engines | 214 |
| Mass Production | 216 |
| How Aircraft Fly | 218 |
| Aircraft Propulsion | 222 |
| Aircraft Instruments | 224 |
| Supersonic Flight | 225 |
| Aircraft Construction | 226 |
| Helicopters | 230 |
| Balloons and Airships | 231 |
| Milestones in Aviation | 232 |
| Airports | 234 |
| Modern Farming | 236 |
| Staple Foods | 238 |
| Food Preservation | 240 |
| Inventions | 242 |
| Weights and Measures | 244 |
| Index | 247 |
| Acknowledgements | 256 |

# Rockets Into Space

Today we live in what is often called the Space Age. Man has not only ventured into space and back, but has at last set foot on another world – the Moon. Since the first artificial satellite *Sputnik 1* went into orbit around the Earth in 1957, hundreds of spacecraft, both manned and unmanned, have been launched into space.

Just how is it done? What is involved in getting a spacecraft off the Earth and into space, and how indeed does it stay up there at all?

A satellite stays up in orbit by virtue of its speed. The minimum speed a satellite must have to remain in orbit is 17,500 miles an hour, which is very high indeed by normal standards. Even the fastest jet-engined fighter planes travel at little more than 2,000 miles an hour.

So how do we achieve such a fantastic speed as 17,500 mph? The answer lies in the *rocket engine*. This is the only engine so far developed that is powerful enough to launch a satellite. And, what is more important, it is the only engine we know that can work in space.

**Step Rockets**

A tremendous amount of power is required to lift any object through the atmosphere against the pull of gravity and give it sufficient speed to send it into orbit. No single rocket is adequate for the task. What must be done is to use a combination of rockets linked together one on top of another. The bottom one gives the others a 'piggy back' and gets them moving fast before their own engines fire. In this way the top rocket can be made to travel fast enough to enter orbit. Such a rocket combination is called a *multi-stage,* or *step* rocket. Each individual rocket in the chain is called a *stage*. It generally is made up of a number of rocket engines clustered together to provide extra power.

Most spacecraft are launched into space on top of a three-stage rocket. The first stage, or *booster,* is the biggest because it has to lift itself, the other rockets, and the spacecraft through the densest part of the atmosphere. The booster fires for only one or two minutes before its fuel supply is exhausted. Then it separates from the rest of the vehicle and falls back to Earth. The second stage ignites for a few minutes and thrusts the now lighter vehicle higher and faster. When its fuel is exhausted the second stage

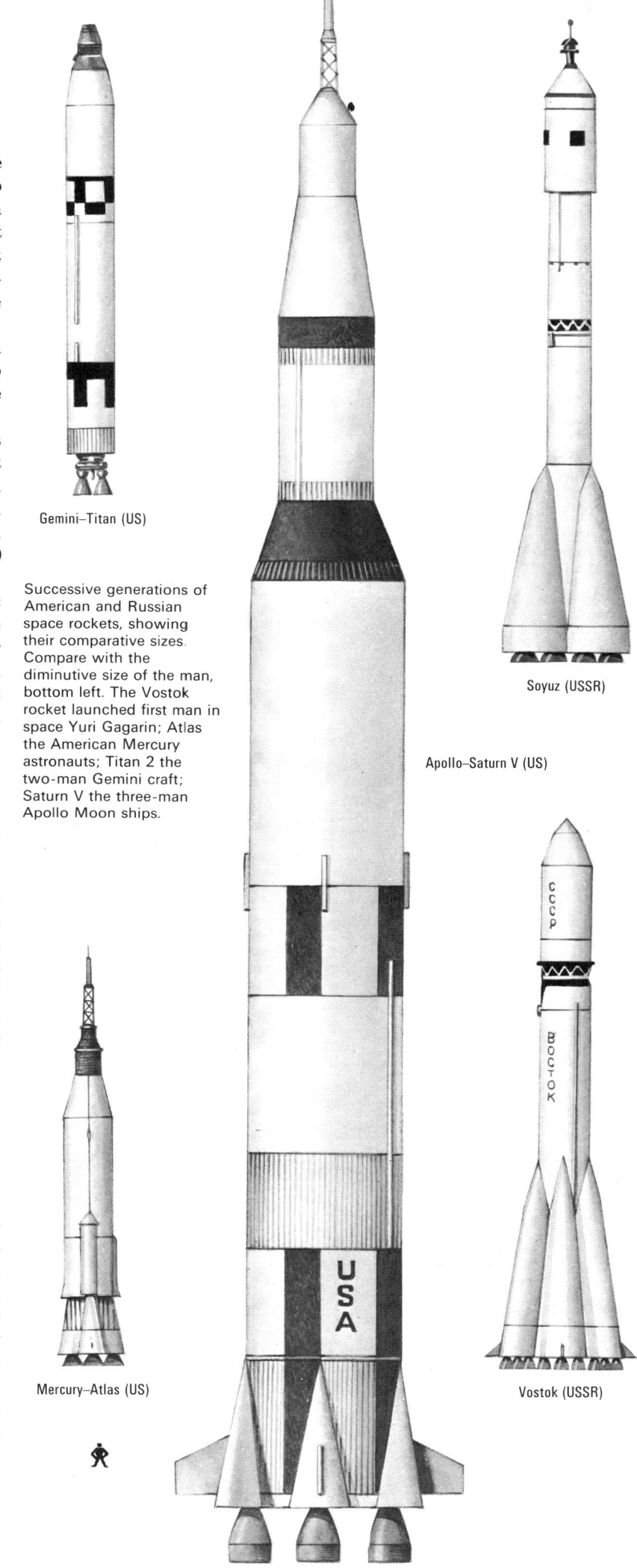

Successive generations of American and Russian space rockets, showing their comparative sizes. Compare with the diminutive size of the man, bottom left. The Vostok rocket launched first man in space Yuri Gagarin; Atlas the American Mercury astronauts; Titan 2 the two-man Gemini craft; Saturn V the three-man Apollo Moon ships.

Gemini–Titan (US)

Soyuz (USSR)

Apollo–Saturn V (US)

Mercury–Atlas (US)

Vostok (USSR)

is discarded, and the third stage fires to boost the spacecraft into orbit. In less than 15 minutes the spacecraft has been accelerated from rest to a speed of 17,500 mph!

The booster launches the rocket straight upwards, but the later stages are made to curve over gradually in order to put the spacecraft into orbit parallel with the Earth. These stages may be steered by small jets on their sides or by swivelling the nozzles on the rocket engines.

**Liquid-Propellant Rockets**

The rocket, like the jet engine, works by *jet propulsion*. The power comes from the thrust of a jet of hot gases formed by burning fuel inside the engine. As the jet of gases shoots backwards, the engine is propelled forwards.

The rocket engine can work in space because, unlike the jet engine, it does not have to rely on the oxygen in the atmosphere to burn its fuel. It carries its own oxygen supply. The substance that provides the oxygen is called an *oxidant*. The massive rockets used to launch spacecraft into orbit and beyond carry *liquid* fuel and oxidant. They are called *liquid-propellant* rockets. Two kinds of fuels are commonly used for these rockets – kerosene, also called paraffin, and liquid hydrogen. Liquid hydrogen is hydrogen gas that has been cooled to a very low temperature (about –255°C) until it turns into a liquid. The most common oxidant used for spacecraft is liquid oxygen, which again is very cold (about –185°C).

The rocket engine itself is relatively simple

A Saturn V three-stage rocket accelerating rapidly away from its launch pad. The cluster of five engines comprising the first stage combine to deliver an incredible 7,500,000 pounds of thrust to lift the massive rocket off the ground. The first-stage engines burn kerosene and liquid oxygen. Subsequent stages burn liquid hydrogen and liquid oxygen.

**Jet Propulsion**

Rockets work on the principle of jet propulsion, as do aircraft jet engines. The thrust *forwards* produced by a rocket or jet engine comes from the reaction to a high-speed stream, or *jet* of hot gases going *backwards*. You can easily demonstrate the effect of jet propulsion for yourself by blowing up a balloon and then letting it go. The air shoots backwards from the neck of the balloon when you release it, and this propels the balloon forwards.

To understand just how this jet propulsion works, think of what happens inside the balloon. When you blow it up and pinch the neck to prevent the air coming out, the balloon does not move. Inside, the pressure of the air is the same in all directions. The pressures on opposite sides are the same. And so are the pressures on the front and back (at the neck). When you release the neck, air rushes out of it. Inside the balloon, the side pressures are still the same, and the forward pressure is still the same. But there is no longer any backward pressure, because the air is escaping. This means that there is a net forward pressure on the balloon, which propels it forwards.

And so it is with the rocket and the jet engine. The net forward pressure of the expanding hot gases produced inside the engine thrusts the engine forwards. Jet propulsion has nothing to do with the jet of gases 'pushing' against the air as some people mistakenly believe.

In fact the presence of an external atmosphere inhibits the jet action and reduces the engine's efficiency. The jet engine, of course, must have an atmosphere to operate in, since it relies on that atmosphere for the oxygen to burn its fuel. The rocket on the other hand carries its own oxygen.

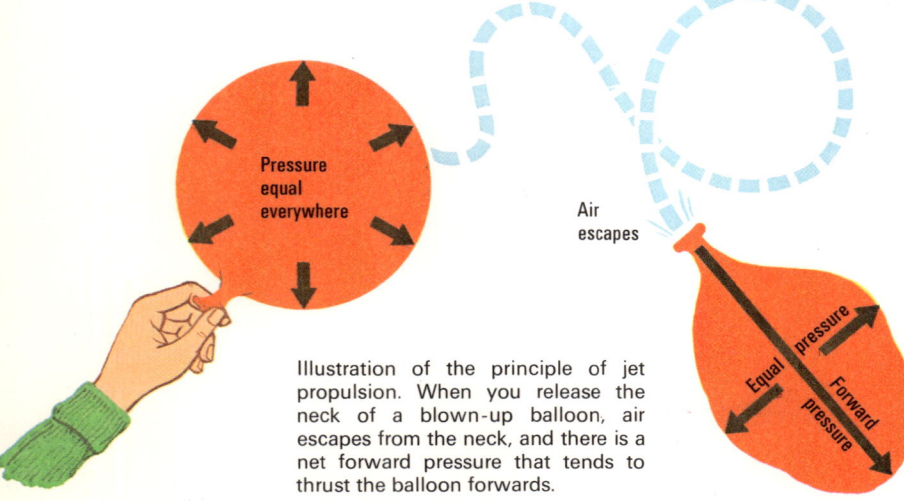

Illustration of the principle of jet propulsion. When you release the neck of a blown-up balloon, air escapes from the neck, and there is a net forward pressure that tends to thrust the balloon forwards.

in operation and really quite small. Rockets are so huge because of the massive tanks of fuel and oxidant they must carry. Only by burning fuel at a truly fantastic rate, two tons or more every *second*, can the engines develop sufficient thrust to lift a heavy load into space. The rocket engine consists essentially of a large open-ended chamber called the *combustion chamber,* which opens out to form a wide nozzle. Fuel and oxidant from the storage tanks flow through valves to pumps that force them through tiny nozzles, or *injectors,* into the combustion chamber.

The propellants come from the injectors in the form of very fine mists, which mix easily together to form an explosive mixture. Then the mixture is ignited, or made to burn by a spark from a kind of sparking plug. (This is just what happens in an automobile engine, too). The expanding hot gases produced leave the combustion chamber through the nozzle at very high speed. Reaction to these high-speed exhaust gases thrusts the rocket forwards. Thereafter the rocket will continue to work until its supply of propellants runs out or is deliberately shut off.

The temperature developed in the combustion chamber by the burning fuel is as high as 3,000°C. This is high enough to melt any suitable materials of construction for the chamber and the exhaust nozzle. These parts must therefore be cooled. This is done by circulating the very cold, incoming fuel around the chamber and nozzle walls. This is similar to the cooling system of an automobile engine, in which water circulates through the engine walls.

### Solid-Propellant Rockets

A few space rockets are powered by *solid,* not liquid propellants. They are not as widely used for space launchings as liquid types because they are nowhere near as powerful and they are difficult to control. However, they are much simpler than the liquid types. Their engines consist of not much more than a combustion chamber and an exhaust nozzle. The common firework rocket is an example of a solid-fuel rocket. It has gunpowder as a propellant. Most guided missiles used by the armed services have solid propellants, too, as have scientific sounding rockets.

Rockets powered by solid propellants have the advantage over liquid-fuel rockets that they are ready for instant use. (This is obviously necessary for a defensive guided missile which might be needed at any time!) Practically all liquid-propellant rockets have to be filled with fuel and oxidant immediately before the rocket is due to take off.

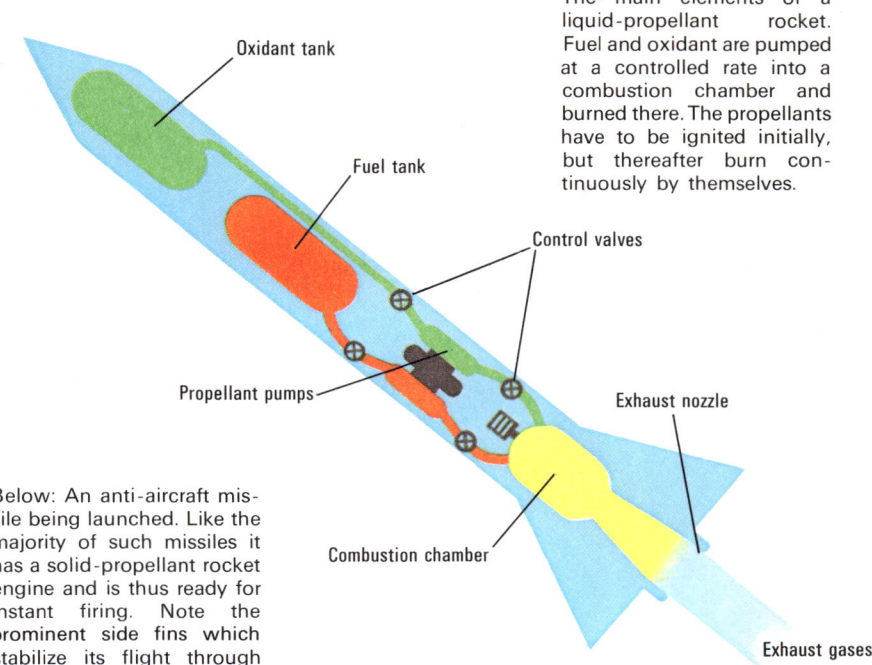

The main elements of a liquid-propellant rocket. Fuel and oxidant are pumped at a controlled rate into a combustion chamber and burned there. The propellants have to be ignited initially, but thereafter burn continuously by themselves.

Below: An anti-aircraft missile being launched. Like the majority of such missiles it has a solid-propellant rocket engine and is thus ready for instant firing. Note the prominent side fins which stabilize its flight through the atmosphere.

# Satellites into Orbit

In our usual experience we know that 'what goes up must come down'. If you throw an object through the air, it travels a certain distance and then falls to the ground. The 'pull' of the Earth's gravity brings it down. One thing you do find, however, is that the harder you throw the object, the faster it travels and the farther it goes before falling back to the ground. For a while you are overcoming gravity by speed. And speed is the secret behind space flight.

Imagine that you throw that object harder and harder so that it goes faster and faster and farther and farther before falling. At a certain speed the rate at which the object falls towards the Earth exactly matches the curvature of the Earth's surface. At this speed the object therefore remains at the same height above the Earth. In other words it goes into orbit as an artificial satellite. The same kind of explanation holds for Earth's natural satellite, the Moon, too.

The speed at which a body will go into orbit is called the *orbital velocity*. At a height about 100 miles above the Earth, the orbital velocity is about 17,500 miles an hour. It falls off progressively with increased height because there is less gravity to overcome. At 1,000 miles high, for example, it is only 15,800 mph. For a satellite to remain in orbit, it must maintain its orbital velocity. If its speed drops too low, it will begin to fall towards the Earth.

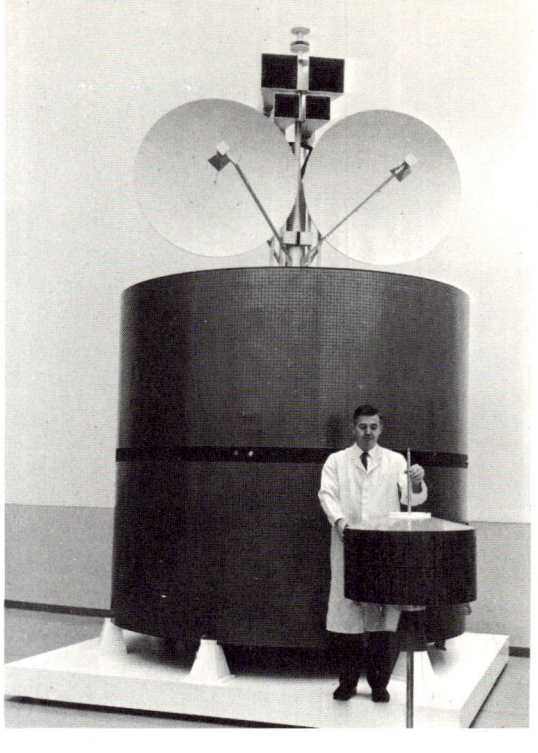

Satellites have brought about astonishing improvements in communications. The picture shows a full-size model of one of the big Intelsat satellites at present in a stationary orbit above the Atlantic Ocean. Note the directional antennae at the top and the mosaic of solar cells covering the body of the satellite.

Weather satellites also perform a very useful function, beaming information back to Earth about the global weather situation. The cloud-cover pictures they take are particularly interesting and storm regions can readily be identified. The mosaic of cloud-cover pictures below taken by a weather satellite show the presence of several hurricanes, which have been named.

## Atmospheric Drag

Most satellites are launched to orbit well over 100 miles above the Earth. This is done to avoid the resistance, or *drag* of the atmosphere. If the satellites were launched much lower, the drag would quickly slow down the satellite to below its orbital velocity. Then it would fall lower and lower. And as it fell, it would meet greater and greater resistance from the air. The friction, or rubbing of the air against the satellite would produce great heat which might eventually cause the satellite to burn up. This is exactly what a meteor does when it enters the Earth's atmosphere at high speed from outer space. It burns itself up in a fiery streak which we see from the ground as a 'shooting star'.

The atmosphere does not suddenly end at a certain point above the Earth, of course. It just gradually becomes thinner and thinner until it merges into space. Even at 500 miles up there is still a faint atmosphere, although it hardly exerts any appreciable effect on spacecraft.

## Stationary Orbits

The time a satellite takes to circle once around the Earth is called its *period*. The period for a satellite 100 miles up is about $1\frac{1}{2}$ hours. 1,000 miles up, it is about 2 hours. And 22,300 miles up, it is 24 hours. This is very interesting, of course, because it is the same length of time as our day. If we launch a satellite into an orbit 22,300 miles up over the equator in the same direction as the Earth is moving, it will circle once around the Earth in the same time as it takes the Earth

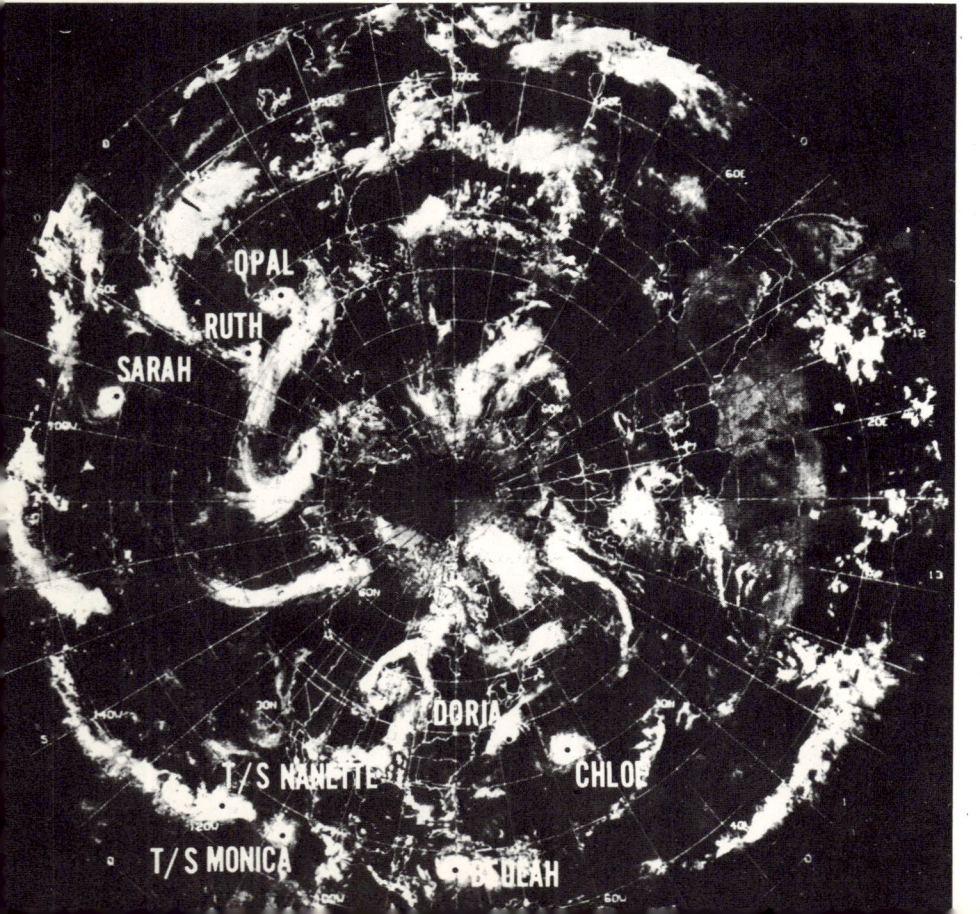

to rotate once. In other words, to an observer on the Earth's surface, the satellite appears fixed in the sky. Accordingly, the 22,300 mile-high orbit is known as a *stationary orbit*. This orbit has a special importance for communications satellites.

We have talked about satellites 'circling' the Earth, but this is not strictly true. The orbit of a satellite is seldom circular in shape. It usually has the shape of an *ellipse,* which is a kind of oval. Because of this shape, the satellite is not always the same distance away from the Earth. The point of the orbit closest to Earth is known as the *perigee,* and the point farthest from the Earth is known as the *apogee*. In an elliptical orbit the satellite does not always travel at the same speed, either. It travels faster at perigee and then gradually slows down as it swings away from the Earth. It drops to its lowest speed at apogee, and then speeds up once more as it travels closer to the Earth.

## Kinds of Satellites

We have seen how a satellite can be launched into orbit and why it can stay there without falling back to Earth. Now let us look at the kinds of things satellites do in space.

The most useful of all the satellites are the *communications satellites*. They are used to relay telephone and television signals around the globe. The signals are transmitted to the satellite from one ground station. The satellite receives and *amplifies,* or strengthens the signal and transmits the now stronger signals automatically to another ground station.

The latest satellites, which include the *Intelsat* series, can handle several different television channels and hundreds of telephone calls all at once. They can not only handle much more 'traffic' than earlier communications satellites, but they are also 'fixed' in the sky in a stationary orbit 22,300 miles above the Earth. This means that transmitting and receiving stations on Earth can 'lock' onto the satellite permanently.

*Weather satellites* have also proved extremely useful to us. They orbit several hundred miles up, taking a variety of measurements of the state of the atmosphere and relaying the information back to Earth. Most important, they take television pictures of cloud formations over a wide area of the Earth's surface. Meteorologists can trace from these pictures the paths of storms and hurricanes and predicts where they will strike. In this way they can warn people in advance of possible danger. The weather satellites are launched into an orbit that takes them constantly over different parts of the Earth's surface. In this way a global picture of the weather situation can be built up. This has revolutionized the whole science of meteorology and greatly improved the accuracy of long-range weather-forecasting.

Satellites have also a great future in the field of *navigation*. A system of *Transit* navigation satellites, which has been used by the US Navy for some time, is now available to commercial shipping. Cunard's *Queen Elizabeth 2* was one of the first passenger liners to use it. By using special equipment, which tunes into and processes the signals coming from the satellites, the position of

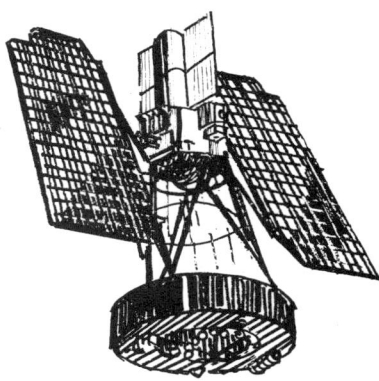

One of the Nimbus weather satellites launched by the United States during the 1960s. Note the "wing" panels of solar cells which provide energy to power the spacecraft's electrical equipment.

The giant dish antenna of one of the United States deep space tracking stations. Huge antennae like this are needed to receive the faint signals from spacecraft deep in space. They are highly directional so that they can precisely locate the spacecraft and issue instructions o it.

### Tracking

Scientists must always know the positions in space of the spacecraft they have launched. They often have to send signals to it to switch on radio equipment or fire a rocket motor. And they can only do this if they know exactly where the spacecraft is. Following the path of a spacecraft is called *tracking*. Tracking stations are situated at various places around the world so that at least one of them can maintain contact with the spacecraft for most of the time. Not all of these stations are on land. The United States, for example, operates quite a number of floating tracking stations situated in the middle of the oceans.

Tracking is usually done by *radio* or by *radar*. In radio tracking, stations on the ground follow the path of a spacecraft by picking up signals it sends out from a radio beacon on board. Radar tracking of a spacecraft is much the same as the radar tracking of aircraft. Radio signals sent out by a ground station and reflected from the spacecraft indicate its position in space. Of course, the radar 'echoes' from spacecraft hundreds or thousands of miles away are much weaker than those from aircraft only a few miles up. Therefore all the tracking equipment has to be that much more sensitive. The transmitting and receiving aerials are shaped like a huge dish. They are capable of transmitting a powerful beam deep into space and of receiving very faint signals. They are completely steerable so that they can be made to follow the path of a spacecraft across the sky. Some of the largest 'dishes' used are more than 200 feet in diameter.

the ship can be pinpointed to within 200 yards. And the system works in all weathers, too.

Communication, navigation, and meteorology are the fields that have benefited most from satellite launchings. But not all satellites are immediately useful to mankind. Many are launched in the interests of scientific research. They contain a variety of instruments to measure the conditions in near and outer space. They send back streams of information about such things as radiation, magnetism, and meteors, which increases our fundamental knowledge of the universe. Early on in the Space Age scientists discovered from satellite measurements the presence of intense belts of radiation high above the Earth – the so-called *Van Allen belts*. They could well prove a danger to manned space flights in the long term.

### Telemetry

The process of measuring things from a distance, as is done in spacecraft, is known as *telemetry*. The telemetric equipment in a spacecraft consists of the measuring instrument itself, a tape-recorder, and a radio transmitter. The instrument gathers information in the form of tiny electrical signals, which it feeds into the tape-recorder as the spacecraft travels through space. When a scientist on Earth wishes to recover the information, he sends a signal to the spacecraft. The signal switches on the transmitter, which plays back the tape-recorded information. The scientist records this information himself and converts it later into actual measurements or photographs.

### Solar Cells

All the instruments, tape-recorder, radio equipment, and so on need electrical power to operate. It is no use putting an ordinary battery into a spacecraft because it would soon run down and there would be no way of recharging it in space. And so most spacecraft are powered by batteries that are recharged by the Sun, which always shines in space. These *solar batteries* are made up of many tiny cells made of silicon, each of which produces a tiny electric current when sunlight falls on them. The 'paddles' you often see on spacecraft are made up of hundreds of these cells.

Other sources of power that are sometimes used in space are the *fuel cell* and the *radioactive generator*. The fuel cell produces electricity by making hydrogen and oxygen gas combine to form water. The radioactive generator is powered by a rod of radioactive material such as plutonium. It has a very long life indeed.

Above: The Earth as seen from an orbiting satellite, proof, if proof were needed, that the Earth is indeed spherical. Space as we can see is inky black. The atmosphere scatters the sunlight that strikes it and appears blue.

Opposite: A superb photograph of the Gulf of California, taken from about 150 miles up. (North is to the right of the picture.) Such photographs are naturally of immense benefit to mapmakers, or cartographers. But many other people, from farmers to oil prospectors, can and have also benefited from orbital photography. This particular photograph was taken in infra-red light, which shows healthy green foliage as red. Diseased foliage can often be distinguished by infra-red photography of this sort.

# Space Probes

So far we have considered only spacecraft that remain in orbit around the Earth as artificial satellites. But many spacecraft are launched much farther afield – to the Moon and to the planets. These long-range spacecraft are generally called *probes*. The Russian name for a probe is *zond*. Probes are launched full of instruments which measure and record such things as temperature, radiation, and magnetism and send the information back to Earth. Many carry television cameras to relay pictures.

## Lunar Probes

The early lunar probes simply flew past the Moon and took photographs of it. Later ones, such as the *Ranger* series, crash-landed after taking close-up television pictures of the surface. Then came those, such as the *Surveyors,* which actually made a soft landing. The *Surveyor* series were outstandingly successful. They all sent back thousands of close-up pictures of the lunar landscape. Some even carried out simple experiments to scoop up and analyse the soil. A detailed photographic survey of the surface was carried out by another series of probes which went into lunar orbit.

Just what are the problems involved in launching a spacecraft to the Moon? Obviously, it is going to need a lot more power to rocket a probe 239,000 miles to the Moon than to put a satellite into orbit.

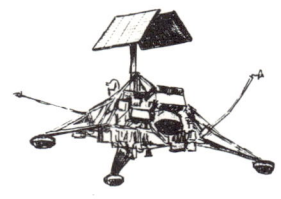

A Surveyor lunar probe. A craft of this kind made the first true soft landing on the Moon in 1966. It proved that the lunar surface was firm for it sank only a little when it landed. Surveyor probes sent back hundreds of excellent close-up pictures of the lunar surface, and tested its consistency and chemical composition.

Below right: Model of a Mariner space probe. Several of these craft have been sent to take photographs of the planet Mars. Note the sail-like panels of solar cells. The picture below left shows one of the closest pictures taken by Mariner 6 in 1969 from about 2,130 miles up. It covers an area about 45 miles square. The large crater at the bottom is about 15 miles across.

The quickest way to reach the Moon would be to use a powerful rocket that burned for the whole journey. But that would be a great waste of fuel. There is another method, which, although much slower, enables the journey to be accomplished with the minimum of fuel. This is a great advantage, of course, because with less fuel on board, more useful instruments can be carried for the same weight of probe. The idea of this method is to give the probe an initial boost and then let it 'coast' towards its target. This technique is called 'free-fall'.

What happens in practice is this. The probe is launched towards the Moon with a speed of 24,500 miles an hour. Then its rocket engines are cut off. Being under the attraction of the Earth's gravity, it gradually slows down. When it is about 200,000 miles away from Earth, it is hardly moving at all. From that point on, however, the Moon's gravity becomes progressively stronger than the Earth's and begins to pull the probe towards the Moon.

The probe moves faster and faster until, a few miles above the Moon's surface, it is travelling at about 5,000 mph. If the probe is to go into orbit or make a soft landing on the Moon, it must be slowed down. Otherwise it will crash and be smashed to pieces. On the Moon there is no atmosphere, like there is on Earth, to slow down, or brake, an object travelling near the surface. The only way braking can be achieved is by turning the probe back to front and firing a blast on the rocket engines in the direction in which the probe is travelling. Slowing the probe down

to about 3,000 mph takes it into a lunar orbit. Slowing it down further takes it out of orbit and makes it drop towards the surface. As it does so, it accelerates under the action of the Moon's gravity. To land softly, the probe must again be slowed down by retro-braking. This must happen all the way down until the probe settles gently on the surface.

## Planetary Probes

The problems of sending probes to the planets are much greater, of course, than those of sending probes to the Moon. The distances involved make aiming, control, and communication with the probes incredibly difficult. The 'closest' planets are Venus and Mars, which, even at their minimum distances, lie respectively about 24 million and 34 million miles from Earth. Venus and Mars bear the closest similarity to Earth in terms of size and conditions among the planets of the solar system. It has long been thought possible that life of some kind might exist on one of them. But the information gathered

The lunar crater Copernicus taken from 28 miles up by Lunar Orbiter 2. The mountains rising from the floor of the crater are about 1,000 feet high. Central mountain peaks of this nature are typical of many of the large lunar craters. Described as "the picture of the century" when it was first released in 1967, it is probably the finest pre-Apollo lunar photograph.

by a number of probes sent to both planets has indicated that there is little likelihood even of primitive life there. As for life on the other planets, this is even less likely. Mercury is far too hot, and Jupiter and the outer planets far too cold for life as we know it.

To travel from Earth to another planet, a probe must escape completely from the Earth's gravity. By so doing, it can then move away from the Earth's orbit around the Sun towards the orbit of one of the other planets. The minimum speed a probe must be given to escape is about 25,000 mph. This speed is called the *escape velocity*. If we launch a probe with a velocity of about 26,000 mph in the same direction as the Earth is travelling in its orbit, the probe will travel out from the Earth's orbit towards the orbit of Mars. If we launch the probe with the same speed but in the direction opposite to that in which the Earth is travelling, it will travel in towards the orbit of Venus. The path the probe takes from the Earth's orbit to that of another planet is called a *transfer orbit*.

# Man in Space

In April 1961 Yuri Gagarin made the first pioneering manned flight into space in *Vostok 1*. He was the first *cosmonaut*, as the Russians call their space pilots. We call space pilots *astronauts*. Both names mean something like 'sailing to the stars'. The astronauts have not yet gone that far, but they have reached the Moon. The American civilian astronaut Neil Armstrong made the first human step on the Moon in July 1969.

Just what are the extra problems involved in manned space flight? Obviously we must not only launch man safely, but we must also protect and sustain him in space and bring him safely back to Earth. To satisfy all these necessary requirements, a manned spacecraft must be a lot bigger than unmanned craft. They are made in a number of separate sections, which in the giant *Apollo* Moon rocket, are called *modules*.

## Spacecraft and Life-Support

The spacecraft proper consists of a crew section and an equipment section. (In Apollo these sections are called respectively *command module* and *service module*. In addition Apollo has a *lunar excursion module,* which is a 'moonship', or separate spacecraft designed to land on the Moon.)

The crew section is where the astronauts live in space. It contains couches for the astronauts to lie on and all the controls, instruments, the computer, and communications gear which enable them to control the spacecraft. It also has several little manoeuvring engines which can change the attitude of the craft in space. This part of the spacecraft is the only one that returns to Earth.

The equipment section contains the main power supply, rocket engine and fuel, and *life-support system*. This is the system that supplies fresh air to the crew section, removes stale air and odours, maintains the correct humidity and temperature, and so on. In flight generally the crew breathe the air provided by the life-support system. In space, of course, there is no air.

When they leave their spacecraft, the astronauts wear a *spacesuit* connected to a portable life-support unit which carries out the same functions as the main on-board unit. The spacesuit, in fact, takes the place of the Earth's atmosphere, which presses down on us, gives us air to breathe, helps to keep us warm, and protects us from harmful radiation. The spacesuit is designed to do all these things. It is made up of several layers,

American astronaut entering centrifuge. Inside, he will be strapped in a seat and whirled round at high speed. He will be subjected to centrifugal forces that simulate the high "G" forces the astronaut will experience during launching and re-entry. During these periods they may experience forces approaching 10 G, or ten times the force of gravity.

from inner, water-cooled 'combinations' to a protective, outer garment.

## Launching

Launching a man into space is not as simple as launching a lump of metal. Metal can, within reason, be made to withstand tremendous stresses without breaking. But a flesh-and-blood man is a lot more delicate. The acceleration at the launching of a rocket are really tremendous. The rocket's powerful engines thrust the spacecraft from rest to a speed of 17,500 miles an hour in only about 12 minutes! If you have been in a high-speed lift, you have probably experienced the unpleasant sensation when it accelerates upwards quickly from rest. You 'leave your stomach behind'. We call the forces due to such acceleration *G-forces* – G stands for the acceleration due to gravity.

You can imagine how great the G-forces are in an accelerating space rocket. To withstand such forces the astronauts have to lie flat on their backs during take-off. They strap themselves firmly to padded couches. Certain aspects of their training help to prepare their bodies for such forces.

As a great contrast to launching, when they weigh ten times what they do on Earth, up in orbit they are *weightless!* In orbit the inward pull of gravity has effectively been cancelled out by the resultant forces that are set up by the spacecraft's motion in a circle (or rather, an ellipse). Up in space then nothing can 'fall'. The astronauts can float in the air, 'lie' on the ceiling, and turn continual somersaults if they want to, all with little effort. While the state of weightlessness seems ideal, it does have drawbacks.

Eating and drinking are quite a problem. The astronauts cannot drink water from a glass because the glass just will not pour. The water stays where it is in the glass when it is tipped up! The astronauts therefore drink by squeezing liquid right into their mouth from a tube like a toothpaste tube. They generally eat from a tube, too. Most of their food is dehydrated. So they mix it with water into a paste which they can easily squeeze from a tube.

### Re-Entry and Landing

Astronauts returning, say, from the Moon approach the Earth at a speed of nearly 25,000 miles an hour. Somehow they must be slowed down to a speed of only a few miles an hour so that they can land safely. Just as there was tremendous acceleration at launching, there is now tremendous deceleration. And to withstand it, the astronauts must lie flat on their back. As we mentioned earlier, the crew section is the only part of the spacecraft to return to Earth. The other part or parts are jettisoned before the crew section re-enters the atmosphere.

The atmosphere acts as a powerful brake on the spacecraft. The drag on a body travelling at 25,000 miles an hour is fantastic. The friction between the craft and the air generates great heat which threatens to burn it up like a shooting star, or meteor. But the base of the craft, which enters the atmosphere first, has a very special coating called a *heat shield*. It is a special kind of refractory material made from a combina-

American astronaut Edwin Aldrin photographed by first man on the Moon Neil Armstrong. Photographer and the lunar module can be seen reflected in Aldrin's gold-covered sun visor. Note the bulky space suit and back pack. The multi-layer spacesuit is necessary to protect the astronaut in the hostile lunar environment. The back pack comprises portable life-support system and communications equipment and is controlled from the box on the astronaut's chest.

tion of plastic, asbestos, and several other materials. And it protects the astronauts as the atmosphere slows their craft down.

When the craft is travelling slowly enough parachutes open above it and carry it gently down to Earth. The American astronauts always land in the sea, Russian cosmonauts invariably land on land. They use a different technique, of course. The cosmonauts often descend inside their spacecraft, using retro-rockets beneath their craft to slow it down. Or they may parachute from the craft just before it touches down.

Left: Astronauts lie flat on their back at launch and at re-entry to withstand the high G-forces developed at those times.

Right: The crew module of an Apollo spacecraft just after splashdown. The flotation collar is in position, and the astronauts have transferred to the life raft to await pick-up by helicopter. Note the scorched appearance of the command module.

# Steps Into Space

**1232** Chinese soldiers used rockets during their war against the Mongols.

**1600s** Johannes Kepler developed his laws of planetary motion, which correctly described planetary orbits as ellipses.

**1687** Sir Isaac Newton published his *Principia* in which he described his law of universal gravitation and his laws of motion. His law of gravitation states that every particle of matter attracts every other particle with a force that varies with the masses of the particles and the distance between them. Doubling the mass doubles the attraction. Doubling the distance *reduces* the attraction to a quarter. This theory explained Kepler's laws and enabled astronomers to account for the movements of the heavenly bodies. Newton's third law of motion states that to every action there is an equal and opposite reaction'. This is the principle on which the rocket works.

**1903** Konstantin Tsiolkovsky, a Russian schoolmaster, published a paper suggesting the use of rockets for space flight, and propounded the step-rocket principle.

**1923** Hermann Oberth in Germany, published his famous book *The Rocket into Interplanetary Space*.

**1926** Robert Hutchings Goddard fired the first liquid-propelled rocket in the United States on March 16. It used petrol and liquid oxygen as propellants and rose almost 200 feet in the air.

**1927** A Society for Space Travel (VfR, or Verein fur Raumschiff-fahrt) was formed in Germany which carried out research on rockets.

**1942** The Germans carried out the first successful launching of the liquid-propelled V-2 rocket (at first called the A-4) at Peenemunde on an island in the Baltic Sea. It used alcohol and liquid oxygen as propellants. Later, Hitler used the V-2 with an explosive warhead to bombard London. The V-2 development was under the direction of Wernher von Braun, an ex-member of the VfR. After the war von Braun went to the United States and later became director of the American space programme.

**1949** USA launched their first two-stage rocket using a V-2 as the first stage. The upper stage, a *WAC Corporal,* reached an altitude of almost 250 miles.

**1957** USSR launched *Sputnik 1*, the first artificial satellite, on October 4. It was a metal globe 23 inches in diameter with aerials projecting from it for its radio transmitter. The transmitter sent back 'bleeps' of information as it orbited the Earth every 96 minutes, between 145 and 560 miles high. It remained in orbit for 92 days, by which time (January 4, 1968) the drag of the very thin atmosphere even at that height above the Earth, had slowed it down and caused it to burn up like a shooting star. The satellite weighed 184 pounds, which at the time was considered surprisingly large. But it was nothing compared with *Sputnik 2* which was launched on November 3. This one weighed over 1,100 pounds, almost half a ton! It carried the world's first space traveller, a dog called Laika, who survived the ascent and remained well until its oxygen supply gave out. The orbit of *Sputnik 2* varied from 150 miles to more than 1,000 miles. It remained aloft for 162 days.

**1958** USA launched its first satellite, *Explorer 1*, weighing only 31 pounds, on January 31. It transmitted information that showed the presence of intense 'belts' of radiation girdling the Earth. These became known as the *Van Allen belts* after the scientist in charge of the experiment.

(*Explorer 1* remained in orbit until April 1970, by which time it had travelled 58,376 times around the Earth.)

**1959** *Lunik 1* (USSR) and *Pioneer IV* (USA) were the first space probes launched to the Moon. *Lunik 2* crashed on the Moon in September. *Lunik 3* sent back a picture of the far side of the Moon which we never see from Earth.

**1960** USA launched the first weather satellite, *Tiros 1*, and the first passive communications satellite *Echo 1*. In August USSR launched two dogs, Strelka and Belka, into space and recovered them alive and well after 18 orbits of the Earth.

Left: The infamous rocket V-2, *Vergeltungswaffe zwei,* or 'vengeance weapon two', used by Hitler to bombard London in the closing months of World War 2. Wernher von Braun directed the V-2 project at Peenemünde. After the war he went to the United States and developed successive generations of rockets for the American space program, below. But the Russians beat the Americans into space. Their second orbiting spacecraft Sputnik 2 contained the dog Laika, right.

**1961** Yuri Gagarin (USSR) in *Vostok 1* (meaning 'East') became the first man in space on April 12, making one orbit of the Earth. USSR launched a probe to Venus. For the USA, Alan Shepard (May) and Virgil Grissom (July) each made sub-orbital flights.
**Yuri Alekseyevich Gagarin** was a major in the Russian air force. At the highest point of the orbit, he was more than 200 miles above the Earth. He made a perfect landing inside the re-entry capsule of the craft. He had travelled more than 25,000 miles in the 1 hour 48 minute flight. Born at Gzhatsk, west of Moscow, in 1934, Gagarin learnt to fly at the Ovenburg Air School. It was ironic that, after facing the immense hazards of the first-ever space flight, he should be killed, in 1968, in an air crash here on Earth.

**1962** John Glenn in *Friendship 7* made 3 orbits of the Earth on February 20 to become the first American in space. USA launched *Telstar*, the first active communications satellite. USSR launched *Mars 1* probe to Mars, and USA *Mariner 2* probe to Venus.

**1963** Valentina Tereshkova (USSR) became the first woman cosmonaut.

**1964** USSR launched *Voshkod 1* (meaning 'Sunrise') containing three cosmonauts. USA launched *Ranger 7* Moon probe which sent back close-up television pictures of the Moon before crashing on the surface.

**1965** Alexei Leonov (USSR) made the first 'walk' in space on March 18, when he spent just over 20 minutes outside his two-man spacecraft *Voshkod 2*. USA made the first two-man flights in their *Gemini* programme. Edward White, from *Gemini 4*, made the first American 'walk' in space. *Gemini 6* made a successful rendezvous with *Gemini 7*, which spent almost a fortnight in orbit. USA launched *Early Bird* the first commercial communications satellite in a stationary orbit. *Mariner 4* (USA) sent back pictures of the crater-scarred surface of Mars over a distance of 135 million miles. It was a fantastic feat of communications. On November 26 France became the world's third space power when it launched its first satellite, *A-1*, with a *Diamant* rocket from Hammaguir base in the Sahara.

**1966** *Luna 9* (USSR) made an instrument landing on the Moon in February and transmitted pictures of the surface back to Earth. *Surveyor 1* (USA) made the first true soft Moon landing in June and sent back a series of close-up pictures. Both USSR and USA put *Luna* and *Lunar Orbiter* probes into orbit around the Moon which sent back thousands of detailed pictures of the surface. *Gemini 8* made the first successful docking, or link-up manoeuvre with another vehicle in space, an essential step for subsequent space flights.

**1967** In January three American astronauts Virgil Grissom, Edward White, and Roger Chaffee, died in a flash-fire in an *Apollo* spacecraft during a practice countdown on the ground, delaying the Apollo programme aimed at putting astronauts on the Moon by 1970. Vladimir Komarov (USSR) became the first in-flight space casualty in April when he was killed returning to Earth in *Soyuz 1* (meaning 'Union'). In September USSR achieved the first automatic docking in space by linking two *Cosmos* satellites. In November an unmanned craft, *Apollo 4*, made a successful high-speed (25,000 mph) re-entry after its launch on top of the *Saturn 5* rocket designed to carry American astronauts to the Moon.

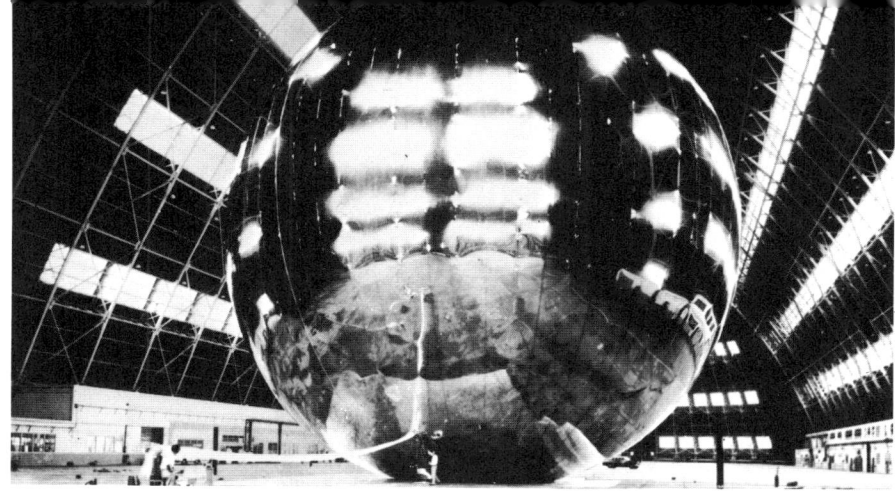

Top: The first communications satellite, Echo 1, an aluminum-coated plastic balloon 100 feet in diameter.

Three famous names in space history; above, Yuri Gagarin; above right, Valentina Tereshkova; right, John Glenn.

Below, American astronaut Edward White made the first American walk in space. Less than two years later he was tragically killed during a practice "countdown".

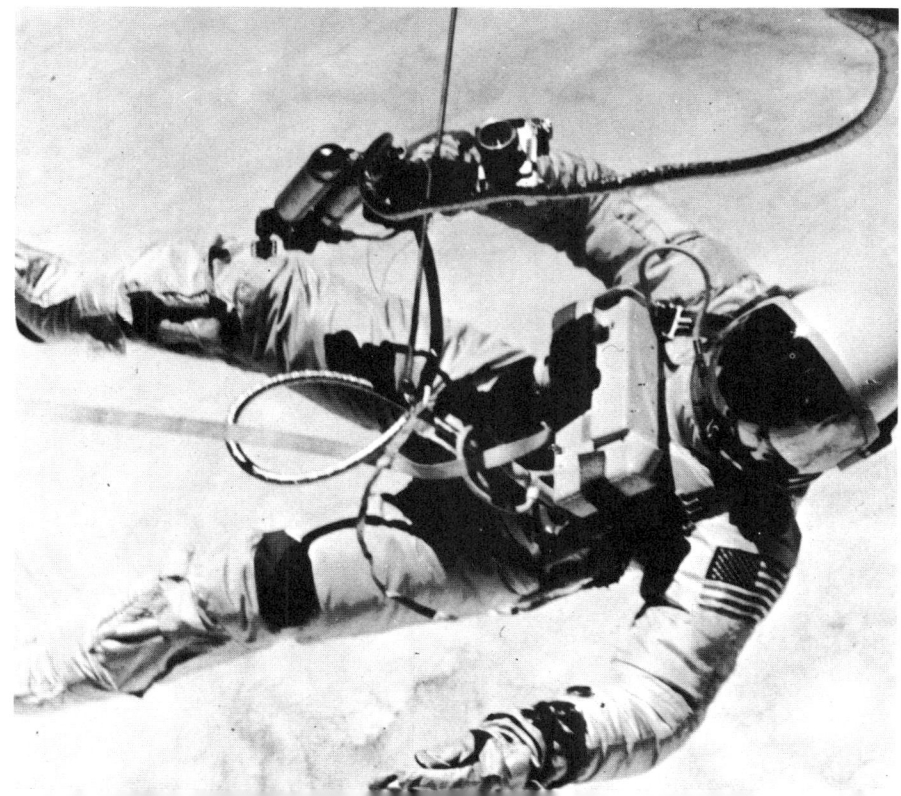

Astronauts James Lovell, William Anders and Frank Borman circled the Moon in Apollo 8 in December 1968.

**1968** USSR recovered unmanned spacecraft *Zond 5* (September) and *Zond 6* (November) after they had flown round the Moon. (*Zond* simply means 'probe.') In October, a manned *Apollo 7* made a successful 10-day flight in orbit, testing all the systems of the *Apollo* craft. In December, Frank Borman, James Lovell and William Anders in *Apollo 8*, became the first men to travel to the Moon and back. During the spectacular six-day flight they made ten orbits of the Moon, about 70 miles above the surface. Everything functioned perfectly. People throughout the world saw live television pictures of Earth and the Moon taken by *Apollo 8* from space.

### 1969 January
*Soyuz 4* with one cosmonaut on board met in orbit and docked successfully with *Soyuz 5*, containing three cosmonauts. Two of the cosmonauts transferred to *Soyuz 4* before both vehicles returned to Earth.

### July
Michael Collins, Neil Armstrong, and Edwin Aldrin in *Apollo 11* repeated the Moon trip, but this time Armstrong and Aldrin in the lunar module, descended to the Moon's surface. They landed at 8.18 p.m. GMT on July 20. On *July 21 at 2.56 a.m. GMT* Armstrong stepped from the lunar module onto the Moon's surface and became the first human being to set foot on another world. Later Aldrin joined him in the 'Moon walk'. They reported that movement on the Moon was much easier than they had expected, despite their heavy life-support packs on their backs. They set up a television camera which transmitted pictures of the historic event, remarkably clear in the circumstances, back to Earth. They collected rock samples for subsequent analysis on Earth and carried out a number of experiments. Armstrong spent 2 hours 47 minutes walking on the surface. Aldrin's stay was somewhat shorter. In all they remained on the Moon 21 hours 36 minutes, before they took off to rejoin Collins in the parent craft. They left behind on the lunar surface a seismometer, which is a device for measuring tremors in the ground.

### November
Astronauts Charles ('Pete') Conrad, Alan Bean, and Richard Gordon in *Apollo 12* blasted off from Cape Kennedy in a blaze of lightning at the start of their trip to the Moon on November 14. On November 19 Conrad and Bean made a pin-point landing in the lunar module only 600 feet away from the *Surveyor 3* lunar probe that had landed $2\frac{1}{2}$ years previously.

### 1970 February
Japan became the world's fourth space power when it launched its first satellite, *Osumi*, from the Kagoshima Space Centre.

### April
*Apollo 13* blasted off en route for a third Moon landing with James Lovell, Fred Haise and John Swigert on board. But an explosion in the service module almost caused disaster 200,000 miles out. The astronauts returned safely, having used the life-support system of the lunar module to survive. On April 24 China became the world's fifth space power. It launched a 380-pound satellite, designated *Chincom 1*, from a rocket center near Shuang Cheng Tsu.

### November
USSR soft-landed a robot space probe, *Luna 16*, which scooped up some Moon dust and returned it to Earth. *Luna 17* soft-landed on the Moon, disgorging an eight-wheeled vehicle, *Lunokhod 1*.

### 1971 February
*Apollo 14* crewed by Alan Shepard, Edgar Mitchell and Stuart Roosa made another Moon trip, visiting a site near the crater Fra Mauro.

### April
USSR launched *Salyut* space station into Earth orbit. *Soyuz 10*, with a three-man crew, docked briefly with *Salyut*.

### May
Cosmonauts Georgy Dobrovolsky, Viktor Patsayev and Vladislav Volkov in *Soyuz 11* again docked with Salyut and remained linked for a record 24 days. The spacecraft made a perfect return to Earth. But when the hatch was opened the three cosmonauts were found to be dead, killed by de-pressurization of the re-entry capsule.

### August
Astronauts David Scott, James Irwin and Al Worden made the fourth successful Moon trip, to Hadley Rill in the Sea of Showers. They drove a battery-powered 'car', the lunar rover.

### November
US space probe *Mariner 9* became the first spacecraft to go into orbit around Mars. Later a Russian probe followed *Mariner* into orbit and released a capsule that soft-landed on the surface of the planet.

### 1972
In March US launched space probe *Pioneer 10* on a trajectory towards Jupiter to photograph the Red Planet from a distance of 100,000 miles. In April US lunar mission *Apollo 16* visited the Cayley Plains region of the lunar highlands, crewed by John Young, Charles Duke, and Thomas Mattingly. US launched Apollo 17, last in the Apollo series, in December. Crewed by Ronald Evans, Eugene Cernan and Harrison Schmitt, Apollo 17 landed at a site between the Taurus mountains and Littrow crater.

### 1973
In January, USSR landed the second remote-controlled Moon vehicle, Lunokhod 2, on the Sea of Serenity. In May, US launched their Space Station Skylab into a near-earth orbit.

Opposite: Apollo 12 astronaut setting up an experiment on the Moon's surface. The bleak, flat landscape is typical of the Moon's maria, or "seas". Below: Apollo 16 astronaut putting the battery-powered lunar roving vehicle through its paces.

# Destination Moon

The most complicated, the most exciting and, of course, the most dangerous space flights attempted so far have been the *Apollo* series of manned missions to the Moon and back. What kind of equipment and what kind of maneuvers are employed on such a trip?

Let us start at the beginning and look at the giant rocket standing on the launching pad ready to take off to the Moon. The launching vehicle is the giant *Saturn V*. The first stage, or booster alone stands no less than 186 feet tall. It burns kerosene and liquid oxygen in its cluster of five engines to develop a thrust of more than 7½ million pounds. The second stage also has five engines, with liquid hydrogen and liquid oxygen as propellants, developing 5 million pounds thrust. The third stage has a single engine also burning liquid hydrogen and liquid oxygen.

Perched high on top of the third stage is the *Apollo* spacecraft, which is made up of three sections. First comes the *lunar excursion module*, the only part that will set down on the Moon. Then comes the *service module*, the part containing all the main control systems, the computer, and so on. Linked to this is the *command module*, or crew compartment that houses the astronauts. Topping the massive structure is the *escape tower* which could, during a faulty launching, separate the command module from the rest of the rocket to rescue the astronauts. From the exhaust nozzles of the booster's engines to the tip of the escape tower, the vehicle measures 364 feet. It weighs more than 3,000 tons!

Five days before the launch is due, the 'count-down' begins. All the thousands of parts and systems are painstakingly checked and rechecked. A day or so before zero hour the propellants are pumped into the massive tanks. A few hours to go, and the astronauts ride by lift up the gantry surrounding the rocket and enter the command module.

A human footprint in the firm lunar soil. It was made by one of the pioneering Apollo 11 astronauts on July 21, 1969, and will probably remain undisturbed for centuries, there being little apart from chance meteoric impact to obliterate it.

Some scientists believe there is a continuous but gradual turn over of lunar dust due to temperature difference between lunar day and night.

The diagram shows the elements of the Moon trip, as carried out by the Apollo spacecraft. The technique is termed lunar orbital rendezvous. A section of the spacecraft descends to the Moon's surface, while the main part of the craft continues in lunar orbit. Then part of the lunar landing craft ascends to rendezvous with the main craft orbiting above it.

They strap themselves in and make final checks. The gantry rolls away, the last seconds tick away, then 'Ignition!'. Flame and smoke billow from beneath the monster rocket, and, as the thrust builds up, it begins to lift off the launching pad. Soon, after burning propellants at the rate of 15 tons a second, the booster burns out and falls away. Altitude about 40 miles; speed about 6,000 mph; time about 2 minutes after lift-off. The second stage fires for about 7 minutes and boosts the craft to 15,000 mph, 100 miles up. The second stage and the escape tower are then jettisoned, and the third stage fires for 2 minutes or so to thrust the craft into a so-called 'parking' orbit 115 miles up with a speed of about 17,500 mph.

All the essential systems are rechecked and a precise flight plan is worked out for the trip to the Moon. If all is well, the third stage is re-ignited and burns for some 5 minutes to increase the spacecraft's speed to more than 24,000 mph. This pushes the craft into an orbit towards the Moon.

Next comes an important docking manoeuvre to align the three modules for their respective roles in the Moon-landing. The combined command and service unit separates from the lunar module. It does a 'U' turn and then docks with the lunar module. The upper part of the command module is now linked with the lunar module. The whole unit then pulls away from the third stage and heads for the Moon. Later, the course may have to be corrected by firing the engine of the service module.

As *Apollo* approaches the Moon, the astronauts turn the spacecraft round and fire the service module's engine as retrobraking to slow them down to about 3,600 mph. At this

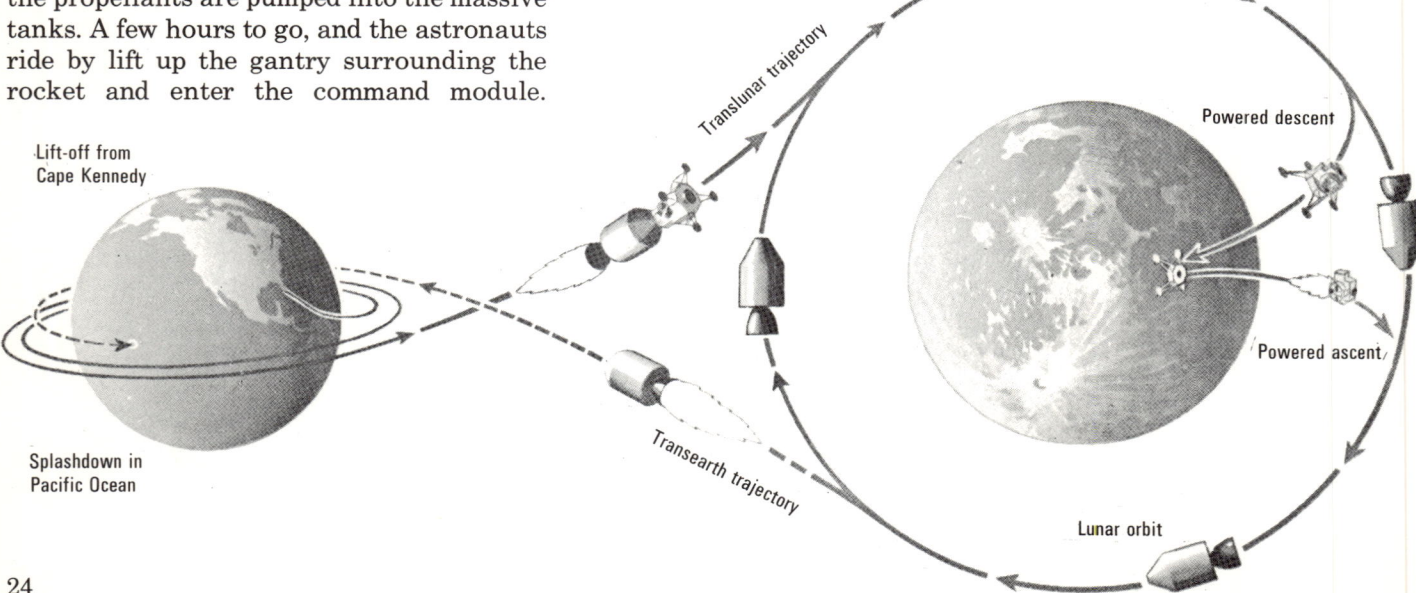

speed they go into lunar orbit. Two of the astronauts enter the lunar module, check its systems, and then separate from the parent craft. They use the lunar module's descent engine as a retrobrake and descend to the Moon's surface.

After spending some time on the surface exploring and carrying out experiments, the astronauts return to their craft and prepare to leave the Moon. The lunar module is actually made up of two sections, an upper and a lower. So far the two have been attached together. Now they are detached. And the lower section becomes a launching pad for the upper, which naturally houses the crew. The crew fire the ascent engine to blast the upper section into lunar orbit once more. There they dock with the parent craft that has been orbiting above while they were on the Moon. The lunar excursion crew rejoin the other astronaut and then jettison the lunar module.

The Apollo trips provided us with some stunning photographs of the lunar landscape. This one shows part of the area explored by the Apollo 15 astronauts at the edge of the Sea of Showers. In the background are the Apennines, a range of mountains reaching to 12,000 feet.

To return to Earth, the crew fire the service module's engine for a few seconds to thrust them to a speed of about 5,500 mph. This will take them away from the Moon and back towards the Earth. As the spacecraft approaches the Earth, its speed increases to almost 25,000 mph under the influence of gravity. Just before it begins to re-enter the Earth's atmosphere, the astronauts jettison the service module. They aim the command module base-first through the atmosphere so that the heat shield bears the brunt of the frictional heating. The atmosphere slows them down rapidly and then, at about 24,000 feet, small *drogue* parachutes open, followed at 10,000 feet by the main parachutes. By the time they splash down in the sea they are travelling at only a few miles an hour.

The scorched, blackened command module is all that is left of the giant rocket which roared into space a few days and half a million miles before.

# Atoms and Elements

All matter in the universe is made up of certain basic substances called the *chemical elements*. These elements are composed of tiny particles called atoms, a word derived from the Greek and meaning indivisible.

The atom is not a solid 'lump' of matter as was once thought, but consists mainly of empty space. Nor is it the smallest particle of matter. But it is the smallest particle that retains distinct chemical properties.

You can picture the atom as a miniature solar system. In the center of the 'system' is the *nucleus*, where practically all the mass of the atom is concentrated. The nucleus has a positive charge. Circling in 'orbit' around the nucleus are tiny, lightweight particles called *electrons*, which have a negative

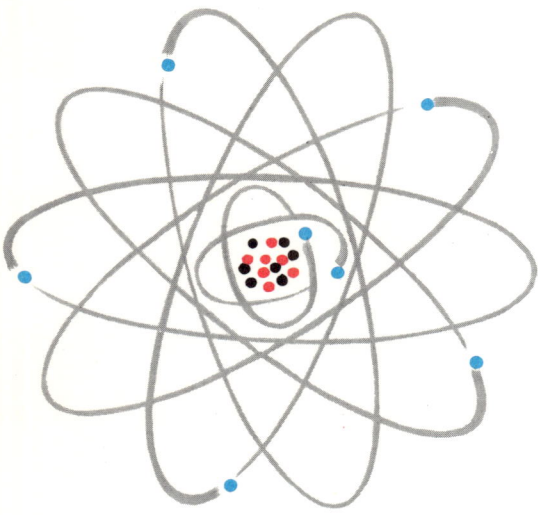

Diagram of the nitrogen atom, showing 7 electrons orbiting around a central nucleus containing 7 protons and 7 neutrons. The electrons have a negative electric charge, the protons a positive charge, and the neutrons no charge. The atom is thus electrically neutral.

● Electron
● Proton
● Neutron

charge. The attraction between the nucleus and the electrons holds the atom together. Atoms are very, very tiny. Hundreds of millions of atoms side by side would measure less than an inch!

The nucleus is made up of two more fundamental particles – protons and neutrons. The *proton* has a positive electric change; the *neutron* is neutral. A proton has the same sized electric charge as an electron, though opposite in sign. And an atom has equal numbers of protons and electrons. Thus, it is electrically neutral. The mass of the proton is about 1,800 times that of the electron. (It would take 160 million million million million protons to weigh one gram!)

Neutrons have more or less the same mass as protons, and are held within the nucleus by powerful forces. They cannot exist for long outside the nucleus and disintegrate to form protons and electrons. All atoms except ordinary hydrogen have neutrons in the nucleus. Helium, the simplest atom after hydrogen, has two neutrons with two protons in the nucleus. The atoms of an element are not always identical. The number of neutrons in their nucleus may vary. Because the neutron has no charge, the presence of a different number of neutrons in the nucleus of an atom makes no difference to the chemical properties of that atom. All that the presence of greater or fewer numbers of neutrons affects is the mass of the atom.

Atoms of an element which differ in weight in this way are known as *isotopes*. Generally one isotope of the element predominates in Nature. Take iron, for example. The common isotope has 30 neutrons with 26 protons in its nucleus. But there are other stable isotopes of iron with 28, 31 and 32 neutrons in the nucleus.

## Atomic Number and Atomic Weight

We call the number of protons in the atoms of an element the *atomic number* of that element. The number naturally tells us the number of planetary electrons the atoms have, too. By arranging the elements in order of their atomic number, they can be placed in groups with similar properties. Such a classification is known as the *periodic table*.

Every kind of atom has a definite and characteristic weight, and when elements combine, they do so in proportion to the weights of their atoms. For convenience the weight of the carbon atom is fixed at 12 and is called its atomic weight. The atomic weights of the other elements are expressed in comparison with the atomic weight of carbon. Thus hydrogen becomes 1·008, oxygen 15·999, chlorine 35·453.

## The Elements

There exist in nature 90 chemical elements, of which two (mercury and bromine) are liquids, eleven are gases, and the remainder solids. The simplest and lightest atom and element is hydrogen with 1 proton only (atomic number 1) in the nucleus and one electron circling around it. The most complex and heaviest atom is uranium, which has 92 protons (atomic number 92) and 146 neutrons in the nucleus and 92 electrons circling around it. In addition to these 90 natural elements, man has made 13 'artificial' elements which have no counterpart in nature, making 103 in all.

As soon as the idea of atomic weights for the elements became established in the 1800s, scientists began searching for relationships between the chemical properties of the elements and their atomic weights.

# Periodic Table

The elements arranged in order of their atomic number, or the number of protons in the nucleus, giving the so-called periodic table. Each vertical division is termed a group and each horizontal division a period, Nitrogen is thus in Group V, Period 1. The end group comprises the unreactive inert gases, which have a very stable electronic structure. The shaded squares in the diagram represent the so-called transition elements which have closely related properties. The small red squares represent the rare earth elements, which can be separated chemically only with great difficulty. The small blue squares represent the actinide series of elements and the artificial, or man-made elements from plutonium to lawrencium.

Hydrogen *H*
Helium *He*
Lithium *Li*
Beryllium *Be*
Boron *B*
Carbon *C*
Nitrogen *N*
Oxygen *O*
Fluorine *F*
Neon *Ne*
Sodium *Na*
Magnesium *Mg*
Aluminum *Al*
Silicon *Si*
Phosphorus *P*
Sulphur *S*
Chlorine *Cl*
Argon *Ar*
Potassium *K*
Calcium *Ca*
Scandium *Sc*
Titanium *Ti*
Vanadium *V*
Chromium *Cr*
Manganese *Mn*
Iron *Fe*
Cobalt *Co*
Nickel *Ni*
Copper *Cu*
Zinc *Zn*
Gallium *Ga*
Germanium *Ge*
Arsenic *As*
Selenium *Se*
Bromine *Br*
Krypton *Kr*
Rubidium *Rb*
Strontium *Sr*
Yttrium *Y*
Zirconium *Zr*
Niobium *Nb*
Molybdenum *Mo*
(Technetium *Tc*)
Ruthenium *Ru*
Rhodium *Rh*
Palladium *Pd*
Silver *Ag*
Cadmium *Cd*
Indium *In*
Tin *Sn*
Antimony *Sb*
Tellurium *Te*

Iodine *I*
Xenon *Xe*
Caesium *Cs*
Barium *Ba*
Lanthanum *La*
Cerium *Ce*
Praseodymium *Pr*
Neodymium *Nd*
(Promethium *Pm*)
Samarium *Sm*
Europium *Eu*
Gadolinium *Gd*
Terbium *Rb*
Dysprosium *Dy*
Holmium *Ho*
Erbium *Er*
Thulium *Tm*
Ytterbium *Yb*
Lutetium *Lu*
Hafnium *Hf*
Tantalum *Ta*
Tungsten *W*
Rhenium *Re*
Osmium *Os*
Iridium *Ir*
Platinum *Pt*
Gold *Au*
Mercury *Hg*
Thallium *Tl*
Lead *Pb*
Bismuth *Bi*
(Polonium *Po*)
(Astatine *At*)
(Radon *Rn*)
(Francium *Fr*)
(Radium *Ra*)
(Actinium *Ac*)
Thorium *Th*
(Protactinium *Pa*)
Uranium *U*
(Neptunium *Np*)
(Plutonium *Pu*)
(Americium *Am*)
(Curium *Cm*)
(Berkelium *Bk*)
(Californium *Cf*)
(Einsteinium *Es*)
(Fermium *Fm*)
(Mendelevium *Md*)
(Nobelium *No*)
(Lawrencium *Lw*)

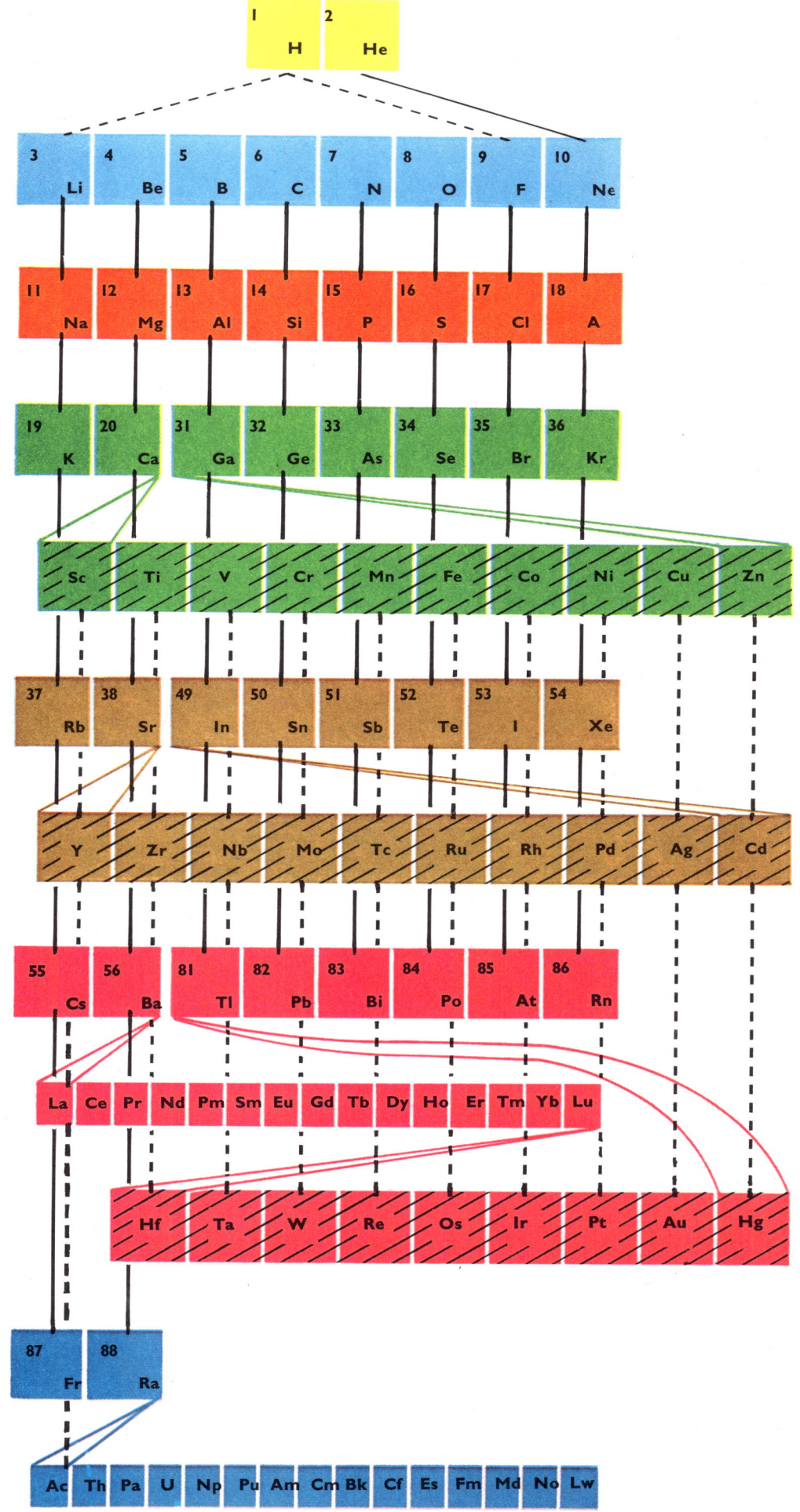

Emission of alpha-particles from radioactive thorium atoms. The cross track near the top of the picture is that of a proton ejected during the disintegration of a nitrogen atom by one of the alpha-particles.

The Russian chemist Dimitri Mendeléef (1834-1907) published in 1869 a table which showed that if the elements are written down in order of increasing atomic weights, elements with similar properties can be grouped together in an orderly way.

## The Periodic Table

In the modern version of the periodic table the elements are listed not by atomic weight, but by atomic number. The table has eight vertical *groups* and seven horizontal *periods*. The last group of each period (that is, before the 'pattern' of properties begins to repeat itself) is made up of the very unreactive, so-called *inert* gases from helium to radon. The table immediately brings out both a group and a period relationship.

Elements in the same group have similar chemical and physical properties and give rise to compounds with similar structure and properties. Group 1 shows this admirably. The elements in it are all soft, highly reactive metals. They decompose water and give rise to strong alkalis (such as sodium hydroxide, or caustic soda), which form salts (such as sodium chloride) with acids. Group 7 – the so-called *halogens* from chlorine onwards – are also a closely related family.

The first period contains only hydrogen and helium. Hydrogen is unique in that it has characteristics of both Group 1 and Group 7. But for the other periods, a relationship can be seen. The first members of a period (that is, Group 1) are elements (the alkali metals) which are unmistakably metallic in character. The members of Group 7 are halogens which display the character of a non-metal. In between these extremes, the elements show a gradual loss of metallic and a gradual gain in non-metallic character.

Periods 2 and 3, the so-called short periods, are quite straightforward. Then in Period 4, things apparently begin to go wrong. This, the so-called first long period, has not eight but eighteen members. Eight of them still have the characteristic properties of members of the previous period. But sandwiched between calcium and gallium (that is, between Groups 2 and 3) are the extra ten elements from scandium to zinc. These ten elements are known as the *transition* elements and show only slight differences in properties one with another. As you can see, with the exception of scandium, they are among the most important of our metals. There is another transition series in Period 5, the second long period. And each of the elements in this transition series resembles the appropriate member of the first transition series: zirconium resembles titanium, silver resembles copper.

A similar situation exists in Period 6, which also has a transition series, containing our most precious metals, platinum and gold. But the first 'member' of the series is not a single element but a family of elements! They are known as the *rare earths* and are so closely related that it is extremely difficult to distinguish between them and separate them. There is an equivalent closely related series of elements in Period 7, too. The first one is actinium, and the series is called the *actinide* series.

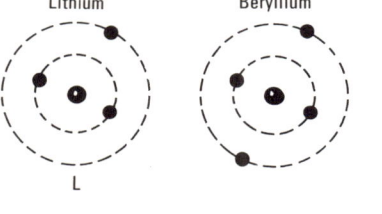
Hydrogen  Helium  Lithium
K
L

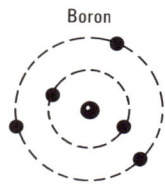
Beryllium

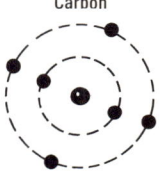
Boron

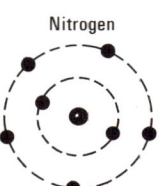
Carbon

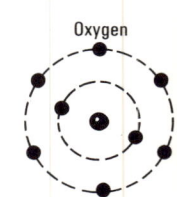
Nitrogen  Oxygen

### Electron Structure

As we have already mentioned, the number of 'planetary' electrons in an atom to a large extent determine its properties. Not only the number, but also the grouping of the electrons is important. By studying the electron structure of the atoms of chemical elements, we can explain regularities and oddities in the periodic table.

We can consider the electrons as occupying a number of concentric 'shells' around the nucleus. There are seven shells, usually denoted K L M N O P Q, the K shell being closest to the nucleus. Each shell can hold a certain maximum number of electrons: K 2, L 8, M 18, N 32.

The simplest atom, hydrogen, has one electron circling the nucleus, in the K shell. The next, helium, has two electrons circling the nucleus, in the K shell. You can imagine the electron structure of other atoms as being built up like this in a series of stages, adding an electron at each stage. (This is, of course, equivalent to adding a proton to the nucleus each time, thus increasing the atomic number by 1 and advancing along the periodic table by 1 step.)

With the inert gas helium, the K shell has its full complement of 2 electrons. When you add another electron to give lithium, the additional electron goes into the L shell. So we say that lithium has an electronic configuration of K 2, L 1, or simply 2, 1. Continuing in a similar way, beryllium has a configuration of 2,2; boron 2,3; carbon 2,4; and so on until the inert gas neon 2,8.

Likewise in Period 3 the next shell begins to build up from sodium K 2, L 8, M 1 to the inert gas argon 2, 8, 8. Period 4 begins with potassium K 2, L 8, M 8, N 1 and calcium 2, 8, 8, 2. But then instead of the outer N shell beginning to fill up, the M shell does until it has reached its full complement of 18 with zinc 2, 8, 18, 2. Then the outer N shell begins to fill up again to become 2, 8, 18, 8 with the rare gas krypton. This explains how the transition series fits into the periodic table in the way it does. The positions of the second transition series, the rare earths, the third transition series, and the actinide series in the table can be readily understood by considering their electronic structure in the above way.

When atoms combine together to form molecules they do so in definite, simple proportions. It is the electrons in the outer shells of the atoms which provide the means for combination (see page 166).

# Radioactivity

We have already seen that the atoms of an element are made up of a nucleus of positively charged protons and uncharged neutrons, with negatively charged electrons 'orbiting' around it. And we have noted that many of the elements consist of two or more different isotopes – atoms with the same number of protons but with a varying number of neutrons. In all, there are among the 90 natural elements more than 320 isotopes.

In the majority of isotopes, the protons and neutrons are bound strongly together by nuclear forces and are stable. In about 50 of the isotopes, however, the nucleus is unstable and spontaneously emits a stream of high-energy particles or radiation or both. This phenomenon is known as *radioactivity*, and the isotopes are called *radioactive isotopes*.

One of the best-known radioactive elements is uranium. Natural uranium is made up of three isotopes with atomic weights of 238, 235 and 234. It is mainly U-238 (as we can write it), with only very slight traces of the other two isotopes. The French scientist Henri Becquerel first discovered the natural radioactivity of uranium in 1896. His discovery that the uranium ore pitchblende gives out an even more penetrating radiation than uranium led the Curies to their discovery of radium. (See page 32.)

The nature of radioactivity was first elaborated by Lord Rutherford, who also identified two of the kinds of rays radioactive elements emit. He called them *alpha-rays* and *beta-rays*. *Gamma-rays* are another kind of radiation emitted by radioactive elements.

## Alpha-, Beta-, and Gamma-Radiation

*Alpha-rays*, or *alpha-particles* as they are often called, consist of two protons and two neutrons. Because they contain protons, they have a positive electric charge. Alpha-particles are, in fact exactly the same as the nuclei of helium atoms. And helium gas is often present in 'pockets' in radioactive rocks. Alpha-particles are emitted with high energies from the radioactive atom but, being relatively big and heavy, they are easily slowed down. A few thicknesses of paper will stop them.

*Beta-rays* are moving streams of electrons. Most natural radioactive atoms emit ordinary electrons with a negative charge. But some emit positrons, or electrons with a positive charge. They can be emitted from a radioactive atom with a wide range of energies. Having little mass, and a smaller electric charge, electrons are more penetrating than alpha-particles.

*Gamma-rays* have no electric charge. They are electromagnetic waves like X-rays but with an even shorter wavelength (see page 43). Like all electromagnetic waves, they travel at the speed of light. Gamma-rays have very high energies and are very penetrating. They constitute the main danger when radioactive material is being handled. Quite often a radioactive atom emits gamma-rays along with either alpha- or beta-rays. But it is seldom that alpha- and beta-rays are emitted together.

## Radioactive Disintegration

The only place that radiation can come from in a radioactive atom is the nucleus. Thus, when an atom emits an alpha-particle, it loses two protons and two neutrons from its nucleus. But it is the number of protons in the nucleus which distinguishes the atoms of the element from those of another. Therefore, when a radioactive atom disintegrates by alpha-radiation, it turns into the atom of another element. This change into another element is known as *transmutation*.

The mechanism behind the radioactive emission of beta-radiation, or electrons is a little more complicated. There are, of course, no electrons in the nucleus of an atom. What happens is this. In the case of negative-electron radiation, one of the neutrons in the nucleus spontaneously changes into a proton and an electron, which is ejected at high speed. In the case of positive-electron radiation, one of the protons changes into a

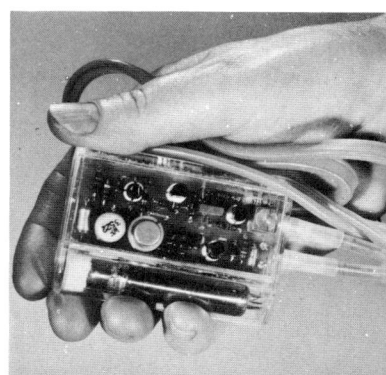

An isotope-powered battery for a cardiac pacemaker implanted in the body. The battery unit has a design life of 10 years, which is substantially greater than that of a conventional battery. Thus replacement surgery is required much less often.

Isotopes are widely used in industry. The picture shows an operator taking a gamma-ray photograph of a welded seam in 3 inch thick steel. The photograph will show up any flaws in the welding.

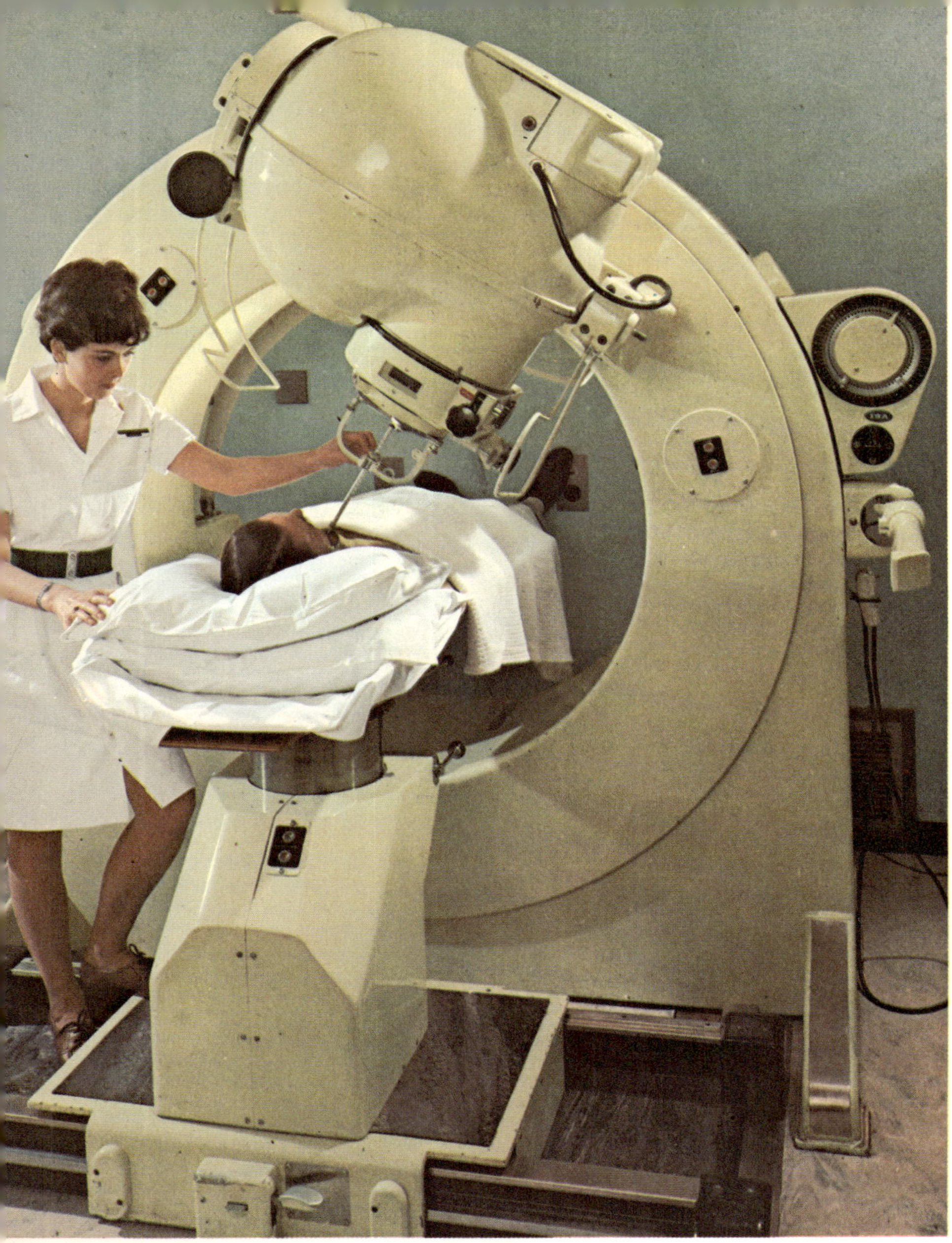

Above: An airfield radio marker beacon in the Outer Hebrides operated by an isotope-powered generator. Termed Ripple (radioisotope powered prolonged life equipment), it is capable of supplying 4 watts electrical power for at least 5 years. Left: Treating a patient in a radiotherapy unit with radiation from radioisotope cobalt-60. Tumerous conditions can often be successfully treated by radiotherapy.

neutron and ejects a positron. Again in both cases the number of protons in the nucleus has changed, although the overall number of particles remains the same. Thus beta-radiation, too, results in the atom changing into the atom of another element.

Gamma-radiation, however, has no effect on the number of particles present in the nucleus. Thus no transmutation takes place. The emission of gamma-radiation merely carries off excess energy from the atom, thereby making it more stable.

What happens after the emission of alpha- or beta-radiation when the original atom has changed into another atom? The new atom may well be stable. But, on the other hand, it may also be radioactive itself. In which case it will disintegrate by alpha- or beta-radiation. This goes on with atoms changing into other atoms until a stable atom is reached. Then the radioactive process ceases. For example, uranium-238 emits an alpha-particle and becomes thorium-234, which in turn emits a beta-particle and becomes protactinium-234, and so on. After more disintegrations the atom turns into radium-226, and after even more into lead-206. And there the radioactive series ends because this is a stable isotope of lead.

The length of time it takes for a given radioactive isotope to change, or *decay* into a different isotope varies. The rate of decay is expressed in terms of the length of time it takes half of a given amount of the isotope to change. This is known as its *half-life*. Some radio-isotopes are long-lived and some are short-lived. The half-life of uranium-238 is about $4\frac{1}{2}$ thousand million years! The half-lives of some other radio-isotopes are only a few thousandths of a second.

### Changing the Elements

Scientists have at last achieved what the alchemists strove to do for centuries. They can bring about the 'transmutation of the elements', the changing of one element into another. Such transmutation occurs naturally as the result of radioactivity. But scientists do it by bombarding an element with atomic particles. The bombardment changes the internal structure of its atoms so that they become another kind of atom.

Lord Rutherford achieved this first artificial disintegration of an atom in 1919. He bombarded nitrogen gas with alpha-particles from the radioactive element uranium. When this happens, some of the nitrogen atoms (7 protons) capture an alpha-particle (two protons and two neutrons) and at the same time eject a proton. Thus, the atom is left with an extra proton, which means that it is now an atom of oxygen (8 protons), not nitrogen.

Similar changes can be brought about in other elements, too. Bombarding beryllium (4 protons, 5 neutrons) with alpha-particles produces carbon (6 protons, 6 neutrons), and a neutron is ejected. Such reactions are a valuable source of neutrons, which can themselves be used to bombard elements. Neutrons are very effective at bringing about nuclear disintegrations because they have no electric charge.

There are good reasons for transmuting the elements in this way. It provides us with an insight into the structure and organization of the nuclei of the atoms. It enables us to produce artificial radioactive isotopes which we can use in many ways. Also it allows us to build up elements with heavier atoms than uranium that do not exist on Earth at all. Neutron bombardment of uranium will produce neptunium, which decays into plutonium. Neutron bombardment of plutonium produces americium, and so on.

### Atom-Smashers

Using the radiation from natural radioactive materials such as uranium and radium is obviously very limiting. Natural radioactive elements are scarce, and the speed and energy of the particles cannot be varied. We would be in a much better position if we could produce our own particles and accelerate them to whatever speed and energy we require. And we can in fact do just that in machines popularly called 'atom-smashers', although their proper name is *particle accelerators*.

The principle of the accelerators is simple really: produce charged particles – protons (+ve), alpha-particles (+ve), electrons (–ve) – and apply a high voltage to them so that they are electrically repelled.

The early atom-smashers were called *linear* accelerators and this type is often used today. The charged particles are accelerated down a long, straight tube. The most powerful machines however, are *circular* accelerators, which are known by such names as cyclotron, betatron and synchrocyclotron. The principle of these doughnut-shaped accelerators is to make the particles travel round and round in a circular path. A high-frequency electric voltage is applied to the machine so that the particles get a 'kick' twice every revolution.

A Van de Graaf electrostatic generator. These machines are used to generate voltages of several million volts for powering atom-smashing machines, or particle accelerators.

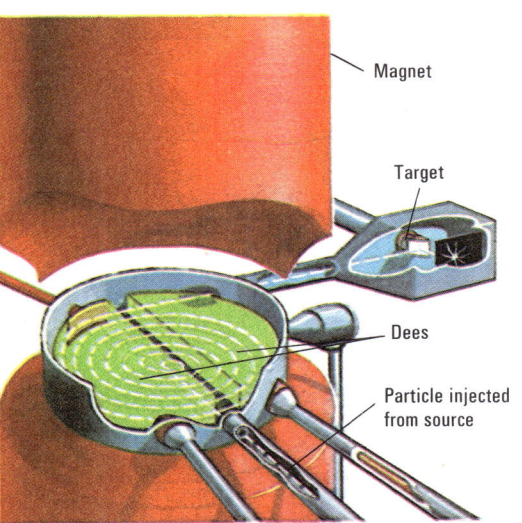

The principle of the cyclotron. It consists of two D-shaped electrodes (dees) which are connected to a rapidly alternating electric voltage. The dees are situated between the poles of a powerful magnet. A charged particle is injected into the dees and is constrained by the magnetic field to move in a circle. The electrodes are made alternately positive and negative at the appropriate times to repel the injected particle as it circles through the dees, hence accelerating it.

# Pioneers of the Atomic Age

### Antoine Henri Becquerel

Some substances, when they have been exposed to bright sunlight, glow in the dark for some time afterwards. This effect is called *phosphorescence*. In 1895 a Frenchman named Henri Poincaré, who later became famous as a brilliant mathematician, suggested that phosphorescent substances might give off X-rays. X-rays had only just been discovered. In 1896 a friend and fellow-countryman of Poincaré's, Antoine Henri Becquerel (1852-1908), decided to put the matter to the test. He exposed crystals of a uranium compound to sunlight to make them phosphorescent and placed them on thickly wrapped photographic plates. On developing the plates he found that the plates had been blackened by rays from the uranium compounds. It looked as if Poincaré had been correct.

A few days afterwards Becquerel happened to be developing some old photographic plates to see if they were still good enough to use. The plates had been kept in a drawer in which there were traces of the uranium compounds. He found to his astonishment that the plates had been blackened by the uranium, even though they had been kept in the dark. The blackening had nothing to do with phosphorescence after all. Becquerel had discovered what came to be called *radioactivity*. He found that the uranium ore pitchblende was more intensely radioactive than even uranium itself.

### The Curies

Becquerel was friendly with Pierre Curie, a lecturer in physics at the university of the Sorbonne in Paris, and his wife Marie, who was a chemist. He mentioned his discovery regarding pitchblende to them, and they resolved to try to isolate the unknown, powerfully radioactive element that must be present in the ore.

The Curies were fortunate in obtaining from the Austrian government loads of pitchblende 'waste', from which the uranium had already been extracted. They sweated and slaved purifying and separating the ore in an old wood shed in the Sorbonne which they used as their laboratory. In July 1898 they announced that there were not one, but two unknown radioactive elements in pitchblende. They isolated one, which Madame Curie named polonium, after Poland, where she was born. They did not succeed in isolating the second one, which they called radium, until 1902.

For their pioneering work on radioactivity, Becquerel and the Curies were jointly awarded the 1903 Nobel physics prize. The radium the Curies had prepared so far was in the form of a compound. They therefore set about producing pure radium. Pierre, however, was tragically killed in a street accident in 1906, and Marie continued alone. Breaking with all tradition, the Sorbonne offered Marie Pierre's position as professor of physics, which she accepted. In 1910 Marie succeeded at last in preparing a minute amount of pure radium. This brought her a second Nobel prize in 1911. Soon afterwards she helped to found, and became the first director of the Radium Institute in Paris.

Marie Curie died of leukaemia, a disease of the white blood cells. It has since been shown that radiation can cause leukaemia, so it is more than likely that Madame Curie died as a result of excessive exposure to the intense radiation from the elements she had discovered.

### Lord Rutherford

After Becquerel's discovery of radioactivity scientists began to investigate the nature of the radiation, or rays from uranium. Among them was Ernest Rutherford (1871-1937), a New Zealand-born physicist. At the turn of the century he was professor of physics at McGill University in Canada. There he discovered and named two kinds of rays — alpha-rays and beta-rays.

Assisted by the British physicist Frederick Soddy (1877-1956), Rutherford developed an explanation of radioactivity. This he published in 1902 as the 'theory of atomic transmutation'. The atoms of radioactive elements break down from time to time and give off alpha- and beta-rays. In so doing they change into atoms of other chemical elements. For this theory Rutherford won the 1908 Nobel chemistry prize.

Three years afterwards Rutherford proposed his revolutionary new theory concerning the structure of matter on which present atomic theory is based (see page 26). The Danish physicist Niels Bohr (1885-1962) later developed Rutherford's theory in a more detailed and quantitative way.

Rutherford followed J. J. Thomson as director of the famous Cavendish Laboratory at Cambridge in 1919. By then he had succeeded in artificially changing, or transmuting one element into another. He bombarded nitrogen gas with the alpha-rays from a piece of radium. A new kind of atom was produced, oxygen, and particles were given off, which proved to be charged atoms of hydrogen. In subsequent experiments he bombarded other elements and always found that charged hydrogen atoms were produced. He thus concluded that these particles, which he called *protons*, were basic to all atoms.

Because of these fundamental discoveries about the nature of the atom, Rutherford has been called the 'father of nuclear science'. His period as director of the Cavendish Laboratory was one of the most spectacular in its history. In 1932 James Chadwick (b. 1891) discovered the neutron, the other main fundamental nuclear particle, and John Douglas Cockcroft (b. 1897) and Ernest Thomas Sinton Walton (b. 1903) split the atom using artificially accelerated particles.

Madame Curie pictured in her laboratory.

Lord Rutherford

Albert Einstein

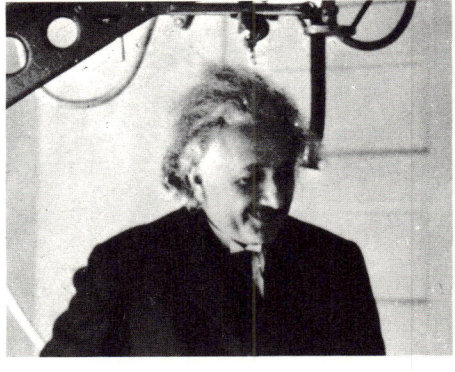

### Max Planck

Until 1900 everyone has assumed that light, heat, and other kinds of radiation are emitted in a continuous 'stream'. However, in that year the German physicist Max Planck (1858-1947) introduced his *quantum theory*, which revolutionized scientific thinking. He showed that radiation is given off in the form of tiny 'packets' of energy which he called *quanta* (singular *quantum*), just like bullets being fired from a machine-gun. Albert Einstein extended Planck's theory to explain the photo-electric effect. Niels Bohr applied it to the theory of atomic structure. Planck received the 1918 Nobel physics prize.

### Albert Einstein

Albert Einstein (1879-1955) published his explanation of the photo-electric effect in 1905 for which he received the 1921 Nobel physics prize. In 1905, too, he published his revolutionary *special theory of relativity*. By this theory he demonstrated, among other things, that mass is not an invariable, separate quantity after all. It is related to energy. The relationship is given by his famous equation: $E = mc^2$, where $E$ is energy, $m$ is mass, and $c$ is the velocity of light.

This equation forms the basis of the theory behind the atomic bomb (see opposite). Einstein, who was an extraordinary kind and gentle man, was immensely saddened that his work on the equivalence of matter and energy should lead to the development of such a terrible weapon of mass destruction.

Einstein also pointed out that time and space are related, too. In trying to fix a point, say, in the sky we must specify not only its longitude and latitude and height above ground (the 'three dimensions'), but also the time at which it is there. For this reason, time is sometimes called the *fourth dimension*. However, Einstein's theory of relativity is really far too complicated to explain in simple terms.

Einstein was born in Ulm in Germany of German-Jewish parents. In 1933 he was virtually outlawed by the Nazis and spent the remainder of his life in the United States, as director of the Institute of Advanced Study at Princeton, New Jersey.

# Atomic Energy

## Fission

We can change one element into another by bombarding it with atomic particles such as alpha-particles, electrons, protons, and neutrons. Normally when an atom is bombarded, the nucleus absorbs the bombarding particle and ejects another different particle. But the bombardment of uranium with neutrons brings about a different and particularly interesting kind of nuclear disintegration. When the nucleus of the uranium atom absorbs a neutron, it becomes completely unstable and breaks into two fragments. This splitting process is known as nuclear *fission*.

When fission occurs, a tremendous amount of energy is released. Heat and powerful radiation are given out and two or more high-energy neutrons are emitted.

The total mass of the fragments and the neutrons produced during fission is slightly less than the original mass of the bombarded uranium atom and the neutron which hit it. In other words some mass or matter has been 'destroyed' during fission.

The destruction of matter is, of course, contrary to the law of conservation of matter, which says that matter can be neither created nor destroyed. But the German genius Albert Einstein demonstrated that mass and energy are equivalent (see opposite). He summed this up in his famous equation $E = mc^2$, where $E$ is the energy, $m$ is the mass, and $c$ is the velocity of light. The velocity of light is very great – 186,000 miles a second, or 300,000 kilometers a second. In any units, therefore, $c^2$ will be a very big quantity. In other words the energy $E$ equivalent to even a tiny quantity of mass $m$ will be enormous. This explains the tremendous energy released in fission, produced by the conversion of the 'lost' mass.

Natural uranium is made up mainly of the isotope uranium-238, together with a tiny amount of uranium-235. U-235 is *fissile*, or able to undergo fission, whereas U-238 is not, though it will absorb high-energy neutrons. Slow neutrons will split U-235. A few other heavy elements besides uranium are potential fissile materials. They include thorium and protactinium. The isotope plutonium-239, an artificial element produced by bombarding uranium, is fissile, too.

## The Chain Reaction

We have seen that the fission of a uranium atom as a result of neutron bombardment produces two or more high-energy neutrons. Under suitable conditions these neutrons will themselves split other atoms, releasing more neutrons, which in turn will split other atoms, releasing still more neutrons. Thus, a tremendous *chain reaction* occurs, with more and more atoms being split as more and more neutrons are released. In view of what has been said about the energy released during fission, you can imagine what a build-up in energy occurs as a result of such a chain reaction. In fact the fission of

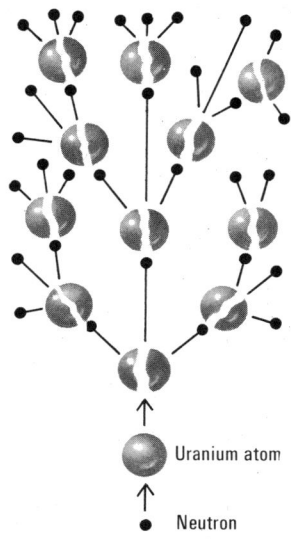

Chain reaction with uranium atoms. Each splitting, or fission of a uranium atom by a neutron produces three more neutrons, which can themselves cause fission in other atoms.

The gigantic fireball developing after the explosion of a hydrogen bomb. Hydrogen bombs can be made that have the explosive force of more than 50 million tons of TNT. The hydrogen bomb works by nuclear fusion— heavy hydrogen nuclei are made to combine, or fuse together to form helium. Since the reaction can take place only at extremely high temperatures, an atomic, fission bomb is used as a triggering device.

## The A-Bomb

Uranium-235 and plutonium-239 are the only suitable materials for an uncontrolled chain reaction because they undergo fission so readily. The principle of the A-bomb is, unfortunately, very simple: bring together enough U-235 or Pu-239 to make a critical mass.

In one type of A-bomb a wedge of the fissile material is forced by a conventional high-explosive into another lump of the fissile material, the two parts together having the required critical mass. The chain reaction begins, and in a millionth of a second the whole mass has undergone fission, releasing incredible energy. In another type of bomb, a hollow sphere of fissile material is blasted inwards to form a critical mass. Atomic bombs are used now as devices to 'trigger off' the even more powerful *hydrogen bombs,* which work on the different principle of fusion. The H-bombs imitate the energy-producing process going on inside the Sun (see page 180).

The atomic bomb utilizes the principle of nuclear fission. An uncontrolled chain reaction results in the release of enormous amounts of energy in the form of heat, light, blast and radiation. The photograph left shows the complete devastation of the Japanese city of Hiroshima after the explosion of an atomic bomb with an explosive power of 20,000 tons of TNT.

1 lb of fissile material produces as much heat as is released by burning 3,000,000 lb of coal!

The chain reaction will occur 'under suitable conditions'. If a lump of fissile material is too small, too many of the neutrons produced by fission are able to escape, and no chain reaction occurs. But as the size of the lump is increased, fewer neutrons escape. Eventually, at a so-called *critical mass,* sufficient neutrons remain to support a chain reaction.

Scientists utilize the energy released by the chain reaction in two ways. The first is for destruction in the shape of the atomic bomb, whose explosive violence results from an *uncontrolled* chain reaction. The second is for peaceful purposes in the shape of the nuclear reactor, which produces heat from a *controlled* chain reaction.

A nuclear reactor utilizes a controlled chain reaction so as to liberate the energy of nuclear fission gradually. The diagram shows a typical reactor set-up. Fission takes place in the reactor core, at a rate determined by the position of the control rods. Coolant circulates through the reactor core to extract the heat developed. It gives up its heat in the heat exchanger before recirculating. The heat turns water to steam, which is then fed to a conventional turbine. Extensive concrete and steel shielding surrounds the core to prevent the leakage of deadly radiation.

## Nuclear Reactors

Nuclear reactors are designed to allow a chain reaction to occur at a controlled rate. A critical mass of nuclear 'fuel' is brought together, only so-called *control* rods are inserted into it. These rods are made of materials like the metal cadmium which absorb neutrons. When the control rods are fully home, they absorb so many neutrons that no chain reaction occurs. As they are withdrawn, there comes a point when enough neutrons remain to sustain a chain reaction.

Natural uranium (mainly U-238), or uranium enriched with either U-235 or Pu-239 may be used in reactors. Reactors using natural uranium have to incorporate what is called a *moderator,* such as carbon or heavy water (water formed from the heavy isotope of hydrogen called deuterium). The function of the moderator is to slow down the neutrons produced by the fission of U-235. Normally they would be absorbed by U-238, which would prevent a chain reaction occurring. But U-238 does not absorb slow neutrons so readily.

Nuclear reactors used to produce heat for power purposes have a coolant circulating through them. This gives up its heat in a heat exchanger to turn water into steam, which is then used to drive turbo-generators in the normal way (see page 85).

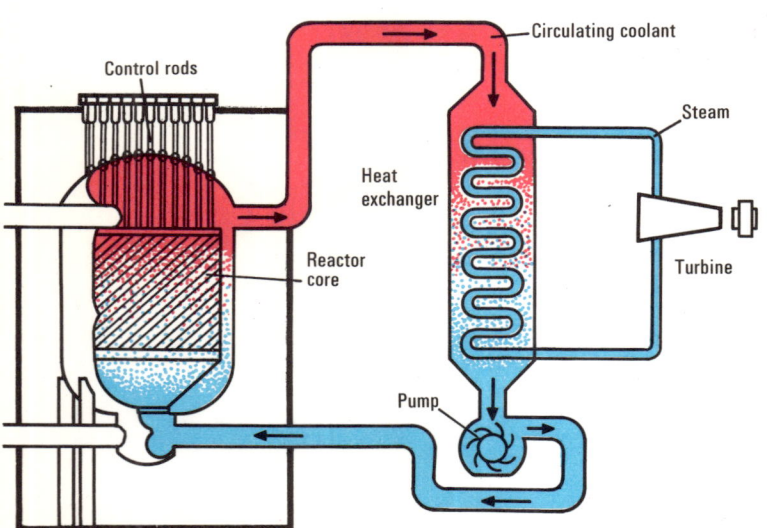

The fast reactor at Dounreay in Scotland. It is a so-called breeder reactor for it 'breeds' new fuel as it is operating. Fast reactors have no moderator and hence must use enriched uranium fuel.

# Common Plastics

**Acrylics** There are quite a few acrylic plastics, but only two types are well-known. One type we know better as the transparent, glass-like plastic, Perspex. The other type is the one from which the acrylic synthetic fibers are made. Both types are thermoplastics. Acrylic fibers appear under trade names such as Acrilan, Courtelle, and Orlon. They are suitable for all kinds of clothing, from underwear to sweaters, and for most other fabrics, too. Such fabrics are light and very soft and warm.

**Bakelite** was the first plastic that was made entirely from chemicals. We probably know it best as the black, heat-proof handles on saucepans and electric irons. But it has very many other uses, from ash-trays and electric-light fittings to heat-proof laminates for kitchen working surfaces. Despite its dark color, its use is widespread because it is so cheap. Bakelite is named after Dr. Leo Henrik Baekeland, the Belgian-born chemist who first made it in 1909. It is made from phenol and a solution of the gas formaldehyde. Both of these substances are plentiful and cheap. Phenol, or carbolic acid, is obtained from coal tar. (See also Melamine-formaldehyde and Urea-formaldehyde.)

**Casein** Milk is one of Nature's natural foods. We drink it and make butter, cheese, and yoghurt from it. The chemist, however, makes plastics from it! The substance in milk that forms the plastic is called *casein*. Casein plastic is not as widely used today as in the past, but buckles, buttons, knitting needles and umbrella handles are some of the common objects still made from it.

**Celluloid** was the first man-made plastic. It has the great disadvantage that it burns very readily, and other plastics are now replacing it. Celluloid is still used, however, for making table-tennis balls. In World War I, it was used in the form of a solution, called *dope*, to coat the fabric on aircraft wings. It made the fabric more taut and more resistant to the weather.

Celluloid, as its name suggests, is made from cellulose. Cotton linters, a pure form of cellulose, is treated with a mixture of acids and made into cellulose nitrate, or *nitrocellulose*. As a plastic material, nitrocellulose by itself is too brittle. A substance called camphor is therefore added to it to plasticise it.

**Cellulose Acetate** There are two main forms—acetate and triacetate. Both can be made into fibers for weaving into cloth. We call the acetate form *acetate rayon*. We know the triacetate form better under the trade names Tricel and Arnel. Another widespread use of cellulose acetate is in the form of a crystal-clear film. Vast amounts are used to make photographic film. Cellulose acetate does not burn as readily as the celluloid it replaced. The clear 'windows' in display cartons for all kinds of goods are made of cellulose acetate film. In solid form cellulose acetate is tough and has a high melting point. One of its most obvious uses is for the handles of cutlery and toothbrushes. It is also used for moulding toys. Cellulose aceto-butyrate is a modified form of cellulose acetate which has greater strength and heat resistance than the acetate alone.

**Epoxy Resins** are a kind of thermosetting plastic. They can be dissolved in solvents to make very powerful adhesives, which will join practically any two surfaces together permanently, even metals. They are mostly used in industry, but you can buy them at 'do-it-yourself' shops for use in the home. They come in two tubes. One contains the resin solution, the other a chemical to make the resin set hard. You mix them just before use. Epoxy resins are also used in lacquers for coating the insides of cans.

**Lucite** is an acrylic plastic that was developed in the U.S. by Du Pont shortly before World War II. Its features include optical clarity and resistance to shattering and sunlight degradation—all important properties in such applications as lenses, aircraft canopies and windows, automotive parts, glazing material, and even jewelry. As an acrylic resin it exists today as a popular housepaint.

Its proper name is *methyl methacrylate monomer*.

**Melamine–Formaldehyde (M.F.)** plastics are thermosets which are widely used for making high-quality plastic tableware, such as Melaware. The properties of such articles are similar to those of articles made of Bakelite. But articles in melamine-formaldehyde can be made in very much cleaner, brighter colors.

**Nylon** is a thermoplastic which is perhaps best known in the form of fibre. It was the first synthetic fibre and is still the most important. It is used in all kinds of clothing, from stockings and shirts to anoraks and furs. Like all synthetic fibres, nylon resists creasing and, being waterproof, 'drip dries' quickly. A great deal of nylon is used in making carpets. It is ideal for this purpose because it is hardwearing and resists bending. Actually, there is more than one kind of nylon, but they all have similar properties. They are all *polyamides*.

Nylon is exceptionally strong and rot-proof. These properties make it ideal for fishing nets, which have got to take the strain of a heavy catch and resist the rotting action of the sea water. For these nets nylon is used in the form of a thick, solid thread. In this form, too, it is used for stringing tennis rackets.

Nylon is also widely used in the form of a solid plastic. Many machines, including home food-mixers and beaters, have nylon gears and bearings. Moving nylon parts need no oiling because they slide over each other easily. They are also much quieter than moving metal parts.

**Perspex** One of the commonest uses for this crystal-clear, tough thermoplastic is for unbreakable watch glasses. But it is used for contact lenses, dentures, and artificial eyes.

Being hard, transparent, and tough,

Left: Polyvinyl chloride being used for protective clothing at the Royal Botanic Gardens, Kew. The suit is air-conditioned.

Right: The effectiveness of silicone treatment of fabrics for water repellency is clearly shown here.

Below: The launching of the Royal Navy's first fiberglass boat, HMS Wilton. Fiberglass reinforced polyester is being used increasingly as a moulding material.

Perspex has many industrial uses: safety goggles, machine safety guards, corrugated (ridged) roofing sheets, and so on.

Another reason why Perspex is so widely used is that it can be very easily moulded into shape. Unlike other plastics, Perspex is made directly into sheet during manufacture. Its proper name is *polymethyl methacrylate*, or *PMMA*.

**Polyesters** We know the polyesters best in the form of synthetic fibers. Terylene, Dacron, and Fortel are well-known trade names of polyester fibers. They are used to make most kinds of textiles. Such fabrics keep their shape well because the fibers do not stretch. Polyesters in this form are thermoplastics, like all synthetic fibers.

Some polyesters, however, can be treated with certain chemicals to make them into a so-called resin that can be heat set. This thermosetting resin is used with glass fibers to make car bodies, boat hulls, fishing rods, suitcases, and many other domestic and industrial products. Most of these products are made by laminating. Alternate layers of fiberglass and resin are applied to a mould and built up to the required thickness. Heating sets the laminate hard.

**Polypropylene** is very similar to high-density polythene. It is often difficult to distinguish between the two plastics. Both are so light that they float on water. But polypropylene has a higher softening point and is more shiny than polythene.

**Polystyrene** is one of the commonest plastics. We find the most familiar examples of polystyrene in the kitchen. Pure polystyrene is crystal-clear and ideal for see-through storage containers and measuring jugs. But it is very brittle. Its most distinctive feature is the metallic ring it makes when it is dropped. Other plastics land with a muffled thud. Inside the refrigerator we find another form of polystyrene. The inner lining is made of polystyrene toughened with synthetic rubber. It is not clear, but opaque.

Polystyrene is found in yet another, totally different form. Very lightweight, white ceiling tiles are made from it. In this form it is called *expanded* polystyrene. It is made by heating tiny polystyrene beads with a foaming agent. A gas is produced which 'fluffs up' the polystyrene into its expanded form. Expanded polystyrene is really a solid foam. The gas trapped inside makes it an excellent heat insulator. Expanded polystyrene is also widely used for packaging. Throw-away cups and pots for plant bulbs are other familiar uses.

Polystyrene is one of our cheapest and most useful plastics. Being a thermoplastic, it is very easy to shape. It is made by polymerizing styrene, a chemical readily obtainable from petroleum. Styrene is also an important ingredient in synthetic rubber.

**Polytetrafluorethylene**, or P.T.F.E. for short, is a plastic related to polythene. This is the coating on non-stick frying pans and saucepans. 'Teflon' is a familiar trade name for this plastic.

**Polythene** is probably the most familiar of all plastics. Washing-up bowls, 'squeeze' detergent bottles, and plastic bags are some of the most obvious examples of the use of polythene. Practically everything we buy these days, large or small, seems to be wrapped in a polythene bag: clothes, mattresses, furniture, toys, and, of course, food of all kinds. Polythene is so widely used in film form for packaging because it is cheap, transparent, tough, and flexible. It is not as transparent as the Cellophane used for wrapping chocolate boxes and cigarette packets, but it is very much tougher. Polythene bags keep food fresh because they are airtight.

In solid form, polythene appears in two forms. One is soft and flexible and is called *low-density* (light) polythene. The other is hard and rigid and is called *high-density* (heavy) polythene. They are made from the same raw material, ethylene, but under different conditions. Ethylene is a gas obtained by processing petroleum, or crude oil. When heated under pressure, the gas forms the plastic polyethylene, which is the correct chemical name for polythene. Polythene can be shaped easily because it is one of the thermoplastics.

Polythene in solid form looks very much like the paraffin wax of a candle, and the low-density form feels very waxy, too. This is not really surprising because chemically polythene and wax are very similar.

**Polyurethane** is the plastic which is widely used in 'foam' form for upholstery, car seats, draught-excluders, the lightweight, insulating linings for raincoats, and so on. The foam is formed by a reaction in which bubbles of gas become trapped in the setting plastic. Paints and varnishes based on polyurethane are extremely hard and resistant to scratches.

**PVC** The shiny, brightly coloured raincoats that have become fashionable are made from the plastic we know best by the initials PVC. The initials stand for *polyvinyl chloride*. PVC can be found practically everywhere and in many different forms. In flexible sheet form it is made into shower curtains and tablecloths. It is coated onto fabric to give the so-called *leathercloth* used for making seat covers for motor cars, handbags, dresses, and so on. Some shoes are moulded from PVC; the 'uppers' are embossed to look like real leather. Such shoes are cheap, absolutely watertight, and long-lasting. Washable floor tiles and plastic garden hoses are PVC, too. So is the plastic around electric wire and cables, because PVC is an excellent electrical insulator. 'Unbreakable', long-playing gramophone records are made from PVC. So are tapes for tape recorders. PVC in rigid form is found in guttering and piping.

By itself, PVC is far too stiff and brittle a plastic to be very useful. It is difficult to shape and work. But the addition of a plasticiser makes it soften more easily for shaping and increases its strength in use.

PVC may often be confused with polythene and polypropylene because they are all used for making much the same things. One sure way of telling is to drop the object in question into water. If it sinks, it is PVC, which is heavier than water. Polythene and polypropylene are lighter than water and therefore float.

**Silicones** are very interesting plastic materials. You have probably come across the word 'silicone' in connection with furniture and car polish. Silicone polishes give an excellent shine and a very hard finish. Any water spilt on the polished surface just rolls off. Raincoats are treated with silicones, too, to make them completely waterproof. For both of these uses the silicones are in the form of fluids. In a resin-like form, they are used in heat-resisting paints. They can also be made into a kind of synthetic rubber. Small silicone-rubber balls have become popular toys. They perform the most amazing 'tricks' when bounced.

In the fluid form, silicones are not really plastics at all, because obviously they cannot be shaped. But they do have the long-chain molecules typical of plastics. Silicones are different from all other kinds of plastics because their long-chain structure is built up in a totally different way. It is based on linkages between silicon and oxygen.

**Urea–Formaldehyde (U.F.)** plastics are made in much the same way as Bakelite and have similar properties. They are thermosetting and are heat-resistant. Their advantage over Bakelite is that they are naturally colorless, and can therefore be colored. Electric light fittings and bottle tops are among their most obvious uses.

**PLASTICS IN THE HOME**

# Plastics Technology

An incredible variety of plastic products are available to us today for use in the home and also in industry. At home you find most examples in the kitchen: vinyl tiles cover the floor; hard, heat-resistant Formica sheet covers the working surfaces and the kitchen table; the chairs are upholstered in a wipe-clean PVC leather cloth; the washing-up bowl and 'squeeze' bottles containing detergent are polythene; the linings of the refrigerator is polystyrene; and so on. Outside the kitchen you will find many products made from, or incorporating nylon: drip-dry clothing, carpets, curtain runners, and so on. Practically all electric-light switches and fittings are either phenol-formaldehyde (Bakelite) or urea-formaldehyde plastic.

Those plastics mentioned above account for a great proportion of the plastics in use today. As you have seen for yourself, every plastic looks or feels different. Some are hard, some are soft; some you can see through, some you cannot; some resist heat, some melt easily. Although they may have different properties and be made from different ingredients, plastics do have certain things in common that characterize them as plastics. They are substances with long-chain molecules that can be moulded into shape when they are heated.

### Polymers

Just what do we mean by 'long-chain molecules'? You all know that every substance is made up of tiny units, or molecules. Most substances, like water, for example, have short molecules containing only a few atoms. But the substances we know as plastics have molecules that may contain millions of atoms linked together in the form of a long chain. Take the simplest and best-known plastic polythene, for example. It is made up of atoms of carbon (C) and hydrogen (H), which are linked together like this:

$$\cdots \begin{array}{c} H\ H\ H\ H\ H\ H\ H\ H \\ |\ \ |\ \ |\ \ |\ \ |\ \ |\ \ |\ \ | \\ -C-C-C-C-C-C-C-C- \\ |\ \ |\ \ |\ \ |\ \ |\ \ |\ \ |\ \ | \\ H\ H\ H\ H\ H\ H\ H\ H \end{array} \cdots$$

You notice that the molecule is made up of many repeated units, consisting of two hydrogen atoms to one of carbon ($CH_2$). We call a substance with such long-chain molecules a *polymer*. 'Polymer' simply means 'many parts'. That is why the names of plastics often begin with *poly-*. The last part of the name indicates what the plastic is derived from. For example, 'polyethylene' (the correct name for polythene) indicates that the plastic was made from the chemical ethylene. Polystyrene is made from styrene.

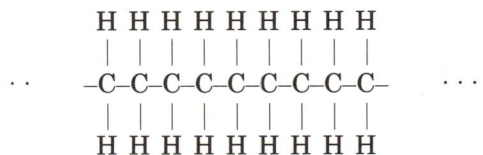

Polyvinyl chloride (PVC) is one of the commonest plastics. It was used to make these lifelike dolls. They were made by the process of slush moulding, which involves pouring a plastic mix into a hollow mould so that a thin skin forms on the inside. Heating completes the forming process.

Polytetrafluorethylene (PTFE) (Teflon) is a great boon to the housewife. As a coating on saucepans, casseroles and frying pans it prevents cooked foods sticking. The pans are cleaned simply by wiping with a wet sponge!

### Polymerization

Polymerization is a way of building up small molecules into the long-chain polymer typical of plastics. If we call the small molecules A, we can represent the polymer as -A-A-A-A-A-A-A-, and so on. The small molecule is called a *monomer* (one part). Polythene, for example, is made by polymerizing the gas ethylene. The monomer is ethylene.

In some plastics the polymer is built up from different monomers. If we call one A and the other B, we can represent the polymer as -A-B-A-B-A-B-A-, and so on. This kind of polymer is called a *copolymer*. Styrene-butadiene synthetic rubber is an example of a copolymer. It has styrene and butadiene as monomers.

The actual methods of polymerization vary enormously, but they generally involve heating the monomers, often under pressure and with a catalyst.

Opposite is an extrusion machine developed to make tubular polythene film. The molten polythene is extruded as a thin pipe, and then air is blown into it to expand it into a huge bubble.

Practically all the plastics are made up, like polythene, of long molecules that have a 'backbone' of carbon atoms. This is because of all the chemical elements only carbon can link with itself in this way. In the silicone plastics, however, the long-chain structure is built up in a different way by means of links between atoms of oxygen and silicon. This is not too surprising because, chemically, carbon and silicon are closely related.

### 'Natural' Plastics

Our own bodies are made up of complicated molecules based upon linked carbon atoms. And so are those of all living things, including plants. It is not surprising, therefore, that we can obtain or make plastic substances from some plants. The best example of this is the latex we tap from the rubber tree and make into rubber (see page 66). Other natural plastics include rosin and shellac. Rosin is made from the sticky gum that oozes out of pine trees. Shellac is made from lac, a resinous substance that the lac insect of India and Burma exudes from its body when it has fed off the sap of certain trees.

But far more valuable than those few natural plastics are the plastics we can make from the cellulose of plants. The most important ones are *celluloid* and *cellulose acetate*. The starting point for these plastics may be either cotton linters or wood pulp.

### Synthetic Plastics

Most of our plastics, however, like nylon and polythene, are *synthetic;* that is, they are made wholly from chemicals. They are manufactured by a process called *polymerization*. The most useful raw materials for making synthetic plastics are crude oil, or petroleum, and coal. They both yield a wide range of valuable chemicals, although coal is a much less important source than it once was.

Many plastics can be softened again by heating after they have been moulded. They are known as *thermoplastics,* or *thermosoftening* plastics. Others remain rigid on reheating. They are called *thermosetting* plastics. The difference between these two kinds of plastics lies in their structure. A thermoplastic is made up of long molecules that are not linked to one another. When it is heated, its molecules can move about in relation to one another, and the plastic softens and eventually melts. A thermosetting plastic on the other hand has long molecules that are chemically cross-linked with one another. They cannot move about when heated, and the plastic remains rigid.

Thermoplastics are by far the most common of the plastics. Polythene, nylon, PVC, Perspex, and polystyrene are all thermoplastics. They can be very easily shaped in the softened state. The best-known examples of thermosetting plastics are Bakelite and the other formaldehyde plastics. By their very nature they are difficult to shape.

There are a large number of ways of shaping plastics. The most important are moulding, extrusion, calendering, laminating and welding.

### Moulding

Moulding is probably the commonest way of shaping plastics. Bowls, 'squeeze' bottles, cups and saucers are all moulded, but in different ways. Bowls are made by injection moulding, squeeze bottles by blow moulding, and cups and saucers by compression moulding.

*Injection moulding* is suitable for a wide range of thermoplastics. The plastic is fed into an injection-moulding machine in the form of a powder or small granules. First it is melted in a heating chamber and then injected into a hollow mould by a plunger. Cold water circulates in the mould, and the plastic sets in the shape of the mould as it cools down. Then the two halves of the mould open, and the plastic object is removed. The mould comes together again, and the process is repeated. All kinds of objects can be made very quickly and very cheaply by this method.

*Blow moulding,* too, can only be used with thermoplastics. Air is blown into a blob of molten plastic inside a cool mould. The plastic expands and takes the shape of the mould. Besides bottles, large numbers of hollow toys are made by this method.

Another common way of making hollow objects from PVC is by *slush moulding*. A mixture of plastic and plasticiser called a plastisol is poured into a heated, hollow mould, which is then rotated. The plastisol starts to set when it touches the mould. When the desired thickness is reached, the excess plastisol is poured out. And the object is 'cooked' into its final plastic form.

*Compression moulding* is used to shape Bakelite and other formaldehyde plastics. Such thermosetting materials cannot be moulded by the other methods because they melt and set firm more or less at the same time. In compression moulding the thermosetting moulding powder is put in the bottom half of a heated mould. As the powder melts and begins to set, the upper half of the mould descends and forces it into shape.

Some plastics, such as PVC, can be moulded without heat. This so-called *cold*

Hollow mould

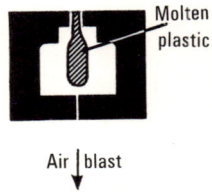
Molten plastic

Air blast

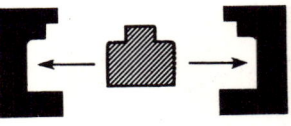

Stages in the blow moulding of a plastic bottle. Compressed air is forced into a sealed tube of molten plastic surrounded by a hollow mould. The plastic expands until it fills the mould and takes its shape. When cool, the bottle is ejected from the split mould.

Calendering is a kind of mangling process used for producing plastic sheet and film, especially PVC. Softened plastic is fed between a series of steam-heated rollers, which squeeze the plastic into thin film.

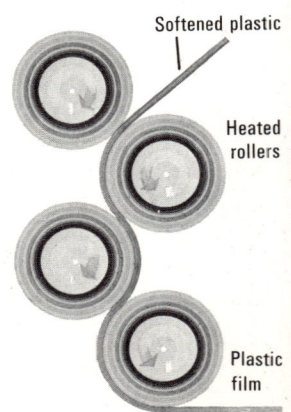
Softened plastic
Heated rollers
Plastic film

Extrusion is another common method of shaping plastics. A screw-like auger delivers plastic pellets to a heated chamber, where they melt. The molten plastic is then forced through the hole of a die. To make piping a plug is inserted in the die hole.

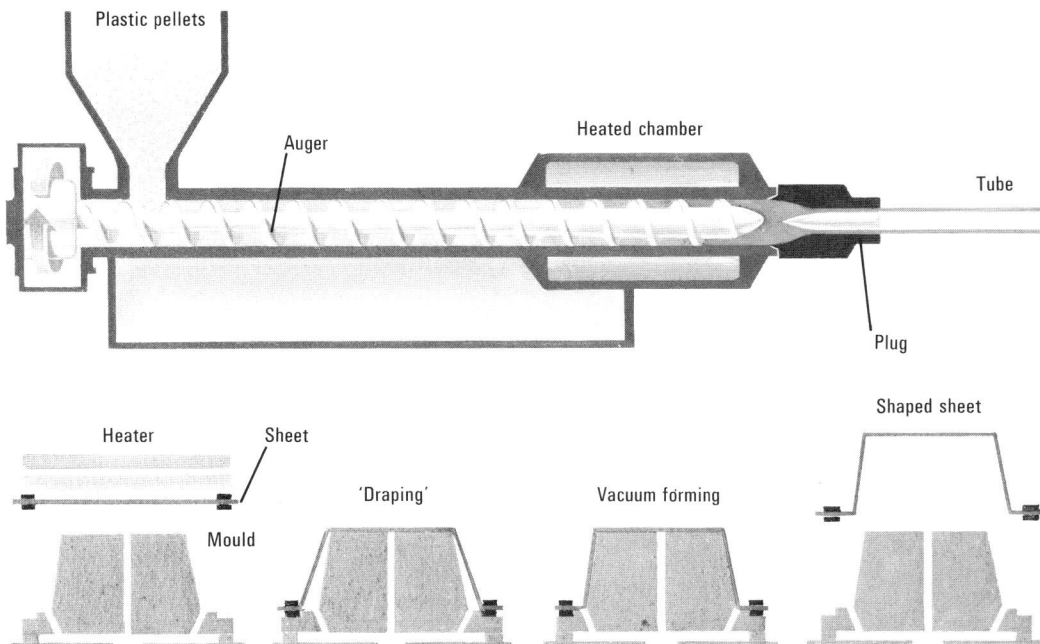

Thermoforming is a technique used to make egg-boxes for example. Plastic sheet is softened by heating and then pressed down over a mould. Vacuum is applied to complete the forming process.

Lamination is the method used to make the tough, heat-proof plastic board used for kitchen work-tops. Several layers of synthetic-resin impregnated paper are sandwiched together in a heated press.

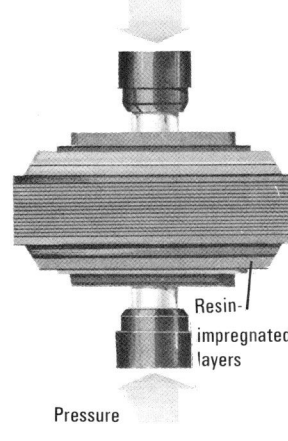

*casting* is one of the simplest moulding methods. The plastic is mixed cold with a plasticiser and placed in a mould. The two react together to form a solid object in the shape of the mould. *Thermoforming* is a method of moulding objects from stiff plastic sheets. Plastic egg-boxes are good examples of this kind of product.

### Extrusion

Plastic rods, pipes, tubes, and film can be made by a process called *extrusion*. Molten plastic is forced through a hole in much the same way that toothpaste is squeezed from a tube. It is a method that can be used for the thermoplastics. Granules of plastic are fed into an extruder, which is a machine somewhat like a kitchen mincer, only much longer. The extruder is heated so that the granules melt. A screw device in the middle of the machine forces the molten plastic out through a shaped die. The plastic takes the shape of the die. To make rod, the die is a simple round hole. To make tubing, the plastic is forced between the die and a rod inside it. To make film, the die is a long slit.

### Calendering

Another way of making plastic sheet and film is *calendering,* which is a kind of rolling or mangling process. It is used for thermoplastics, especially PVC. Massive, steam-heated rollers squeeze softened plastic into a film. The surface may be given a decorative embossed pattern by the final roller. Calenders can also be used to coat other materials with plastic. Playing cards and fabrics can be coated in this way. A film of hot plastic is bonded firmly with the paper or fabric by the heavy pressure of the rollers.

### Laminating

The tough plastic sheet used for kitchen working tops, such as Formica, is made by a process called *laminating*. The sheet is called a *laminate*. Laminating is like making a sandwich. Several layers of material are stacked one on top of the other and pressed together. To make a plastic laminate, layers of cheap material such as paper and cloth are soaked in a solution of formaldehyde resin. They the layers are stacked together, heated, and compressed. The resin sets hard when heated, binding the cloth or paper layers into a rigid sheet.

Another form of laminate is used to make 'fiberglass' car bodies and boat hulls. It is made up of alternate layers of certain kinds of polyester resin and fiberglass. These materials can easily be applied to a shaped mould. No pressure is needed to set them, only heat and a substance called a catalyst, which helps the process along.

### Welding

Most kinds of intricately shaped objects can be produced in one operation by moulding. But there are some that must be made by joining several mouldings together. This may be done by *welding*. It may also be done simply by glueing.

Simple welding involves heating the edges of two pieces of plastic until they melt and then bringing them together. The molten plastics fuse together. A strong joint is formed when the parts cool. Usually molten plastic is added to the joint from a so-called filler rod. This is very similar to the welding of metals (see page 140). Welding is usually done with a gas torch, but only hot gas, not flame is allowed to touch the plastic.

# The Electromagnetic Spectrum

The Earth is a live, warm place because of the energy it receives from the Sun. The Sun obtains its energy by 'burning' hydrogen to make helium in its incredibly hot core (see page 181). It gives off this energy in many forms, but noticeably as light we can see but not feel and heat we can feel but not see. This is true but to a lesser extent for any other hot body as well, for example, a white-hot poker.

The Sun's light and heat travel to us through the empty depths of space in the form of waves. These waves are much like the ripples that spread in ever-increasing circles over the surface of a pond when you throw a stone into the water. In other words, the waves have crests and troughs and vibrate at right-angles to the direction they are travelling. They are called *transverse* waves. They are unlike sound waves which are longitudinal waves that vibrate along the direction of travel (see page 114).

Sound waves need a medium, such as air, to travel through, because the wave travels by means of the disturbances of the medium's molecules. Thus, sound waves cannot travel in space. Light and heat waves can, as we well know. They can because they consist of electric and magnetic vibrations, which are completely independent of the presence of any medium. (The magnetic field around a magnet is always there, no matter whether there is air present or not.) Light and heat, then, are electromagnetic in origin and are forms of what we call *electromagnetic radiation*.

It is reasonable to suppose that there might be other ways in which the Sun can radiate its energy besides light and heat. And we do in fact find that there are many other kinds of radiation. Light and heat, as it were, form the 'tip of the iceberg' with the other radiations remaining below the surface, or rather, beyond our ordinary powers of detection.

## The Visible Spectrum

We can gain an insight into the nature of electromagnetic radiation by passing a beam of white light through a glass prism. We find that the white light, on passing through the prism, is bent and split up into a 'rainbow' of color, which we call the *spectrum*, ranging from red, through yellow, green, and blue to violet. Thus white light is made up of a combination of colors. And we find that these colours cannot be further split up by passing through another prism. We call such light *monochromatic* (single-colored).

How do we interpret this so-called *dispersion* of color in terms of the wave theory? If electromagnetic radiation is a wave motion, we can describe it in terms of the *wave-*

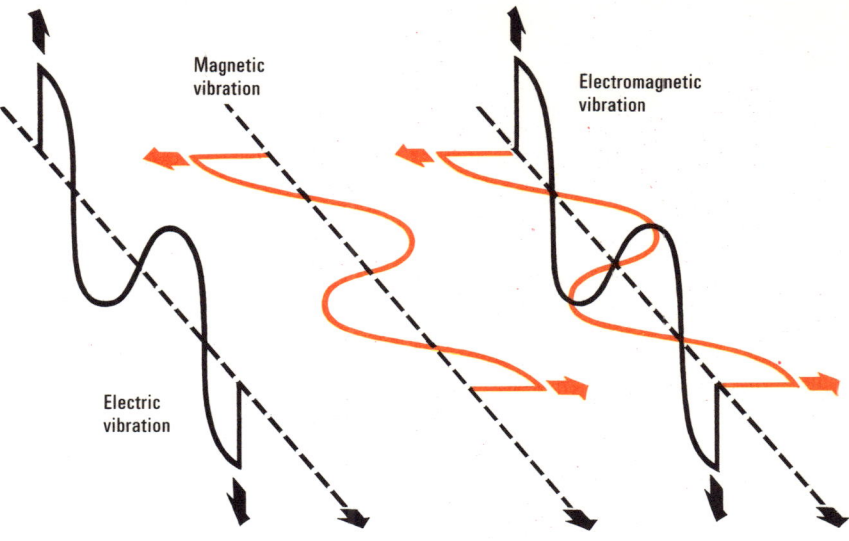

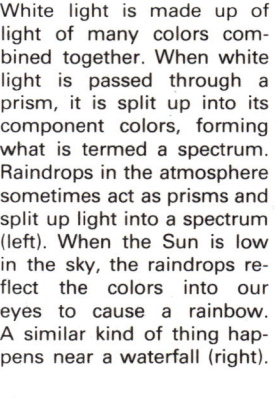

Light, heat, X-rays and radio waves are different forms of the same kind of thing. They are electromagnetic waves, which consist of electric and magnetic vibrations. The diagram gives an impression of how these waves are made up, with the electric and the magnetic vibrations being at right angles to one another.

White light is made up of light of many colors combined together. When white light is passed through a prism, it is split up into its component colors, forming what is termed a spectrum. Raindrops in the atmosphere sometimes act as prisms and split up light into a spectrum (left). When the Sun is low in the sky, the raindrops reflect the colors into our eyes to cause a rainbow. A similar kind of thing happens near a waterfall (right).

*length* and the number of waves passing a given point in a given time, say, a second (the *frequency*). The speed of motion is given by multiplying wavelength by frequency. Light and all electromagnetic radiations travel at the same speed which, in space, or a vacuum, is an incredible 186,000 miles a second ($3 \times 10^{10}$ centimetres a second).

Thus, related characteristics of a wave are wavelength, frequency, and speed. In a given medium the speed is always the same, but there is nothing to stop the wavelength and frequency changing. Thus you can have a small wavelength and proportionately high frequency or a large wavelength and a proportionately low frequency, always obeying the relationship that wavelength × frequency is constant (speed).

Thus, coming back to the many colored spectrum, we can reason that each color represents light of a different wavelength and frequency. And, being different, each color is affected differently during passage through the glass prism.

As light passes through a denser medium (such as glass), it is slowed down slightly, as we would expect, and is therefore bent. The light waves with the most vibrations, or shortest wavelengths are obviously going to be more affected, that is, more bent than those with the least vibrations, or longest wavelengths. Thus, the prism separates light of different wavelengths, which emerge as a spread of colour. From the actual spectrum we can see that red light is deflected least; therefore it has the longest wavelength. And violet light is deflected most; therefore it has the shortest wavelength.

**The Invisible Spectrum**

The spread of color we have described forms the visible part of the electromagnetic spectrum. There are many invisible radiations beyond the red end of the spectrum, that is, with longer wavelengths; and there are many beyond the violet end, that is, with shorter wavelengths.

Let us deal with the shorter ones first. The short-wavelength radiation just beyond the violet is called *ultra-violet*. And it is one form of invisible radiation that we can detect. When ultra-violet light falls on some substances, they fluoresce, or glow. Ultra-violet light also has a strong effect on photographic film. It is the light in sunlight that tans our bodies brown.

Shorter than the ultra-violet rays are the *X-rays* which doctors use to photograph bones and body organs. They are very penetrating and pass through human tissue. But bones being denser, allow fewer rays through and therefore show up on photographs. In

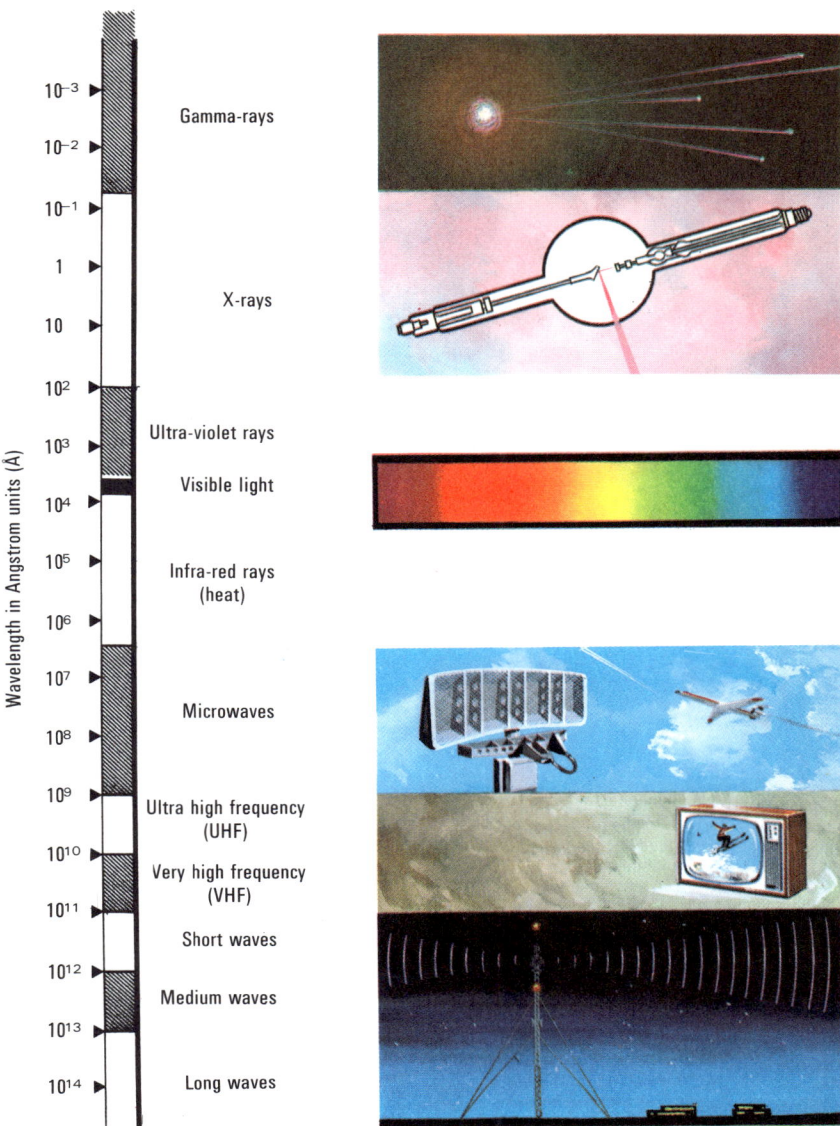

The shortest wavelength waves in the electromagnetic spectrum are gamma rays; the longest are radio waves. The diagram shows the distribution of the various radiations within the spectrum. Wavelengths are measured in terms of Angstrom units, 1 Angstrom unit being equal to a hundred-millionth of a centimetre ($10^{-8}$ cm).

large doses, X-rays can be dangerous. The same goes, but to a much greater extent, for the even shorter-wavelength rays called *gamma-rays*. These are the rays given off by many radioactive materials such as radium (see page 29). And they really are deadly.

What about the radiation beyond the red end of the visible spectrum? We find that the radiation immediately beyond the red is actually that which we feel as heat. We call it *infra-red*. Infra-red radiation is much used now for various forms of heat treatment in medicine.

Beyond the infra-red are the longer waves called *radio waves* that are employed in communications devices. First there are the *microwaves,* which are the ones used in radar equipment for detecting aircraft in the air, ships at sea, speeding motorists, and so on (see page 234). Then come the *ultra-high frequency* (UHF) and *very-high frequency* (VHF) waves that are used for television and high-quality radio transmission. Most of our radio programs, however, are transmitted by longer waves, which we know as *short-*, *medium-*, and *long-wave* radio waves.

# Properties of Light

## Reflection

Light travels in straight lines. We know this by common experience; the edges of the beam of a torch are straight; we cannot see round behind things; and illuminated objects throw shadows.

When light falls onto a surface, a variety of things may happen. It may be reflected, transmitted, or absorbed. In practice, a reflecting surface will also transmit and probably absorb some of the light falling on it. Likewise, a little light will be reflected and absorbed by an object that transmits light, and so on.

Surfaces that reflect light well are called *mirrors*. Most mirrors are made of glass which is silvered on the back to reflect the light. When you shine a beam of light at any angle onto the mirror, you always find that it leaves the mirror at exactly the same angle as it strikes it. In other words the 'angle of incidence is equal to the angle of reflection'. Mirrors have important uses not only in the home but in many optical instruments, including periscopes and telescopes.

## Refraction

An object, such as a pane of glass that allows light through is *transparent*. One that does not, such as a stone, is *opaque*. One that allows a little light through is *translucent*. A grease spot on a piece of paper is translucent, for example.

Light travels through a substance in a straight line, but when it enters another substance it is 'bent'. This happens when a light ray enters glass from the air, or leaves glass for the air. This bending of light at the interface between one medium and another is known as refraction. It occurs because light travels at slightly different speeds in different media. A measure of the extent of refraction in a substance is given by its refractive index ($\mu$), which is expressed by the relationship:

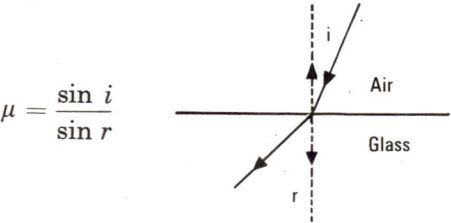

$$\mu = \frac{\sin i}{\sin r}$$

Where $i$ is the angle of incidence of a light beam and $r$ is the angle of refraction.

Light passing from a dense to a less dense medium is not always refracted. When the angle of incidence rises above a certain *critical angle*, the light is totally reflected. This effect is used to advantage in certain optical instruments, such as the prismatic compass and prismatic binoculars, which incorporate totally reflecting prisms.

You are probably quite familiar with one kind of refraction – that between air and water. A swimming pool or any vessel full of water appears much shallower than it really is. A stick half in and half out of the water looks as if it is broken. Both of these phenomena are caused by the water bending the light rays. Refraction can occur in the atmosphere, too. The best-known example of this is a *mirage* in the desert. Light rays from, say, a distant pyramid are progressively refracted and then totally reflected by layers of air of different temperature and therefore densities. An observer sees what appears to be a pyramid standing in a shimmering lake of water.

The principles of refraction are utilized in the manufacture of glass lenses and prisms, with which all kinds of optical instruments are made – cameras, binoculars, telescopes, microscopes, and so on.

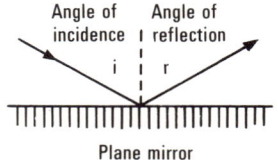

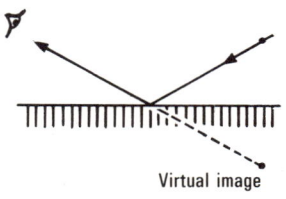

When light falls on a flat, or plane mirror at a certain angle, it is reflected from the mirror at the same angle (above). The image formed by a plane mirror is a virtual one (below). You cannot receive the image on a screen at a point where the image appears to be.

## Mirrors

The simplest mirror is the *plane* mirror, a sheet of plane glass silvered on one side. This is the ordinary mirror we have in the home. When you stand in front of the mirror, your image appears to be as far behind the mirror as you are in front. The light rays reflected into your eyes appear to come from that image. The image does not really exist,

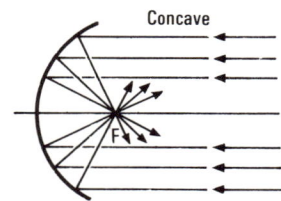
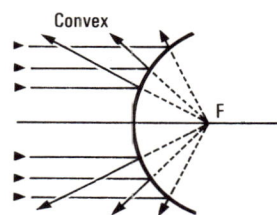

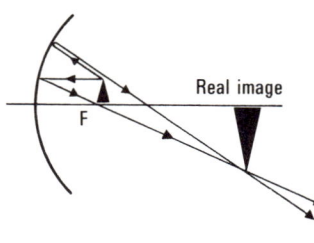
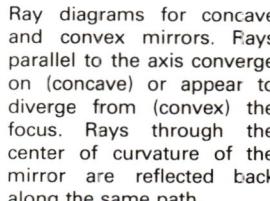

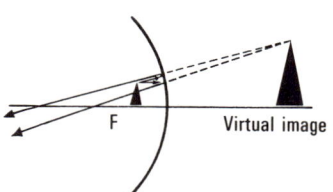
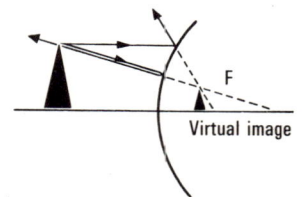

Ray diagrams for concave and convex mirrors. Rays parallel to the axis converge on (concave) or appear to diverge from (convex) the focus. Rays through the center of curvature of the mirror are reflected back along the same path.

however. You cannot form it on a screen placed behind the mirror. We call such an image a *virtual* image. Images that can be produced on a screen are called *real*.

Plane mirrors are the commonest but not the only kind of mirrors. *Curved* mirrors are also widely used. They are generally spherical, or parts of spheres and are either concave or convex. Concave mirrors curve away from you as you face them; convex mirrors curve towards you. Two important points on the axis of the mirror are the *center of curvature* and the *center of focus*, or *principal focus*. The center of curvature is the center of the sphere of which the mirror is a part. The principal focus is the point to which a parallel beam converges or from which it appears to diverge after being reflected.

In a concave mirror objects situated beyond the principal focus will produce real, but inverted images. Objects situated between the focus and the mirror produce magnified erect, but virtual images. Concave mirrors, usually in the shape of a parabola, are used as reflectors in searchlights and motor-vehicle headlamps. Concave mirrors are also used in the powerful astronomical reflecting telescopes to gather the feeble light rays from the stars.

In a convex mirror the images produced are always erect, reduced, and virtual. Convex mirrors are often used for the wing and rearview mirrors of motor vehicles because they have a bigger field of view than plane mirrors.

**Lenses**

Just as there are concave and convex mirrors, there are concave and convex lenses. Lenses are made with any combination of concave, convex, or plane surfaces. The ones in most common use are the double convex and the double concave. The difference between them can be demonstrated by their effect on a parallel beam of light. A convex lens converges the beam and brings it to a focus (the principal focus). A concave lens makes the parallel beam diverge, as if from a focal point (its principal focus).

The convex lens produces a real, inverted image when the object lies beyond the focus. When the object lies between the focus and a point twice the focal length away, the image will be magnified. When the object is between the focus and the lens, the image is erect, magnified, and virtual. This is the kind of image you see when you use a convex lens as a magnifying glass. The concave lens always produces a reduced and erect, virtual image.

The convex lens is by far the most useful of the two, and different combinations of convex lenses are used in microscopes and telescopes, for example. The idea behind both instruments is to use a convex lens (the objective) to produce a real, inverted image of an object. This real image is then viewed with another convex lens (the eyepiece), acting as a simple magnifying glass. Thus the final image will be magnified, virtual, and still inverted. This is alright for astronomical telescopes, but for terrestrial telescopes, an erecting lens is incorporated to produce an erect image.

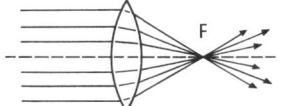
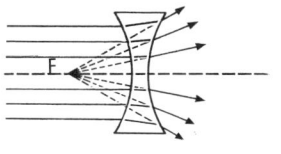
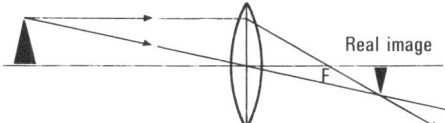
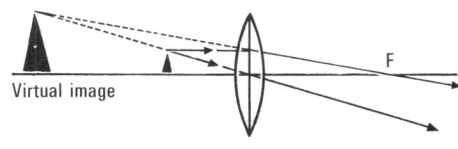
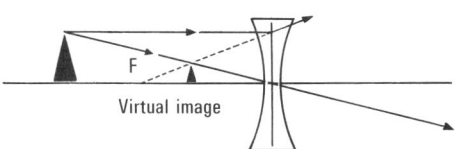

Simple ray diagrams for double-convex and double-concave lenses. Again, it is simple to construct the diagrams and find the position of the image by remembering two simple rules. Rays entering the lens parallel to the lens axis are refracted so that they converge on (double convex) or appear to diverge from (double concave) the focus of the lens. Rays passing through the center of the lens on the axis are not deflected.

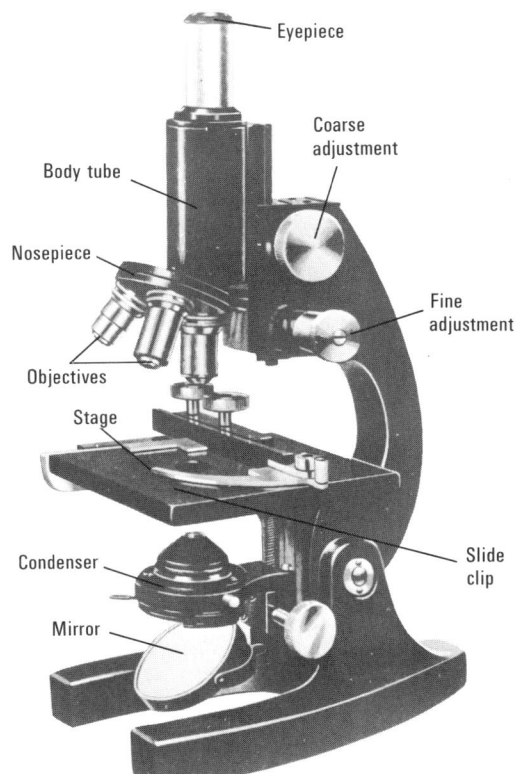
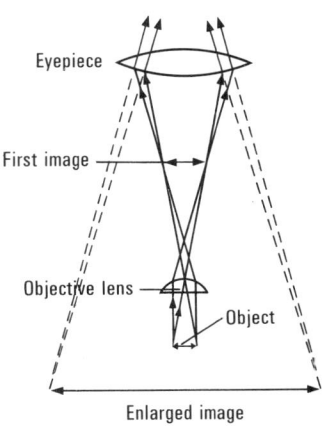

A suitable combination of lenses can produce greatly magnified images, a property used to advantage in such instruments as telescopes and microscopes. The picture left shows a common microscope set-up. The ray diagram above shows how the eyepiece magnifies the image of the object formed by the objective lens.

# Color

## Color in Light

We can demonstrate that white light is made up of many constituent colors by passing it through a prism to form a spectrum (see page 42). It is reasonable to suppose that if we can split white light up into separate colors, we can make white light by combining separate colors. And we find that we can do just that. To the eye the dominant colors in the spectrum are *blue*, *green*, and *red*. Combining these colors together in equal proportions, we can produce white light. We also find that we can make any other color in the spectrum by combining blue, green, and red in certain proportions. Because of their importance we call these colors the three *primary* colors.

A surface that appears blue to us does so because it absorbs all the other colors of the spectrum except blue, which it reflects.

## Color in Paint

Mixing different colored paints together is not the same as mixing different-colored lights. If we mix yellow paint and blue paint, the result is green. But if we mix yellow and blue light, the result is white. The difference lies in the way the color is produced. In light it is by addition, in paint by subtraction.

When ordinary white light falls on an opaque object, such as a paint pigment, the object absorbs some of the constituent colors. The rest are reflected, and combine

These two diagrams show the difference between color in light and color in paint. When you mix together the three primary colors in light, the result is white, left. When you mix together the three primary paint colors, the result is black, right. (Due to the limitations of color printing, the shades of color shown here are not entirely accurate.)

If we combine equal proportions of green and red, we form yellow. Then adding the right proportion of blue to yellow produces white again. We call yellow the *complementary* color of blue. Complementary colors are pairs of colors that produce white light when combined. The other primary colors have their complementary colors, too. The complement of green is a combination of the other primary colors blue and red, called *magenta*. And the complement of red is a combination of blue and green, called *cyan*. Of course, there are many other complementary colors, but the complementaries of the primary colors are the most important.

So far we have discussed the mixing of colors by *addition*. Colors can also be produced by *subtraction*. If we shine white light through a green filter, only green light comes out. What has happened is that the filter glass has absorbed, or subtracted from the white light all the other colors except green. Color by subtraction is the principle behind color photography (see page 51).

to form what we see as the color of the object. Thus, a pure blue pigment absorbs all colors except blue, which it reflects.

If all pigments were pure colors, mixing them together would yield only black. A pure yellow pigment would absorb all colors except yellow and a pure blue pigment all colors except blue. Mixing the two together would therefore produce a mixture that absorbs all colors. But in fact the pigments are not pure colors. Thus, yellow reflects a little red and green as well as yellow. And blue reflects a little green as well as blue. Mixing the two together therefore produces a mixture reflecting green, which is the component common to both of them.

As you have probably found out for yourself, you can produce paint of any color you wish starting with just the three paints yellow, blue, and red. These colors we call the *primary* colors of paint. You notice that they differ from the primary colors of light (blue, green, red). And when you mix the three primary paint colors together you produce black (not white, as with light).

## Analysing the Spectrum

The stars contain many of the chemical elements that exist on Earth, especially hydrogen, helium, calcium, sodium, and iron. How on Earth do we know this when they are so far away? The answer lies in their light, which is the only thing to reach us. Just as the Sun's white light can be split up into a colorful spectrum, so can starlight. By careful observation of a star's spectrum, we can tell not only what elements are present, but also estimate its temperature, density, and the speed at which it is moving!

Practically, it is no good using just a prism to observe a spectrum. We must use an instrument called a *spectroscope*. It consists of three parts – *collimator, prism,* and *telescope*. Light from the source to be studied passes through a narrow slit and a lens in the collimator. The light emerges from the lens as a parallel beam and falls on the prism, which splits it into its component colors, which we view through the telescope.

This simple spectroscope is the kind used in laboratory experiments. The one used for obtaining the spectra of stars is somewhat more complicated and more refined. For example, to gather enough light the spectroscope is attached to the eyepiece end of a powerful astronomical telescope, which focuses an image on the collimator slit. And more than one prism may be used.

## Spectral Types

The spectra produced by luminous bodies are called *emission* spectra. When a solid substance is heated white-hot, the spectrum produced is *continuous*; that is, it extends from red to violet without a break. This is also true of glowing liquids or gases under very great pressure, like those, for example, in the core of the Sun.

When a solid element is heated white-hot, it emits not a continuous spectrum but merely bright lines in certain parts of the spectrum. Gaseous elements subjected to an electric discharge similarly emit such *bright-line* spectra. Each element has a characteristic spectrum, which identifies that particular element.

When we view, say, the Sun's light, through the spectroscope, we find that the spectrum is not continuous, as we would expect, but is crossed by a number of dark lines. And it is found that the positions of these dark lines correspond with the positions of the bright-line spectra of certain elements.

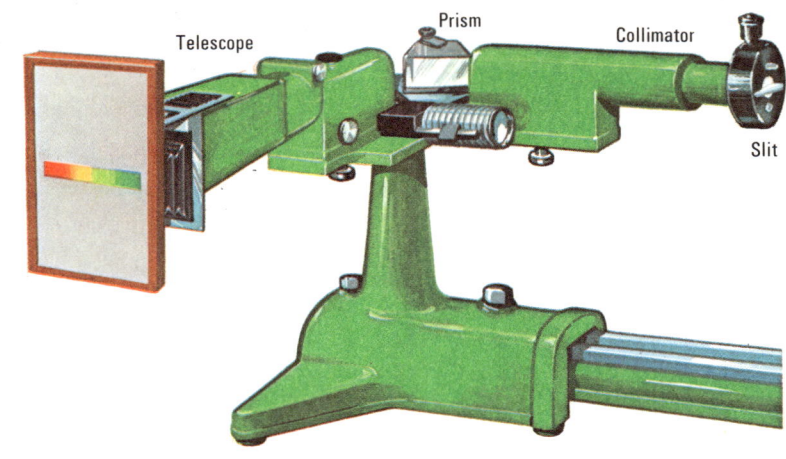

The essential features of a spectroscope. Light enters the instrument through a narrow slit. This thin beam is made parallel by the collimator lens and then passes through the prism, where it is split up into a spectrum of color. The spectrum is then viewed through a telescope or is projected onto a screen.

When metals are heated fiercely, they emit a spectrum of bright lines. Barium and potassium color a bunsen burner flame characteristically. Their spectra are shown below.

This can be explained because when light with a continuous spectrum passes through cooler gas, the gas absorbs the light it would emit if it were luminous. In other words, instead of producing characteristic bright lines, it subtracts light with those lines from the continuous spectrum, thereby giving rise to dark lines. These lines are known as *Fraunhofer* lines after a German optician, Joseph von Fraunhofer (1787-1826), who first observed them. They constitute what are called *absorption* spectra.

Thus, when we see dark lines characteristic of a certain element in the spectrum of a star, we can say that such an element is present in the cooler, gaseous outer atmosphere of the star. As we explain in the section on 'Stars', the spectra also reveal other interesting facts about the stars they come from (see page 186).

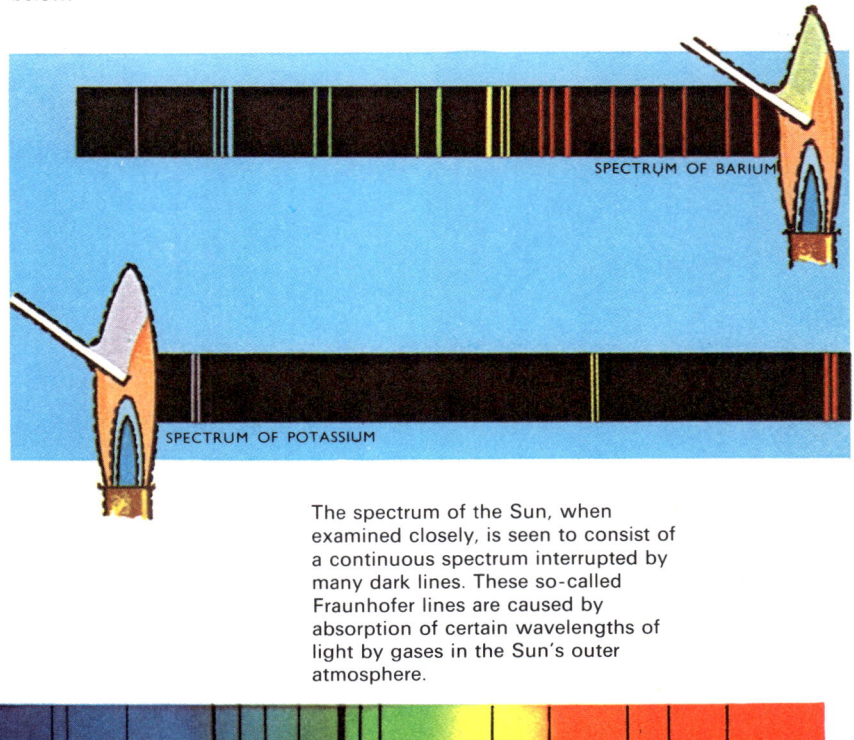

The spectrum of the Sun, when examined closely, is seen to consist of a continuous spectrum interrupted by many dark lines. These so-called Fraunhofer lines are caused by absorption of certain wavelengths of light by gases in the Sun's outer atmosphere.

# Photography

Photography is one of the world's greatest hobbies. There must be few among us who have not taken photographs at one time or another. Taking photographs these days is very simple. You load a roll of film into the back of your camera, point the camera at what you want to photograph, and press a button. Then you wind on the film and 'shoot' again.

The principles of photography are relatively simple. The *camera* is a light-tight box with a hole at one end to let the light through. In the hole is a glass *lens* (actually a combination of lenses cemented together) which gathers and focuses light rays from the object or scene being photographed onto the light-sensitive surface of a *film*. A blind, or *shutter* behind the lens normally prevents light entering the box. Pressing the camera button, or *shutter release* opens the shutter for a fraction of a second to allow light in.

Photography in the early days necessitated a lot of cumbersome equipment and lengthy preparation. The picture above shows a pioneer photographer 'on location'.

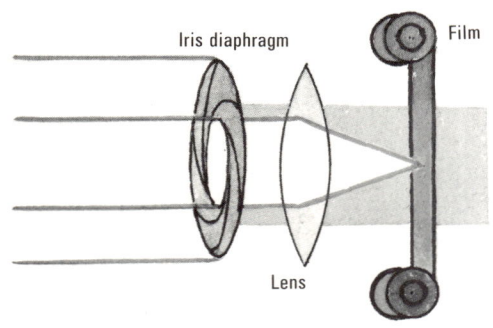

An 'instant' electronic-eye modern camera. Film is loaded into the back of the camera in a cartridge. Exposure is controlled automatically by a photocell. The cube on top of the camera holds flashbulbs for use in the dark.

The light-sensitive surface is altered by the light falling on it and holds an invisible 'image' of the scene photographed. This invisible image can be made visible by treatment with certain chemicals in a process called *developing* (see opposite). *Printing* of the developed image reproduces the scene originally photographed. The photographs may be in black-and-white or in colour, depending on the kind of film used.

**Film**

Roll film is used in most cameras today. Simple black-and-white film is made up of a long, transparent strip of plastic – cellulose acetate. The plastic strip has on one side a kind of jellified coating called an *emulsion,* which consists of minute grains of silver salts (bromide and iodide) suspended in gelatine. These silver salts are sensitive to light and change chemically when light falls on them. The amount they change is dependent on the amount of light that falls upon them. They can thus 'record' a pattern

One of the finest twin-lens reflex cameras. The bottom lens is the camera lens. The scene is viewed through the upper lens. Exposure is adjusted by varying the lens aperture (middle scale) and shutter speed (upper scale). The lower scale indicates film speed.

of light falling upon them. On the other side of the plastic strip from the emulsion in some films is a protective paper backing. Colour films have more than one emulsion (see page 51).

### Box Cameras

The simplest camera is the *box* camera. It has very few controls except for a button to release the shutter and a knob to wind on the film when a picture has been taken to bring a fresh piece of film beneath the lens. The speed at which the shutter moves, or in other words, the time the film is exposed is always the same (usually about 1/30th or 1/50th of a second). The size of the lens opening, or *aperture* is fixed too. This means that the *exposure* – the amount of light reaching the film – cannot be controlled at will and is dependent entirely on the lighting conditions. More light therefore reaches the film on a very bright day than on a very dull day. And the photographs taken may well be too bright (overexposed) or too dull (underexposed). This is obviously not very satisfactory because we want to take good photographs in all kinds of lighting conditions.

Another drawback with box cameras is that the distance between the lens and the film cannot be altered, or in other words, the camera cannot be focused. This means that some parts of the picture we take may be blurred. Blurring may also be caused if the object we are taking is moving. The speed of the box camera's shutter is too slow to 'stop' movement.

### Adjustable Cameras

To overcome these disadvantages, most cameras have adjustable controls. They have a range of shutter speeds and apertures,

The 35 mm camera uses a film only 35 mm wide. Below is one frame of a film, which has been developed into a negative. Enlargement is necessary to obtain a satisfactory size print, above. This flash picture shows the negative-print relationship well.

and a means of focusing. The number of speed and aperture settings varies from camera to camera. Many cameras have six speed settings ranging from 1/15th to 1/500th of a second. 1/25th will stop the movement of a person walking; 1/500th will stop the movement of splashing water. There is also a setting marked *B* or *T*. On this setting the shutter stays open all the time the shutter release is pressed.

The camera usually has six or seven lens-aperture settings, too. The aperture is controlled by means of an *iris diaphragm,* an ingenious device consisting of a number of overlapping metal leaves. The iris is opened or closed to let more or less light onto the film. The aperture settings are described in terms of *f-stops,* or *f-numbers.* The usual range of *f*-numbers is 22, 16, 11, 8, 5·6, 4, 2·8; *f*/22 is the smallest aperture.

The size of the lens aperture also affects the sharpness of the picture taken. With a large aperture, only objects a little closer and a little farther from your subject (what you are photographing) will be in focus. The depth of focus, or field increases as you decrease the size of the aperture.

To focus, the lens is moved towards or away from the film. In its closest position to the film, the lens focuses on far objects from about 30 or 40 feet upwards. In its farthest position from the film, the lens focuses on objects about three feet away. Focusing also affects the picture's depth of focus. The closer the object being photographed is to the camera, the smaller is the depth of focus.

Judging how far an object is away is often difficult. And so some cameras have built-in *rangefinders* to make the task easier. In one kind you see in the camera viewfinder a double image of what you are photographing. You turn the focusing knob until the two images coincide, and then the camera is accurately set. In single-lens and twin-lens reflex cameras, you see on a ground-glass screen exactly the picture that will be recorded on the film. And so you can focus it easily.

Many cameras also have built-in *light meters,* or *exposure meters.* They are 'electric-eye' devices which help you set aperture and speed to give the optimum exposure for any lighting conditions. Many photographers use a separate light-meter. For use indoors, when lighting is poor, practically all cameras are 'synchronized for flash'. In other words they can be used in conjunction with a flash 'gun', which produces a very bright artificial light.

### Developing and Printing

The silver salts in the emulsion of an exposed film carry a latent, invisible image

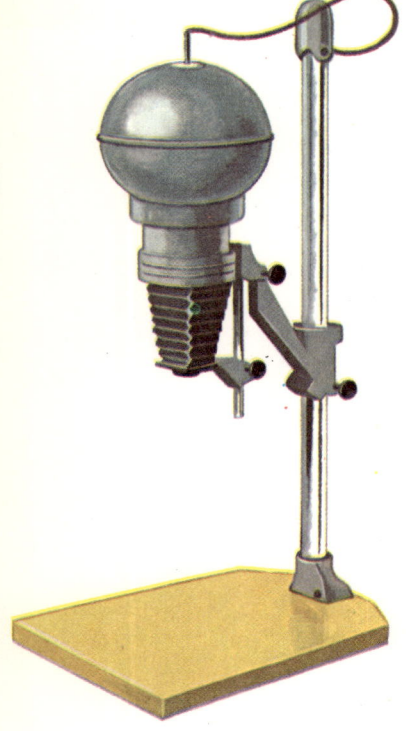

The sequence of developing an exposed film, using a developing tank. Depending on the type of film used, developing is carried out for a certain time at a certain temperature.

A photographic enlarger mounted on a tripod.

of the scene that was photographed. In *developing*, the silver salts that were affected by light are changed into metallic silver by the action of a chemical called the *developer*. The exposed areas therefore become dark. From the developer, the film goes into a *stop bath*, generally acidified water, to remove excess developer. The next stage is *fixing* – making the unexposed silver salts insensitive to light. The fixer is generally hypo (sodium thiosulphate). Thorough washing after fixing removes any traces of silver salts and fixer from the film. Thus the areas where the silver salts were unexposed are transparent.

The exposed dark areas are where light from the object being photographed hit the film. The unexposed, clear areas are where scarcely any light hit the film. Thus, what was light has become dark, and what was dark (that is, reflected little light) has become light. This reversed image is appropriately called a *negative*. Naturally all the stages of developing must take place in the dark, because the film is still sensitive to light until the unexposed silver salts are washed away.

The process of *printing* reverses the negative image on the negative to a positive image on a piece of printing paper. The simplest method of printing is *contact* printing, in which the negative is actually in contact with the printing paper. Enlarged pictures are obtained by projecting the negative image onto the printing paper from a kind of projector called an *enlarger*.

The printing paper itself is much like a piece of film, having a light-sensitive emulsion on a paper backing. In printing light is shone onto the paper through the negative. Transparent areas of the negative let light through to expose the paper. Dark areas prevent light going through. Thus, when the paper is developed (in the same way as before), the exposed areas under the transparent parts of the negative become dark. And the unexposed areas under the dark parts of the negative become light. This is a reversal of the image on the negative and therefore a reproduction of the original picture.

## Color Photography

Taking color pictures these days is just as simple as taking black-and-white. There are two basic kinds of color films: one produces color negatives from which color prints are obtained. The other produces color slides, or *transparencies*, which you view through a projector or by holding up to the light. Both kinds of film work on the principle that any color can be split up into different combinations of the three primary colors red, green, and blue (see page 46).

The color film itself is made up of not one, but several emulsions of silver salts, each sensitive to one of the primary colors. The first emulsion that the light from the

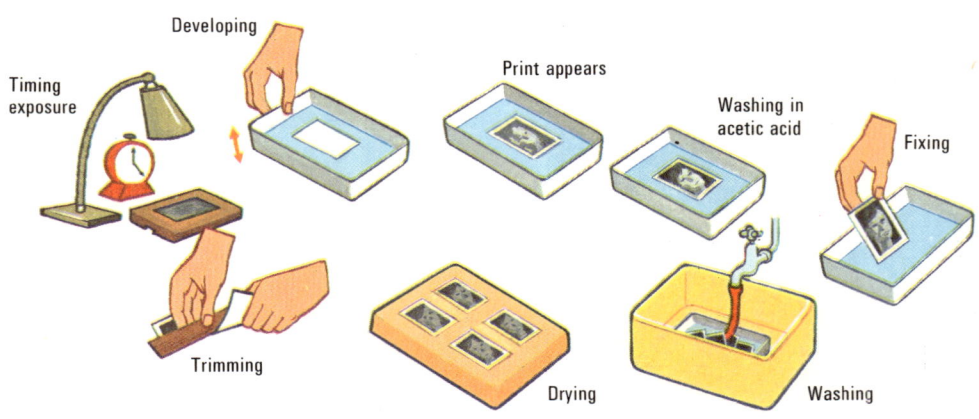

The sequence of contact printing. The negative is placed in a printing frame and exposed for several seconds. Then the printing paper is developed, and the print appears.

lens reaches is sensitive to blue light. Next comes a yellow filter which absorbs any blue light remaining. Then there is a second emulsion, which is sensitive to green light, and a third sensitive to red light. Each emulsion subtracts a certain amount of the light passing through it. That is why this process of color photography is called the *subtractive process*.

Color developing is a complicated process in which the images formed in each layer of emulsion are dyed in colors complementary to the color of the light that formed them: yellow, magenta, and cyan blue. Color printing of negative film takes place on printing paper with three emulsions, just like a color film.

**Motion Pictures**

The pictures we have discussed so far have been 'stills'. We can also take 'movies', too. Actually, the whole conception of 'movies' is a fallacy really. The image we see on a cinema screen is not a continuous one at all. It is made up of a number of separate shots that are projected onto the screen in quick succession. Because of a phenomenon called the persistence of vision, it appears as if the action is continuous. The image of one shot persists in our eyes until the next shot comes along.

To do this, twenty-four separate shots, or frames have to be projected onto the screen every second. If the frames are projected at a slower rate, the film starts to 'flicker' as our eyes begin to detect the separate frames. This happened in early films which were projected at only 16 frames a second. That is why they became known as the 'flicks'.

The film camera, or ciné camera works on the same principles as the still camera, and it takes the same kind of film strip, which is developed in the same way. The difference is that it has a motor mechanism which winds on the film. A claw device holds the film strip still while each frame is shot and then moves the next frame into position. A semi-circular disc shields the film while it is being moved to prevent blurring the image. The sound for the film is recorded in the form of a variable pattern of light on a sound track at the side of the film.

In the film projector, a very bright electric-arc lamp shines through the film strip and throws a magnified image onto the screen. It projects at 24 frames a second and, like the ciné camera, has a device for blocking off the light when the film is actually moving. At the same time, an 'electric-eye', or photo-cell mechanism converts the sound track back into sound. (See 'Sound Recording', page 105).

'Pushkin'. This picture was taken with a relatively large aperture and consequently has a relatively shallow depth of field. The subject is in sharp focus, but the house in the background is blurred.

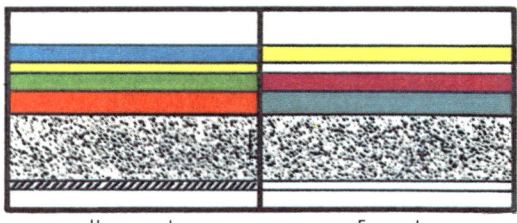

Section of a color film showing pictorially the various emulsion layers. (The raw film is not of course colored until it is processed.)

The principle of the ciné projector. The shutter and Maltese cross device are necessary to allow separate frames to be shown in rapid succession without blurring.

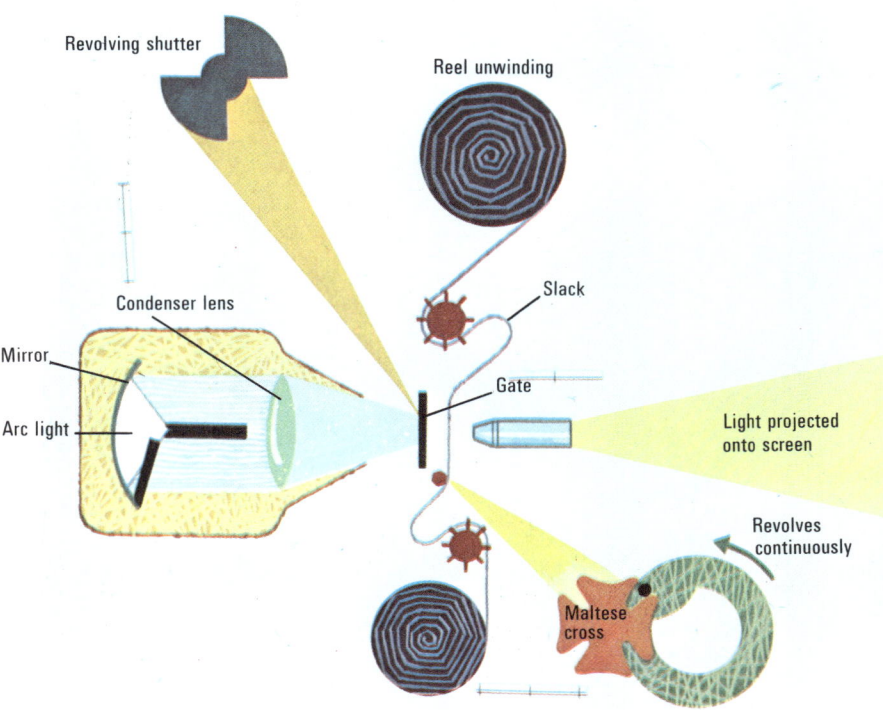

51

# Railways and Locomotives

There is a wide variety of railways and railway trains. Really the only thing they always have in common is that their wheels run on rails. In fact, with some experimental trains using the 'hovercraft' principle, it is not even strictly accurate to say that. Most trains that we see and use do, of course, run on rails and are propelled by a locomotive or motor-coach. The locomotive may be steam (although this is becoming uncommon now), diesel, electric, or gas-turbine. Sometimes the driving motors may be beneath the floor of a passenger coach rather than in a separate locomotive.

The width, or *gauge* of the rails varies in different countries. In Britain and many other countries where British engineers helped to build the early railways, the gauge is 4 feet 8½ inches. This is the gauge in North America, too. But in Russia, for example, the rails are 5 feet apart. In Africa many of the railways have a metre gauge (about 3 feet 6 inches). In Australia there are no less than three different gauges!

*Tramways* are a form of railway. Single or multiple tramcars still run through the streets of many cities. Along the Belgian coast, the trams run along their own stretch of track with stations, just like a railway.

*Narrow-gauge* railways, or light railways, were built in country areas to meet special needs. For example the Ffestiniog Railway (1 foot 11½ inches gauge) was built in North Wales to carry slate from the quarries of Blaenau Ffestiniog to the ships at Portmadoc. There are miniature railways at many seaside holiday resorts and funfairs to give pleasure rides to children. These are often accurate scale models of full-size railways.

## Rack Railways

When railway trains are required to

Contrasting kinds of rail transport. Above right is a miniature railway used for giving pleasure rides to children. Above left, is British Rail's advanced passenger train, which should go into operation in a few years time. Powered by gas turbine, this train will be capable of sustained high speeds. Below is one of French Railways' successful turbo-trains, also powered by gas turbine. In complete contrast is the Swiss rack-operated railway, left. Rack railways have a cog-wheel drive for positive traction on steep gradients.

travel up or down very steep gradients, the normal adhesion, or friction between the wheels and the rails is not sufficient to keep the train going up, or stop it coming down. In such cases, rack railways may be built. Most rack railways have a centre metal rack running between the two rails. The underside of the locomotive carries a gear wheel with teeth that fit into the slots of the rack.

There are quite a few rack railways in the Alps, for example, in Switzerland, Austria, and Italy. An early rack railway that is still operating is the Mount Washington Cog Railway in the United States. This railway first carried passengers to the summit of Mount Washington in 1869. In Britain, the Snowdon Mountain Railway was opened in 1896 up the north-western flank of Snowdon. The one- or two-coach train is pushed $4\frac{3}{4}$ miles to a height of 3,493 feet by a six-wheeled tank locomotive.

## Cable Railways

Cables are used to haul cars up slopes that are too steep even for a rack railway. They usually operate over very short distances, and often two cars are attached to opposite ends of one cable that passes over a large pulley at the top of the slope. In this way, one car ascends as the other descends so that their weights tend to counter-balance one another. This system is known as a *funicular*. Most funiculars are wound up and down by stationary engines. A notable one in Paris carries passengers up and down the slope of Montmartre.

A main-line railway operated by cable is in Brazil. It has four inclines with a gradient steeper than 1 in 10, and the cars are connected in pairs, funicular fashion. Funicular railways may operate over much steeper gradients than this. One in Switzerland climbs an incline of 1 in $1\frac{1}{8}$.

## Monorail Railways

Monorail railways have only one rail. The train may run on the rail or hang below it. A train running above the rail needs a gyroscope to balance it, or else it may hang over either side like a saddle with guide wheels to grip the sides of the rail.

The Alweg monorail system is of this last type. The rail is a concrete beam, and the cars are supported on it by rubber-tired driving wheels. This type of monorail is used in Japan for the $8\frac{1}{2}$-mile line which links Tokyo with its airport.

In a suspended-monorail system the train may hang freely from wheels on the rail, or the wheels may run inside the box formed by two rails close together. This form of construction is used in the Safège system, which has rubber-tired wheels on bogies running inside the box girder. An experimental section of this track was opened in France near Orleans in 1960. The Wuppertal line in Germany is another suspended monorail system. It has been operating since 1901. It is more than 8 miles long and runs along a narrow valley over the River Wupper.

The various types of hover-train now being developed may be considered developments of monorail systems. In the case of the French Aérotrain, the track is shaped like an inverted 'T', and the train is supported on it by a cushion of air. The upright portion guides the vehicles, which may travel at speeds of up to 200 mph.

## Steam Locomotives

For many years steam locomotives were the prime source of power on the railways of the world. Today, they have been replaced in many countries by more modern power units, principally diesel, electric, and gas-turbine. But they nevertheless still retain a glamour and fascination for many people.

A suspended-monorail system in Japan. The passenger cars hang by a bracket from a kind of bogie that travels along the single rail. Japan has been in the forefront of monorail development. The country also has a number of above-rail, saddle-type monorail systems.

Basically, a steam locomotive has a *boiler*, which produces steam that is fed to cylinders to force pistons backwards and forwards. These pistons drive the wheels through *connecting rods*, and cranks. The boiler occupies the bulk of the locomotive. It is a long steel cylinder, with a *firebox* at one end in which is burnt the coal, wood or other fuel. A large number of tubes carry the heat through the boiler to turn the water into steam.

The steam enters the cylinders, one of which is usually mounted on either side of the locomotive. Pistons in the cylinders are linked through piston rods and connecting rods to the driving wheels. Large steam locomotives may have four or more pairs of driving wheels joined by coupling rods.

Exhaust steam from the cylinders is fed through a *blast-pipe* into the chimney, where the draught of its rapid movement helps to draw the fire up more fiercely. This means that when a locomotive works harder, the fire burns more vigorously, and more steam is produced to supply extra power. In a modern locomotive, the steam will reach a pressure of up to 300 pounds a square inch.

## Diesel Locomotives

Diesel locomotives are powered by fuel oil. Like steam locomotives, they are independent of any outside source of power and so can run on the same tracks. Consequently, they have gradually replaced steam locomotives in many countries. Not only can they make long runs without refuelling or servicing, but their power is available quickly. Steam locomotives, on the other hand need time to build up steam pressure. Also, diesels can stop and start more quickly and accelerate more rapidly.

The diesel engine is an internal com-

Left: A beautiful preserved relic of a bygone age, an early Swiss steam locomotive. Right: one of the last generation of British steam locomotives hauling a goods train. Note how dirty the locomotive is and the black smoke belching from its chimney. Dirty, inefficient but fascinating nevertheless, steam locomotives have long since departed from the railway scene.

bustion engine, similar in many respects to the gasoline engine used in automobiles, but using oil, not gasoline as fuel (see page 216).

As in an automobile, the diesel engine cannot be connected directly to the wheels of the locomotive. Various forms of *transmission* are used to transmit the power to the wheels.

In a *diesel-electric* locomotive, the diesel engine is connected to an electric generator. The electricity supplied by this generator is fed to driving, or *traction* motors that are usually mounted on the axles, to which they are coupled through gears. This is the most common form of power transmission in diesels.

Mechanical transmission employs a gearbox, so that the driver has to change gear in a similar way to a truck driver. Such *diesel-mechanical* locomotives are small locomotives used for shunting duties.

The third main type of transmission is *hydraulic*. The power is transmitted through oil in a *torque converter,* which takes the place of a gearbox. This is really a kind of turbine. The engine drives a pump which impels the oil on to turbine blades, causing them to rotate. The shaft of the turbine is linked to the driving wheels of the locomotive. Diesel-hydraulic locomotives are used fairly widely for main-line trains.

## Electric Locomotives

Electric locomotives have many of the advantages of diesel locomotives but are dependent on an external supply of electricity. They can operate, therefore, only on track which is designed for them. The electricity may be supplied as alternating current (A.C.) or direct current (D.C.). If A.C., it is converted to D.C. by devices called *rectifiers* on the locomotive. It may be supplied to the locomotive through overhead wires by means of a hinged, sprung arm called a *pantograph* on the roof, or through a third rail laid alongside or in the middle of the track. Overhead systems operate at 25,000 volts A.C. or more. Some third-rail systems operate at 750 volts D.C.

The traction motors that drive an electric locomotive are similar to those employed in diesel-electric locomotives. They may be mounted on each driving axle. In multiple-unit electric passenger trains, the electric motors are placed at intervals along the length of the train.

Some locomotives, called *electro-diesel* locomotives, can work on both electrified and non-electrified lines. They operate as an ordinary electric locomotive on lines with a third-rail supply, and have a diesel-electric generator set for use on non-electrified lines.

**Gas-Turbine Locomotives**

Gas-turbine locomotives have a power unit similar to an aircraft jet engine. A compressor supplies compressed air to a combustion unit where oil, natural gas or powdered coal is burned. The resultant hot gases operate a turbine. The turbine itself is linked to the driving wheels of the locomotive electrically or mechanically.

Gas-turbine locomotives are economical only if they can be kept working for long periods at full power. Consequently, they have found their greatest applications in countries such as the United States and Russia where there are long runs with heavy gradients. Small gas-turbine units

In contrast to the steam locomotives they replaced, modern locomotives are clean, efficient and have vastly superior performance. On the Continent electric locomotives power the vast majority of the trains (above). In Britain both electric and diesel locomotives are widely used. In the United States, though, diesels are almost invariably used (below).

have recently been developed for intercity travel, notably in France.

Research continues into new forms of power. One promising device is the linear induction motor. This may be visualized as an 'opened-out' electric motor. The motor windings would be in the form of a flat strip in the locomotive, and the track would form the rotor. The windings would move past the rotor, instead of the other way round as in normal electric motors (see page 84).

# Pioneering Days of Rail

The early Greeks and Romans knew the advantages of having some form of track or groove to guide the wheels of vehicles, but the first men to use wheeled wagons on raised tracks were probably miners. These men needed to move coal or ore from the mine face, and they knew that a horse could pull a much larger load when that load was running on rails.

The first rails were of wood, but as metal-working techniques improved the timbers were covered with iron plates. Iron rails were in use in Cumberland as early as 1738. At first, the rails had a raised outer edge to hold the wagon wheels on the track, but later the wheels ran along the edge and themselves had a flange to keep them on the track.

When the first public freight railway, the Surrey Iron Railway, opened in 1803, a horse pulled the 55-ton train of wagons. There were early attempts to design horse-powered locomotives in which the horse rode on a truck that it propelled by turning a treadmill.

## Steam Engines

The first steam locomotive was built by Richard Trevithick in 1804. Twenty-one years later the first public railway to employ steam engines was opened. This was the Stockton and Darlington Railway, whose first passenger train was hauled by Stephenson's *Locomotion No. 1*. A later Stephenson locomotive, *Rocket*, inaugurated the services of the Liverpool and Manchester Railway in 1830.

**Rocket** incorporated the basic ideas of modern steam locomotives. In particular, it had a multi-tube boiler, with a correspondingly more efficient production of steam than in the earlier, single-flue boilers. George Stephenson entered it for a competition to find a suitable engine for the Liverpool and Manchester Railway in 1829. It won easily, pulling 15 tons for more than 60 miles. After the loaded wagons were uncoupled it reached a speed of $29\frac{1}{2}$ mph. The *Rocket* and other engines like it pulled the first trains between Liverpool and Manchester.

This was the first passenger railway to dispense completely with horse traction. In the following half-century many thousands of miles of railway were laid to carry steam locomotives.

Various other means of traction were tried at different times. There were several attempts to operate 'atmospheric' railways. On these lines, the power was supplied by stationary engines. A piston on the underside of the train was sucked along a tube that lay between the rails. Engine-houses that once contained the suction pumps may be seen still standing beside the main line through South Devon. The first railway in Ireland was 'atmospheric'.

Some early locomotive designers were not convinced that a smooth wheel on a smooth rail would give sufficient traction. John Blenkinsop and others built locomotives with large gear wheels that engaged on lugs cast on to the outside edges of the running rail. A similar principle is still employed on rack railways where the train has to be moved up or down a steep slope.

## Battle of the Gauges

The width of the rails on most railways today, 4 feet $8\frac{1}{2}$ inches, was the average wheel spacing of carts, and the early railways were designed so that any ordinary cart with this wheel spacing could use the tracks. Stephenson adopted this width for his locomotives, as did most railway engineers at that time.

One man, however, believed that a greater width would allow bigger, more powerful, and more stable locomotives to be used. Isambard Brunel built much of the Great Western Railway to a gauge of 7 feet. This gauge was very successful and many large locomotives were built, but unfortunately it was in the minority. The rest of the railway system was at the narrower gauge, and the difficulties of moving traffic from one railway to another caused a parliamentary commission to be set up. This came out in favor of 4 feet $8\frac{1}{2}$ inches as the standard gauge The Great Western Railway was gradually converted to the narrower standard. The generous curves and ample clearances that exist on some sections of the main line in the West of England today are a result of the earlier wider standards to which it was built.

## Across America

The railways played a major part in opening up the interior of America and developing the West. Several steam locomotives were imported from England in 1829, but the first regular service in the United States was operated by the American built *Best Friend of Charleston* in South Carolina in 1830. The drive to unite the Atlantic and Pacific coasts came in 1863 when the Central Pacific Railroad in the west, followed by the Union Pacific Railroad in the east, started laying track towards a meeting place in Utah. Tremendous difficulties were overcome in traversing the deserts of Nevada and the mountains of the Sierra Nevada and the Rockies – not to mention hostile Indians! Thousands of Chinese labourers were hired and transported to the States. Towards the end, the construction teams were completing up to ten miles a day. The tracks finally met in May, 1869.

Farther north, the Canadian Pacific Railway built its 2,881-mile line between Montreal and Vancouver. The first passenger train left Montreal in June, 1886, and took $5\frac{1}{2}$ days to cross Canada from east to west.

Elsewhere in the world, railways followed the pioneers. The Trans-Siberian Railway was started in 1891 and the Cape-to-Cairo Railway by Cecil Rhodes in 1899. The Trans-Siberian Railway is the longest railway in the world, stretching for about 5,800 miles. Part of it is now electrified. Trains make the journey between Moscow and Nakhodka in Siberia in about four days. The railway planned to run from Cairo to the Cape of Good Hope was never completed.

## The Development of Steam Power

The growth of the railways has followed the development of more efficient sources of power. The first steam locomotives had a simple boiler consisting of a barrel, partly filled with

Trevithick's first locomotive

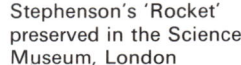

Stephenson's 'Rocket' preserved in the Science Museum, London

George Stephenson

**Richard Trevithick (1771–1833),** a Cornish engineer, designed and built the first steam engines to operate at high pressures. He built a steam road carriage and a railway steam engine. The latter, which ran on a tramway at Penydaren in South Wales in 1807, was the world's first railway locomotive. It weighed five tons and could pull 40 tons. However, the cast-iron plates of the track could not stand up to the pounding they received, and further engines were not taken into general use, although one more was built.

**George Stephenson (1781–1884)** is known as the 'Father of Railways'. The son of a stoker, he learnt to read and write in his spare time. He became a much-sought after railway consultant and designed and built many early locomotives, including the first to be used on public railways. His most famous was *Rocket* which he designed with his son Robert (1803-1859). Robert became as famous as his father as a builder of railways and locomotives. Among his achievements were the Britannia tubular bridge over the Menai Strait.

water and containing a large inner tube that held the fire.

Trevithick improved on this by doubling the tube back into a U-shape, so increasing the heating surface and reducing fuel consumption. He also turned the exhaust steam into the chimney to improve the draught in the firebox. Stephenson raised the efficiency further by replacing the single, large tube with many small ones.

Early boilers were of riveted wrought-iron plates, covered with wood planking for insulation. Later ones were of welded steel, lagged by plaster or asbestos covered with thin steel sheeting. Copper fireboxes and brass tubes were also later replaced by steel units.

Increase in power followed a rise in steam pressures. Early boilers worked at up to 50 pounds a square inch; later ones reached 300 pounds. Efficiency also rose with increase in temperature. Later steam locomotives incorporated superheaters. These were long, narrow, folded tubes carrying steam inside the normal boiler tubes to increase its temperature still further.

Other improvements were made in the operation of the pistons that drove the wheels. It is wasteful to admit steam to the cylinder during the whole stroke of the piston, as the steam can continue to expand beyond its first increase in volume. So 'expansive working' was introduced in which the driver could advance the point at which the steam was cut off. An alternative use for a time was 'compound working', in which a small, high-pressure cylinder was followed by a large, low-pressure one.

**Mallard,** an A4 class Pacific Locomotive, set up an unbeatable record for steam locomotives when in July 1938, it reached a speed of 126 mph. The streamlined Pacific locomotives, with their wedge-shaped front that diverted smoke and steam clear of the driver's cab, were designed by Sir Nigel Gresley.

### Underground Railways

In 1863, the world's first underground railway was opened in London. Worked by steam locomotives, it was the Metropolitan Line between Paddington and Farringdon Street. It was a sub-surface tunnel built by 'cut and cover'. This meant considerable disruption of traffic on the roads above, and engineers investigated the possibility of driving tunnels at a much greater depth.

The first deep tunnel was the Tower Subway, opened in 1870, which carried passengers, by cable traction, under the River Thames. Twenty years later, a true tube railway similar to those of today, was opened between the Monument and Stockwell. This had a major advantage over earlier underground lines — the filth-producing steam locomotives were replaced by tiny, four-wheeled electric locomotives.

Following the example of London, other cities began to build underground railways. Glasgow had a cable railway beneath it by 1896; Paris followed with its Métro in 1900. The New York Subway opened in 1904.

### Electricity

The planners of surface railways soon realized the advantages of electric power. In 1883, Volk's Electric Railway was opened along the seafront at Brighton. Other electric trains ran in Ireland on the Giants' Causeway railway. Then other main-line railways started changing to electric traction.

From the 1950s onwards, railways in many countries embarked upon extensive electrification schemes. Today electric and diesel-electric traction are of prime importance in most countries.

Above: One of the famous locomotives of the Great Northern Line in the early years of the century.

Right: One of the early trains operating on the Circle Line of London's Underground. Note that a steam locomotive was used. You can therefore imagine how dirty it was travelling by underground in those days!

Below: The fastest ever steam locomotive 'Mallard', which reached a speed of 126 mph in July 1938 pulling seven coaches.

# Modern Railway Operation

Most railways carry both passengers and freight. Of the two, freight is usually the more profitable. It may be moved at the time most convenient to the railway company, and requires the minimum of staff for its transportation. Passengers, on the other hand, expect trains to be available wherever and whenever they wish to travel. This means, for example, that some lines are packed during morning and evening peak, or 'rush' hours but almost empty during the rest of the day.

## Freight

*Containerization* – an ugly word meaning the packing of goods into uniform-sized containers – has brought considerable advantages to railway freight operation. Freightliner trains, which consist of flat cars carrying containers or special vehicles

On today's high-speed railways track laying and maintenance are of vital importance. Regular testing of the track by such vehicles as that shown above for misalignment of the rails or defects allows detection of faults before they can cause any accidents.

Left: One of France's crack locomotives "Le Capitol" travelling at speed. Like most modern high-speed locomotives Le Capitol is electric, picking up its electricity supply by pantograph from an overhead wire. Below: To cope with increased high-speed traffic signalling systems have been modernized and streamlined. Signals are now controlled electrically from signal boxes like this.

adaptable to rail or road transport, operate regular, fast services between special terminals at large industrial centers. Coal is carried from pit-head to power station by means of 'merry-go-round' trains. These trains are loaded automatically at the pit while on the move and discharged through bottom-opening cars at the power station, still without stopping.

Large marshalling yards operate automatically to assemble trains. The cars are pushed over a hump where a trackside apparatus 'reads' the load and destination from a coded label on the side of each car. As a car runs down the hump, a computer calculates its weight and speed and applies retarders, which slow it down by the required amount. The car is automatically directed into one of 30 or 40 sidings to join other cars with a similar destination.

## Passengers

Improvements in passenger services are centered mainly on increasing speed. In many countries nowadays electric intercity services operate 100 mph trains. Lightweight, streamlined trains powered by gas turbines, which are capable of travelling up to 200 mph on ordinary tracks, are in an advanced state of development in several countries.

Faster trains need more efficient signals to give earlier warning of hazards. The older style semaphore signals are being replaced by red, yellow and green colored lights. Equipment may be fitted that repeats the signal on an instrument in the engineer's cab, so that he knows when he has reached a signal at danger, even in a dense fog. Train-stop devices are available that automatically apply brakes if there is a danger signal. One signal box can control many miles of track. In the box a large illuminated diagram shows the positions of all the trains in its area.

# Underground Railways —Subways

There is at present a 'boom' in underground railways. Many cities throughout the world are currently planning, building or extending subways under their streets. It is appropriate that London is among the cities involved in this activity, because the first subway trains ran beneath London. Today, the automatic Victoria Line is operating.

There are basically two ways of building underground railways. The earliest lines were 'cut and cover'. This system may be used when a large trench can be dug without too much disruption of surface traffic or buildings. When the tracks have been laid in the trench, walls are built and a roof placed over. The surface amenities may then be replaced.

In most cities today such methods would create chaos, and so engineers tunnel deep underground. The ground under cities such as London is full of obstacles such as gas, water and electricity supplies, sewerage pipes and drains, underground rivers, and older subway lines, and planners of new lines need to go deeper and deeper to find clear ground. These deep tubes are constructed in safety behind *tunnelling shields*.

Of the lines under London, which still has the most extensive underground transit system in the world, 67 miles are in small-diameter, deep-level tunnels and 23 miles are in larger sub-surface tunnels, built by 'cut and cover'. Including surface lines as well, the London Underground system has 254 miles of route.

Coaches on underground trains are usually designed for maximum capacity rather than for comfort, as these systems often have to cope with large crowds of people getting to and from work, and the area devoted to standing passengers is large. Most trains are worked by a driver and a guard, although many of the latest lines operate automatically and need no attendants at all.

In automatic train operation, the trains may pick up instructions by induction from cables between the tracks. Alternatively, programme machines may compute the running programme. The flow of passengers can also be controlled automatically by means of ticket-issuing, inspecting and collecting machines at entrances and exits.

A number of American cities have realized the urgent need of carrying large numbers of commuters in and out of town daily without blocking all the roads. Several 'rapid-transit' systems are planned and under construction. An ambitious one is the San Francisco Bay Area Rapid-Transit System. The townships around the Bay are connected to San Francisco by a 4-mile long underwater tube. This tube, which is earthquake proof, is prefabricated in sections and sunk into position on the floor of the Bay. The complete network comprises 20 miles in tunnels, 24 miles on the surface, and the remaining 31 miles on elevated structures. The trains are aerodynamically designed and are able to run at more than 70 miles an hour.

Subways do not always stay underground. Here the New York City subway is passing Coney Island in Brooklyn. The subway system operates by electricity, which trains pick up from a third rail (lower right). Note that it is protected by insulators.

London's Victoria Line first went into operation in the late 1960s. It was designed for automatic operation with the minimum of staff. Electronic machines check the tickets of passengers entering and leaving the stations.

# Timber

Wood is one of the most valuable of our natural resources. We use it not only in the form of bulk timber for doors, floorboards, furniture, and so on, but also in many other forms. For example, we break it down into pulp, and from this we make paper and rayon, the most widely used of the man-made fibers.

Because timber is so precious we must make sure of a continuing supply by preserving and protecting existing forests and planting new ones. In the past the forests were ruthlessly cut down for fuel and for building, with little thought for the future. Now the trees are treated as a crop, just like cereals or roots. But unlike farm crops, which are planted and harvested within a year, trees take a lifetime or more to mature. Even the quickest-growing conifers do not mature for 40 to 50 years. For oaks, it is more like 150 years.

Tree felling in the vast Canadian softwood forests.

The science of planting and cultivating trees is known as *forestry,* and the people who look after, or 'farm' the forests are called *foresters*. Like the farmer, the forester today makes extensive use of machinery to help him in his work – caterpillar tractors, excavators, power saws, and so on. Also, he may use chemical sprays to control weeds, pests, and diseases.

The prime aim of the forester is to make sure that he plants at least as many trees as are cut down, or *felled*. He sometimes allows the growing trees to re-seed areas of forest that have been felled. But often he raises his own seedlings from seed in a specially prepared forest *nursery*. He plants them, close together, in the forest when they are about 3 to 5 years old. For two or three years afterwards he must weed them regularly or else they will be smothered and killed or stunted. Later on he thins out the trees to allow the others to grow better. Always he must be on the lookout for pests and diseases.

Various methods are used to transport logs from the forest to the sawmills. In winter it may be convenient to carry them by sled, right. Over lakes they are often transported within a floating collar, or boom and hauled by tug, below.

## Felling

Forty, eighty, a hundred years may pass before the trees in a forest reach maturity and are ready to be felled. The trees in virgin forest land may be much older, of course. The forester decides which trees are to be felled and marks them accordingly. Then the felling teams move in with axe and saw to begin their noisy, dangerous work.

In many countries the felling is still done mainly by hand, but in the vast softwood forests of northern Europe and North America, the 'fellers' use a powerful chain saw to cut down the trees. This is probably the most useful saw in the forest. It has a saw-toothed cutting chain, driven by a small gasoline engine, which whirls rapidly around a steel blade. One-man or two-man saws may be used, depending on the size of the tree being felled.

The fellers are so skilled that they can 'drop' a tree exactly where they want to. When you are dealing with a giant of a tree 200 feet tall, 4 feet in diameter, and weighing 50 tons or more, you can imagine the need for accuracy! To 'drop' the tree accurately, the fellers first make a V-shaped notch called the *undercut* low-down on the side of the tree facing the direction in which they want the tree to fall. Then they saw through the trunk on the opposite side and a little above the line of the undercut. Eventually the tree overbalances because of the undercut, and comes crashing down. As the fellers move to another tree, the 'bucker' takes over. With a chain saw, he slices the trunk into shorter logs to make it easier to extract from the forest.

### Extraction

Extraction these days is generally a highly mechanized operation. In some countries, however, horses, oxen, buffalo, and even elephants are still used to drag the logs from the forest. In the simplest method of mechanical extraction, tractors or tractor winches haul the logs over the ground. Often the front end of the log is hoisted up on a tracked or wheeled frame called an *arch* or *sulky*, thus making towing easier.

One of the most spectacular methods of extraction is to use one or more of the tallest trees in the forest as 'spar trees'. The logs to be moved are attached to long cables which pass over pulleys at the top of these trees. A power winch winds in the cable and drags in the logs. But before a tree can be used as a spar, its top must be removed. This is the incredibly dangerous job of the *high-climber*.

Often mechanical spars are used instead of spar trees. They may form part of an extensive system of *aerial ropeways,* or *cableways* linking vast areas of the forest. The logs are bunched together and lifted up on a pulley system that travels along the cable.

The trees are extracted to a clearing in the forest called a *landing* to await transport to the sawmill or pulp-mill. Transporting the logs from the forest is often a headache, for many of the forests are in very cold inaccessible regions of the world. Often new roads and new railways have to be built so that trucks and trains can get through. But quite often it is more convenient to use other methods. One of the commonest is to send the logs by water.

The logs may simply be tipped into a river and allowed to float haphazardly downstream. But often they are bound into rafts which are guided by workers living on them. In Scandinavia and elsewhere great masses of floating logs are towed by tugs across the lakes inside floating collars, or *booms*.

At the sawmill the logs are stored in a vast log pond, where they are protected from fire and disease until they are needed. Then they are fed by chain conveyor into the sawmill itself and sliced by powerful band or frame saws into boards and posts (above).

In some forests the logs are transported down the mountainsides by means of chutes, or *slides*. They may be dry or wet. In wet slides, called *flumes,* the logs may be practically floated down with quite a lot of water or merely wetted to make them slippery. In winter, ice slides provide a very rapid means of transportation. Sledges, too, are often used.

**Through the Sawmill**

The logs from trees felled deep in the forest eventually arrive at the sawmill, where they are sawn up, or *converted* into boards and other familiar forms of lumber. At the sawmill the logs are stored in a huge logpond, or *millpond*. The water protects them from fire and disease until they are needed. Meanwhile, men with long poles and spiked boots walk about on the logs, sorting them for size, quality, and so on.

The logs are fed one by one into a *conveyor* which leads from the pond into the mill. The conveyor is a heavy, spiked chain

Logs are often transported by water in the spring when the streams and rivers are flowing swiftly. From time to time the logs jam together. Then the lumberjacks have to clamber over the treacherously spinning logs and use spiked poles to free them. Sometimes the lumberjacks test their log-riding skill by competing in log-rolling contests.

which moves along the bottom of a wooden chute. High-pressure jets of water remove any dirt and stones remaining on the logs which might damage the saws in the mill. As the log enters the mill, it is clamped on a carriage which moves it towards the first saw. This may be a band saw or a frame saw.

The *vertical band saw* is the one most widely used, especially in North America. The blade consists of an endless steel band with saw teeth on it, often on both edges. It passes around two large wheels, one of which drives it round at high speed. The whirring blade slices quickly through the wood as the carriage presses the log against it. The sliced-off board drops onto rollers and goes for further cutting. The carriage returns to its starting point and repositions itself automatically for the next cut. The action is really much like that of the bacon slicer you see in grocers' shops.

The *frame saw* is favored in northern Europe where the logs are in general smaller. It has a number of parallel saw blades set in a frame, which moves rapidly up and down. The whole log is pushed against it and converted in one operation.

The boards coming from the first saw pass to the *edger,* which consists of two parallel circular saws. The saws square up the edges and remove any bark as the boards pass between them. Next, the boards go to the *trimmer* where they are cut into standard lengths. The trimmer is made up of a number of circular saws in a row, any of which can be lowered to cut the boards passing beneath them.

The trimmed boards are graded according to quality and size and then stacked in the lumberyard. Sometimes they are first passed through a planing machine, where sharp knife blades shave the surfaces very smooth.

But the lumber is by no means ready for use yet. It is still very green. Before it can be used it must be dried, or *seasoned*. This may take anything from a few days to a year or more, depending on the method of seasoning and the thickness and type of wood.

In some mills the logs are converted in an entirely different way. Small logs are sliced into thin sheets, or *veneers,* by a method called *rotary cutting*. The log, with bark removed, is rotated rapidly against a knife blade pressing against it along its length. The thin veneer is peeled from the log in a continuous ribbon and then cut into convenient lengths by a guillotine.

The greatest use for veneer produced in this way is for making *plywood*, which is a kind of wood and glue 'sandwich'. Plywood is used extensively for all kinds of furniture and fittings in the house and in industry. The sheets of veneer are glued together so that the grain, or direction of the wood fibers of one sheet is at right angles to that of adjacent sheets. This makes the resulting plywood almost equally strong in all directions, unlike solid wood, which is stronger across the grain than along the grain.

Veneers of expensive woods, such as mahogany, are often glued on top of cheaper woods to give them the appearance of the solid expensive wood at a much lower cost.

Fire is an ever-present danger in forestry. In the summer a carelessly dropped cigarette end, a smouldering campfire, and often spontaneous combustion can let loose a blazing inferno that can lay waste hundreds of acres of valuable forest land and cause untold damage to wildlife and property.

Crawler tractors are widely used to extract timber from the forest. To make haulage easier the front end of the log is raised on a wheeled frame, or sulky, below.

# Paper

We have the Chinese to thank for the invention (in about A.D. 100) of paper as we know it today. It is said that a Chinese got the idea from watching wasps, which produce a kind of paper to make their nests. This invention was one of the most important inventions in the history of mankind for it made possible mass communications, learning, and knowledge. The first paper was made from the inner bark of the mulberry trees. They are widely cultivated in China because they are the breeding place for silkworms, which gorge themselves on the leaves.

## The Raw Materials

Most paper today, however, is made from a broken-down kind of wood called *woodpulp*. Special qualities are made from other raw materials, including rags, old rope, manila hemp, jute, ramie, esparto grass and even straw. Each kind has its own uses. Rag paper, for example, is used for documents that have to be kept for a long time. Manila is used for envelopes. Quite often paper is made from a mixture of several pulps.

Wood is made into pulp in several ways at the pulp mills, which are generally situated close to the forest regions where the timber is cut. Sometimes the grey, porridge-like pulp is made into paper in the same mill. In this case the pulp is pumped

The majority of paper is made from woodpulp. The pulp arrives at the paper-making plant as dry sheets and must first be converted back into liquid form in a pulping tank, above.

Some of the best quality writing paper is made not from woodpulp but from rags. Here women are sorting rags for quality before they go for pulping.

### Wood—pulp

Three main methods are used to convert wood into wood-pulp. One involves merely grinding the wood mechanically. In the others the wood is broken down chemically.

In the *mechanical* method of making wood-pulp logs are fed to a set of revolving grindstones. Water is sprayed onto them to prevent the wood from charring. The wood leaves the grinder as a fine, wet pulpy mass. The main use for this ground wood-pulp is for making *newsprint*, the relatively coarse yellowish paper on which newspapers are printed. Large quantities are also used for making wallpaper and wrapping paper.

The chemical processes include the *sulphite*, or *acid* process and the *sulphate*, or *alkaline* process. Small, debarked logs are cut into tiny chips and 'cooked' with certain chemicals (the 'cooking liquor') in a sealed vessel called a *digester*. This is like a huge pressure cooker. The chemical action dissolves the mysterious substance called *lignin* which binds the cellulose fibres together in the wood. The sulphite process produces pulp which is very light in colour and which can subsequently be bleached easily to give extreme whiteness. Sulphite pulp is therefore used to produce high-quality writing and printing paper. When purified, it is used to make 'artificial silk', or rayon. The sulphate process produces a pulp of great strength which is used to make such things as wrapping paper and cardboard. It is often called *kraft* pulp, after the German word meaning 'strength'. The pulp is darker than sulphite pulp and is generally used in that state, although it can be bleached to give a fine white paper.

The common softwoods spruce, hemlock and pine are the woods most widely used for making pulp, but hardwoods such as poplar, beech, birch, and elm can also be used. The sulphate process has the advantage that it can use a greater variety of woods than the other processes.

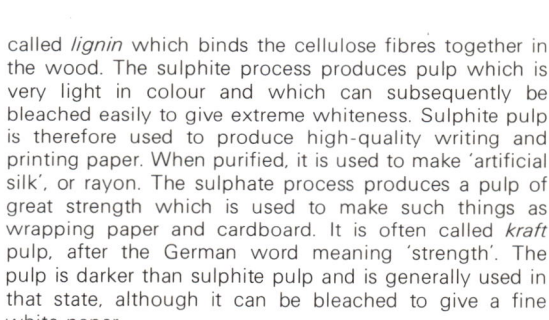

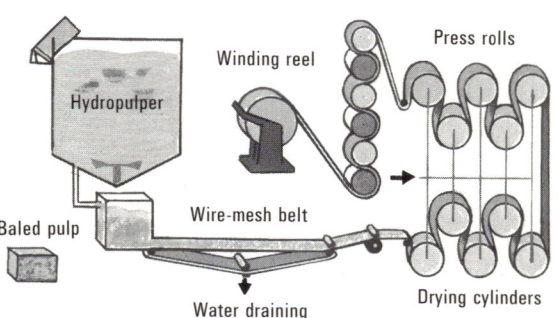

Stages in the production of newsprint, the rather coarse paper on which newspapers are printed. The woodpulp for making this kind of paper is obtained by grinding debarked logs with a grindstone. At the paper factory the pulp is beaten and then mixed with other substances before flowing into the 'wet end' of the paper-making machine. The water drains away, and the damp paper web is dried and pressed into a continuous sheet.

into storage vats and from these directly to the paper-making machines. Often, however, the paper mills are quite separate, and the pulp is transported to them in the form of dry, flat sheets.

## The Paper Mill

The first thing that must be done at the paper mill is to convert the pulp back into liquid form. This is done in a huge, agitated vat called a *hydropulper*. The dry pulp sheets are broken down into separate fibres and mixed with water.

The liquid pulp passes into the *stock preparation machine,* or *beater,* in which revolving bars and knives beat the pulp. This frays the fibers and makes them flexible and cuts them into a suitable length for the paper to be made. The amount and kind of beating determines the nature of the resulting paper. Blotting paper, for example, requires less beating than, say, greaseproof paper, in which the fibers are greatly frayed to produce a very dense paper that is impervious to grease.

The beaten pulp next goes to the *mixer,* There it may be blended with different kinds of pulps, together with a number of other substances which will produce the desired type of paper. Adding *China clay* at this stage gives the resulting paper 'body' and makes it smoother. Adding *size,* a kind of starch or glue, makes the paper easier to print and write on. *Dyes* and *pigments* may also be added at this stage to color the paper. Rosin and alum are generally added to make the paper more water-resistant.

The 'wet end' of the paper-making machine. At this stage the pulp contains over 90% water. It flows over a wire-mesh belt, where most of the water drains, or is sucked away.

The 'dry end' of the paper-making machine. The damp web of paper leaving the wire-mesh belt is led through a succession of drying cylinders and press rollers.

## The Paper-making Machine

The fine, pulpy mass emerging from the mixer passes to the paper-making machine proper. The standard machine is the *Fourdrinier* machine, named after the French brothers who developed it, in England, in about 1803. The largest machines are more than 300 feet long. At the 'wet end' of the machine, the so-called *breast box* allows an even flow of pulp over a swiftly moving wire-mesh belt. The water gradually drains away. Further down the machine, *suction boxes* draw off more moisture.

The soft, moist 'web' of paper produced goes through a number of heavy *press* rollers which squeeze it into a thin, damp sheet. This damp sheet then passes around a large number of steam-heated *drying cylinders* and gradually dries out. Heavy *calender* presses give it a smooth surface finish before it is wound onto massive reels. The paper sheet emerges from the machine at a rate of up to 2,000 feet a minute. Each reel may hold more than 10 miles of paper.

The absorbent paper tissue which is used for toilet rolls, paper handkerchiefs, kitchen towels, and so on, is made by scraping the paper web off the drying cylinders with a knife edge. This produces the typical crepe-like finish. Many papers are surface coated with a sized pigment to prevent absorbency and to give a glossy finish. This is done for 'art' paper, for example.

To give the paper a watermark, or translucent pattern, a roller called the dandy roll with a raised design on it, is gently pressed against the damp paper web early on in the production process. It causes the watermark by displacing the fibers slightly.

# Rubber

Rubber has a remarkable combination of properties that make it one of our most valuable raw materials. Most important of all it is elastic. It is therefore flexible and absorbs shock well. That is one of the reasons why we use it for the soles of our shoes and for auto tires. It is also water-tight and more or less airtight. And, of course, it rubs out pencil marks. That is how it came to be called 'rubber' in the first place. Originally all of our rubber came from a natural source, the 'juice', or *latex* of a certain tree. But much of the rubber used today is made synthetically in chemical factories.

### Natural Rubber

*Natural rubber* comes from the so-called rubber tree which grows best in regions within a few hundred miles of the equator, where the climate is hot and moist. It grows wild in Brazil, in South America. But practically all of the world's natural rubber comes from huge plantations of cultivated trees in Malaysia and Indonesia, in southeast Asia. These cultivated trees are the 'descendants' of seedlings raised in Kew Gardens, in London, from the seeds taken from wild rubber trees in the Brazilian jungle in 1876. Improved strains have since been developed which yield more latex.

.The latex is collected from the trees by *tapping*. Workers cut a narrow, slanting groove about half way around the trunk. It is just deep enough to pierce the living layer of cells beneath the bark. At the bottom of the cut the workers attach a metal spout, and below it, a cup. The milky white latex oozes from the cut and drips into the cup.

The latex is collected and taken to the plantation factory, where it is processed. Only about a third of the latex is pure rubber. The rest is mainly water. The rubber is present as tiny, dispersed drops, or *globules*. To make what is called *crude rubber*, the latex is treated with formic acid. This makes the tiny globules of rubber coagulate, or form larger, solid lumps.

The yellowish-white spongy mass of crude rubber is then rolled into sheets to squeeze out all the water. Afterwards the rolled sheets are hung for several days in a hot smoke-house. The smoke dries and helps to preserve the rubber, which changes in color to an amber-brown. The crude-rubber sheets are then baled for transport to manufacturers for further processing.

The latex is not always processed in the way described above. Sometimes it is shipped to the manufacturers in bulk liquid form.

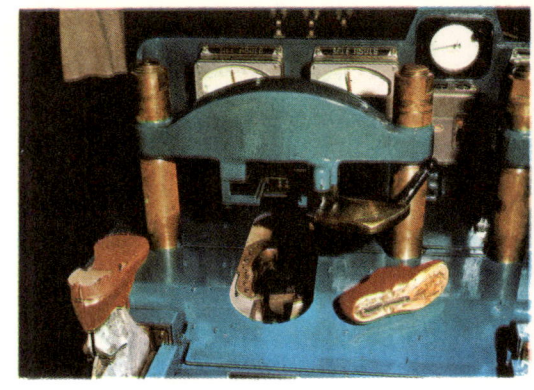

A great deal of synthetic rubber is used these days for all kinds of applications. This picture shows its application in footwear. Both the uppers and the sole and heel are injection moulded from synthetic rubber in this machine.

In this form it is more suitable for making certain products, such as foam rubber and rubber gloves. Before it leaves the plantation, however, a lot of the water it contains is removed. Remember that about two-thirds of the latex is worthless water. The latex is generally concentrated in a centrifuge. This machine rotates at high speed and literally flings much of the heavier water away from the rubbery part. The concentrated latex is then treated with ammonia to preserve it and to keep it from coagulating during transit.

### Synthetic Rubber

*Synthetic rubbers* are man-made products of the chemical industry. They include a wide range of rubber-like materials which are really special kinds of plastics. The most widely used synthetic rubber for general purposes is made from two chemicals obtained from petroleum: styrene, a liquid, and butadiene, a gas. Under certain conditions

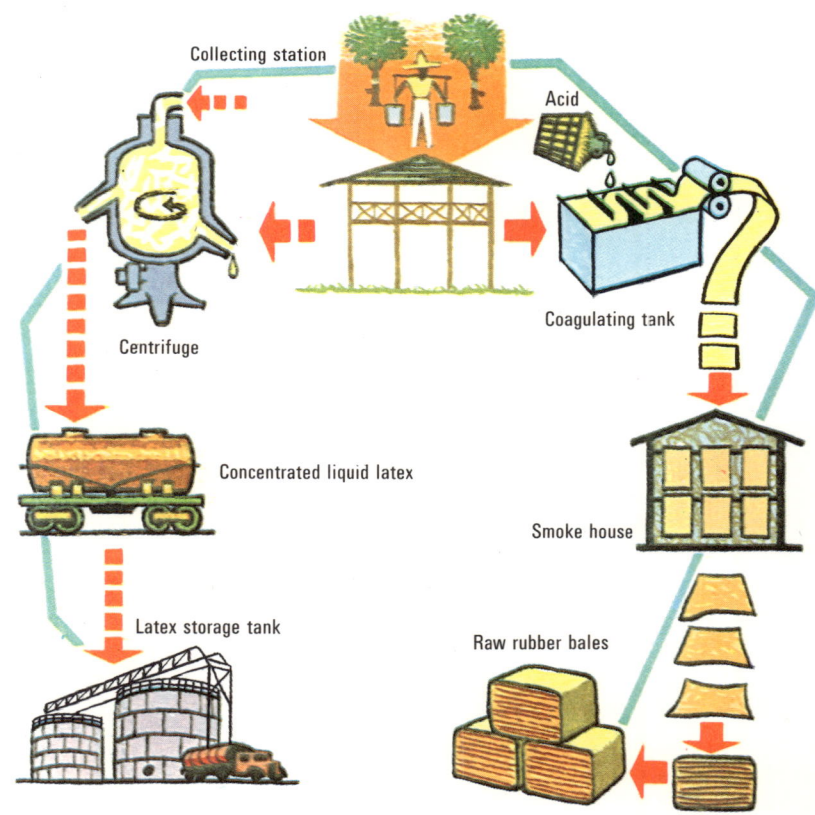

This diagram shows the processing necessary to convert the milky latex from the rubber tree into usable products—either concentrated latex or raw rubber.

styrene and butadiene (in a compressed, liquid form) combine to form a milky white liquid which is also called latex because it resembles natural latex. This synthetic latex is separated and thereafter processed in much the same way as natural latex.

In addition to styrene-butadiene rubber, there are a number of special-purpose rubbers. Among the most interesting are the *silicone* rubbers. They are remarkable in that they remain tough and elastic at very low and at very high temperatures. *Butyl* rubbers are valuable because they hold gases better than most other rubbers. One of their main uses is for the inner tubes and linings of tires. Most synthetic rubbers, and especially *nitrile* rubber, have greater resistance to gasoline, oil, and grease. That is why they are so valuable for making tires, which often come into contact with such substances.

**Processing Rubber**

When dry, raw rubber from the plantation or from a chemical factory arrives at the manufacturers, it first has to be worked into a soft and pliable, dough-like form. This may be done by machines that roll, knead and mangle. Heating and adding softening agents called *plasticizers* quickens the process. In the next stage of manufacturing, called *compounding,* a variety of materials are added to the rubber to improve its properties. Fillers such as clay make it easier to work; pigments such as carbon black increase its strength and durability; sulphur brings about a toughening process called *vulcanization* later on in manufacture.

After compounding, the rubber is ready to be shaped by a variety of methods. Rubber sheet is made by *calendering*. The soft rubber is rolled, or 'mangled' between sets of rollers until it is of the correct thickness. Rubber tubing and strips of various shapes are made by *extrusion* – forcing warm, soft rubber through a shaped hole, or die. Tires, hot-water bottles and toys are made by *moulding*. The rubber is forced into a mould with the shape of the article to be manufactured.

*Dipping* is a method used to fashion products such as rubber gloves, balloons, and 'gum boots,' either from a rubber solution or from liquid latex. A mould of the desired shape is dipped into the liquid. The thin layer that sticks to it is then dipped in an acid bath to coagulate it. The process is repeated until the desired thickness is built up. Foam rubber for upholstery and mattresses is made by beating up the latex into a frothy 'jelly' or by adding a foaming agent to it.

The final stage of manufacture is *vulcanization*. It is often carried out at the same time as shaping. It simply involves heating rubber containing sulphur. The sulphur combines with the rubber and makes it harder, tougher and more elastic; most important, it is less affected by changes in temperature. Natural rubber by itself becomes sticky on a hot day and hard and brittle on a cold day. When a great deal of sulphur is added to vulcanize the rubber, a very hard substance called ebonite is produced. It is used, for example, for the cases of some automobile batteries.

A plantation worker making a fresh cut in the trunk of a rubber tree. The milky latex gradually oozes out and collects in the cup below.

A good proportion of the natural and synthetic rubber produced goes into tire-making. The picture shows newly moulded tires being removed from the moulding press in which they were shaped and vulcanized.

# Printing

The invention of movable type in about 1456 by Johannes Gutenberg of Mainz in Germany was one of the great milestones in the history of human communication and knowledge. Gutenberg's printing press was a very crude device, somewhat like a wine press, and it was operated manually. Today the latest high-speed rotary printing presses are vast electrically driven machines, with bank upon bank of revolving cylinders. They can print five or more colours at the same time. Their output is quite incredible – something like 60,000 printed sheets an hour!

Gutenberg did his printing by inking the surface of the type he had assembled into lines and pages and pressing a sheet of paper against it. The ink was transferred to the paper to form the printed words. This method of printing is called the *letterpress* method and is the one most widely used even today, especially for newspapers. But other methods of printing, such as offset-lithography and gravure have been developed. Offset-lithography particularly has become increasingly important for book and magazine printing in recent years. This book is printed by offset-litho. But no matter what method of printing is used, the words to be printed – the *copy* – must be set in type first.

## Typesetting

The copy is set up in type in the composing room. For most ordinary work, type is set by machine. For special 'display' work, for advertisements or posters, for example, it may be set by hand. The main typesetting machines are the Linotype and Monotype machines, of which the Linotype is by far the most widely used. In both machines the type is cast in hot metal (an alloy known as *type metal,* see page 134) in tiny moulds. As you may guess from their names, the Linotype machine casts the type in lines, the Monotype in individual letters.

On the *Linotype* machine the operator taps out the words on a keyboard like that of a typewriter, only much bigger. As each letter is tapped out, a tiny mould, or *matrix* for this letter is carried into its proper place in the line of type being set. When the line is almost filled, the operator presses a key which pushes wedge-shaped pieces of metal between the words so that they completely fill the line to make an even right-hand edge. This is called 'justifying' the line. The full line is transferred automatically to the casting mechanism where hot metal is forced into the matrices. It hardens to form a single line of type called a *slug.*

On the *Monotype* machine the operator types out the copy on a keyboard. The keys perforate a paper tape to give a pattern of holes. The tape is then fed into the casting mechanism. The tape matrices are set in a 'case', which moves into position over a casting box according to instructions from the tape. Each letter is ejected from the matrix as it is cast to form part of the line of type. One advantage of the Monotype method is that individual letters can be replaced if they are wrong.

A more recent typesetting method which has increased in importance is *film setting,* in which the setting is done photographically. Filmsetting is much faster than hot-metal setting. It is used particularly in conjunction with offset-litho printing, which is itself a photographic method. For letterpress printing the film-set type is used to make the metal printing plate.

After typesetting by whatever method, the type is 'proofed', or printed on long

One of the two main hot-metal typesetting processes is Linotype. Here you see a Linotype operator at the keyboard of his machine typing out the copy to be set. Most newspapers are set by Linotype.

The priciples of the three main printing processes. In letterpress printing the printing surface is raised. In lithography the printing surface is flat. It is treated so that ink adheres only to areas that are to print. In the intaglio process the printing surface is recessed, surplus ink being scraped off the printing cylinder by a doctor knife.

 Letterpress — Ink roller

 Lithography

 Intaglio — Doctor knife

sheets, usually called *galleys* or *galley proofs*. The proofs are checked for accuracy, and corrections are made to the type as necessary.

**Letterpress Printing**

The blocks of corrected type are then *made up*, or assembled into pages. Any photographs have to be converted into metal blocks. This is done by a process called *photoengraving*. Photographs are themselves photographed through a fine-mesh screen which splits up the image into a pattern of dots. An etched plate, or engraving called a *half-tone* is made from the photographic negative. When inked and printed, it reproduces the original photograph as a pattern of lighter and darker dots. You can see them clearly in newspaper photographs.

Color pictures are reproduced in a similar way except that the colours are first split into different combinations of the primary colours blue, green, and red. Separate engravings are then made for each colour, as well as for black.

The made-up pages must then be converted into printing plates. The type and engravings themselves would soon wear away if they were used directly for printing. Two kinds of duplicate metal plates are used for printing – *stereotypes* and *electrotypes*. They are made by pouring hot metal into moulds of the made-up type pages.

Most printing these days is done on high-speed rotary presses. The printing plates are curved to fit round the printing cylinder. The paper to be printed passes between the printing cylinder and an impression cylinder. Most rotary presses are fed with a continuous roll, or 'web' of paper which is cut up into sheets as it passes through the machine.

**Offset-Litho and Photogravure**

Letterpress printing is done from a raised surface and is called *relief* printing. Offset-litho printing is done from a plane surface and is called *planographic* printing. The flat metal printing surface is prepared photographically from a printed copy of the made-up type or from film-set type. The plate is treated so that it attracts the greasy printing ink only where the type image is.

Printing takes place on a rotary press with three cylinders, one above the other. The printing plate is wrapped round the upper plate, or printing cylinder. It transfers, or *offsets* an inked image onto a rubber offset cylinder, which in turn transfers it to paper pressed against it by an impression cylinder.

Gravure, or *photogravure* printing is done from a pitted, or recessed surface and is called *intaglio* printing. It is used mainly for reproducing illustrations. A photographic method is used to etch the pictures and words in a copper cylinder. On the presses, the etched pits retain ink, which is transferred to the paper by pressure. Deeper pits give a greater density of colour than shallow ones.

The majority of printing these days is done on high-speed rotary presses, the printing surface being cylindrical. The picture above shows a photogravure cylinder being lifted onto the press. The picture below shows the printing plate of a letterpress cylinder machine.

# Heat

Heat is a very familiar form of energy. We rely on the heat from the Sun, from burning fuel, or from an electric fire to keep us warm. When fuel burns, heat energy is produced as a result of chemical action, usually the combination of carbon with oxygen. This chemical energy has been changed into heat energy. When we switch on an electric fire, electricity flows through the wires and produces heat. Electrical energy has been changed into heat energy. When we apply the brakes of an automobile, the hard brake linings rub against the whirling brake drums and produce heat by friction. Mechanical energy has been changed into heat energy.

Thus all these kinds of energy are equivalent. Energy can be changed in form but can neither be created nor destroyed. Once it was thought that heat was a substance called *caloric*, present in all objects. But the British scientist James Prescott Joule (1818-1889) demonstrated in a famous experiment that heat is just another form of energy. He determined what is called the *mechanical equivalent* of heat. Joule found that a definite amount of work can be turned into a definite amount of heat. This is one way of stating what is known as the *first law of thermodynamics*. One way of stating the *second law* of thermodynamics is that by itself heat cannot flow from a colder to a hotter body. There are other laws, but they are more difficult to explain.

Of course, it follows that heat energy can be converted into other forms of energy, too. Steam engines and turbines and internal combustion engines are devices for converting the heat energy from burning fuel into mechanical energy of motion. They are therefore classed as *heat engines*.

## Expansion

When you heat a cupful of milk for a bedtime drink, and come to pour it back into the cup afterwards, you find that you have milk left over. The milk has expanded in volume as it has heated. Most things in fact expand when they are heated: liquids expand more than solids, gases more than liquids. We can easily see why this happens. The molecules in the hotter matter have more energy and therefore require more room to move about. Gases expand more than liquids and solids because their molecules are less affected by the attraction of other molecules.

Solids do not expand much, but their expansion cannot always be ignored. Expansion joints have to be left in steel railway lines and bridges and in concrete roads to prevent them buckling in hot weather when they expand. The proportionate increases in length, area, or volume of a material for each degree rise in temperature are expressed in terms of *coefficients* of linear, area, or cubic expansion respectively. The difference between the coefficients of expansion of two materials may be utilized to advantage, as in some kinds of thermostats, which are devices designed to control temperature.

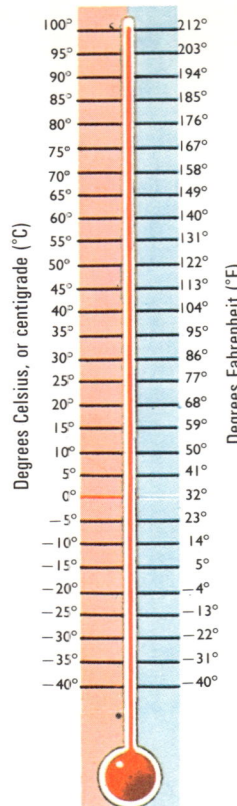

We usually measure degree of hotness, or temperature with a thermometer graduated in degrees centigrade or Fahrenheit. The above gives centigrade-Fahrenheit equivalents for the range between −40°C and 100°C, or −40°F and 212°F. Note that 5 centigrade degrees equal 9 Fahrenheit degrees.

White-hot steel being poured into ingots. Iron, in common with many other materials, changes in color when it is heated. It becomes a dull red, then bright red, and finally white as its temperature rises. The light and the heat are indeed related for they are ways in which the iron loses its surplus energy. They are both forms of electromagnetic radiation. Light is a visible manifestation of energy release. Heat is radiation we can feel but not see. It lies beyond the red end of the visible spectrum in the so-called infra-red region.

## Measuring Temperature and Heat

Some substances expand in a regular way as they become hotter. They may therefore be used in devices to measure degree of hotness, or in other words *temperature*. We call these temperature-measuring devices *thermometers*. The most common kind is the mercury thermometer, in which the length of a thin column of mercury is used to indicate the temperature. Alcohol is another liquid used in thermometers.

The two most common scales for thermometers are the Celsius (or centigrade) and Fahrenheit scales. Temperatures are expressed as so many degrees Celsius (°C) or degrees Fahrenheit (°F). The two fixed points by which most thermometers are calibrated are the freezing point (0°C or 32°F) and the boiling point (100°C or 212°F) of water.

The temperature of a substance is really a measure of the kinetic energy – the energy of movement – of its molecules. The lowest temperature that can exist anywhere is when molecular movement ceases. This is what scientists call *absolute zero*. It is equivalent to a temperature of −273°C or −460°F.

The heat, as opposed to the temperature, of a body is generally measured in units called *calories*. 1 calorie is the heat required to raise the temperature of 1 gram of water by 1 °C. The appropriate unit in the British system in terms of pound and degree Fahrenheit is the British Thermal Unit or BTU. 1 BTU is the heat required to raise the temperature of 1 pound of water by 1°F. All substances vary in the amount of heat they can absorb. Their heat capacity is generally expressed in terms of their *specific heat* – the heat required to raise the temperature of a unit mass by one degree. Thus the specific heat of water is 1.

When materials are heated, they tend to expand. Solids expand less than liquids and gases. The amount of expansion for a given temperature rise varies from material to material. The diagram gives comparative expansions for some common materials. Invar is a special iron-nickel alloy formulated so as to have an extremely low expansion. It is used for example for balance wheels in watches whose performance would otherwise be adversely affected by temperature change.

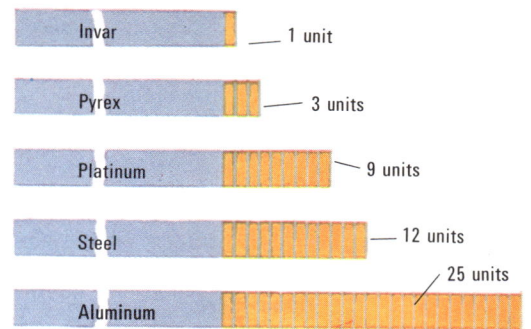

## Heat Transfer

If you put one end of a poker into the fire, you soon notice that the handle becomes hot, too. Heat has travelled along the metal. This form of heat transfer is called *conduction*. The molecules of a hot substance become more energetic and pass on some of their energy to adjacent molecules, which in turn pass on some of their energy to the next molecules, and so on. Metals conduct heat well. Substances such as fiberglass, brick, and cork conduct very little heat. They are called *insulators*.

In a gas, such as air, or a liquid, heat is transferred by a process called *convection*. The heat from an electric fire, say, heats the air. The air expands and becomes lighter than the cold air surrounding it, and therefore rises. Cold air moves in to take its place and in turn expands and rises. Thus, heat is transferred by the movement of the warm air in a so-called convection current.

Heat reaches us from the Sun, but there is no substance in space to conduct or convect its heat to us. The heat energy reaches us by *radiation*. Heat is just one form of electromagnetic radiation; light is another. Heat waves have a longer wavelength than light and lie beyond the red end of the visible spectrum. Therefore they are often called *infra-red rays* (see page 42).

## Latent Heat

Heating a solid usually changes it into a liquid eventually. And heating a liquid eventually changes it into a gas. Thus heat can bring about what is called a *change of state*. On the molecular scale you can visualize the molecules becoming more energetic as heat is added and gradually overcoming the attraction of the other molecules around them to become first a liquid then a gas. At the two great transition points – melting and boiling – the molecules require extra large amounts of energy (heat) to break away from their previous solid or liquid state. The amount of heat required per unit mass for such a change is called the *latent heat*. For melting, it is called the latent heat of *fusion;* for boiling, the latent heat of *vaporization*.

The automobile engine is an internal combustion heat engine which converts the heat obtained by burning fuel into mechanical energy for propulsion. To withstand the heat of combustion the engine is cooled by circulating water through it. Heat is transferred from the engine into the water by conduction. The hot water circulates to the top of the engine, partly by convection but mainly by pumping, and thence to the radiator. It gives up its heat by conduction to air passing over the radiator tubes, and then recirculates.

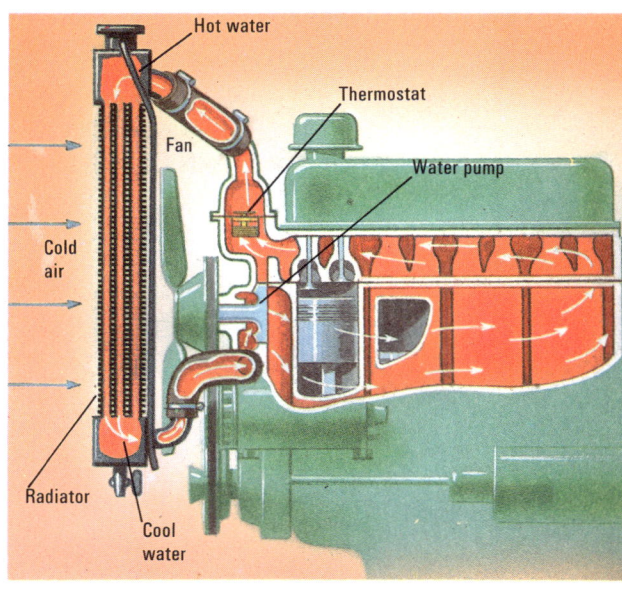

# Steam Engines and Turbines

When you heat water above its boiling point of 100°C (212°F), it turns into steam. For every 1 cubic inch of water that evaporates, you obtain 1 cubic foot of steam. When you cool a closed cylinder full of steam, it condenses, or changes back into water. Since the water occupies only about 1/1700th of the volume it occupied as steam, a partial vacuum will result. These two principles of expansion and condensation form the basis for the operation of steam engines.

The nineteenth century was the great age of steam. Steam engines were the prime power source on land and at sea. Nowadays it is a different story. Internal combustion engines such as the gasoline and diesel engines and gas-turbine, or jet engines have more or less supplanted the steam engine in industry and transport. But steam engines in one form or another still have quite a number of uses.

Basically there are two kinds of steam engines. The traditional steam engine is a *reciprocating* engine (as a gasoline engine is) in which a piston moves up and down. The more modern variant of the steam engine is the *steam turbine,* which is a *rotary* engine. The reciprocating steam engine is little used today because it is very inefficient and wasteful of fuel. A few old ships, for example, still have this kind of engine. So have the few remaining steam locomotives (see page 54). Steam turbines are very much more efficient. They can utilize almost 40 per cent of the energy supplied to them. They are three times as efficient as the reciprocating engine (and almost twice as efficient as the gasoline engine!). Steam turbines power most of the world's ships and the majority of the world's electricity generating stations (see page 85).

A proud relic of the age of steam—a Fowler steam roller. Steam traction engines were once prominent on the transport scene before the internal combustion engine took over. Enthusiasts maintain old engines like this with loving care and display them at rallies.

## Steam Boilers

All steam engines require an efficient boiler to turn water into steam. Modern engines operate at high pressure and so boilers must be able to generate, and be strong enough to withstand, high-pressure steam. There are two main types of boiler, both of them designed so that as much of the water as possible is in contact with the heat from the furnace fire. The *fire-tube* boiler has tubes running through it, through which the hot gases from the furnace pass before they enter the chimney flue. In the *water-tube* boiler it is the water that is held in tubes while the hot furnace gases pass over them.

Part of a 60 megawatt, three-stage steam turbine stripped down so as to show the rotor. Note the progressive increase in the size of the turbine wheels. This is done so as to extract the maximum amount of energy from the steam as its pressure drops on its passage through the turbine.

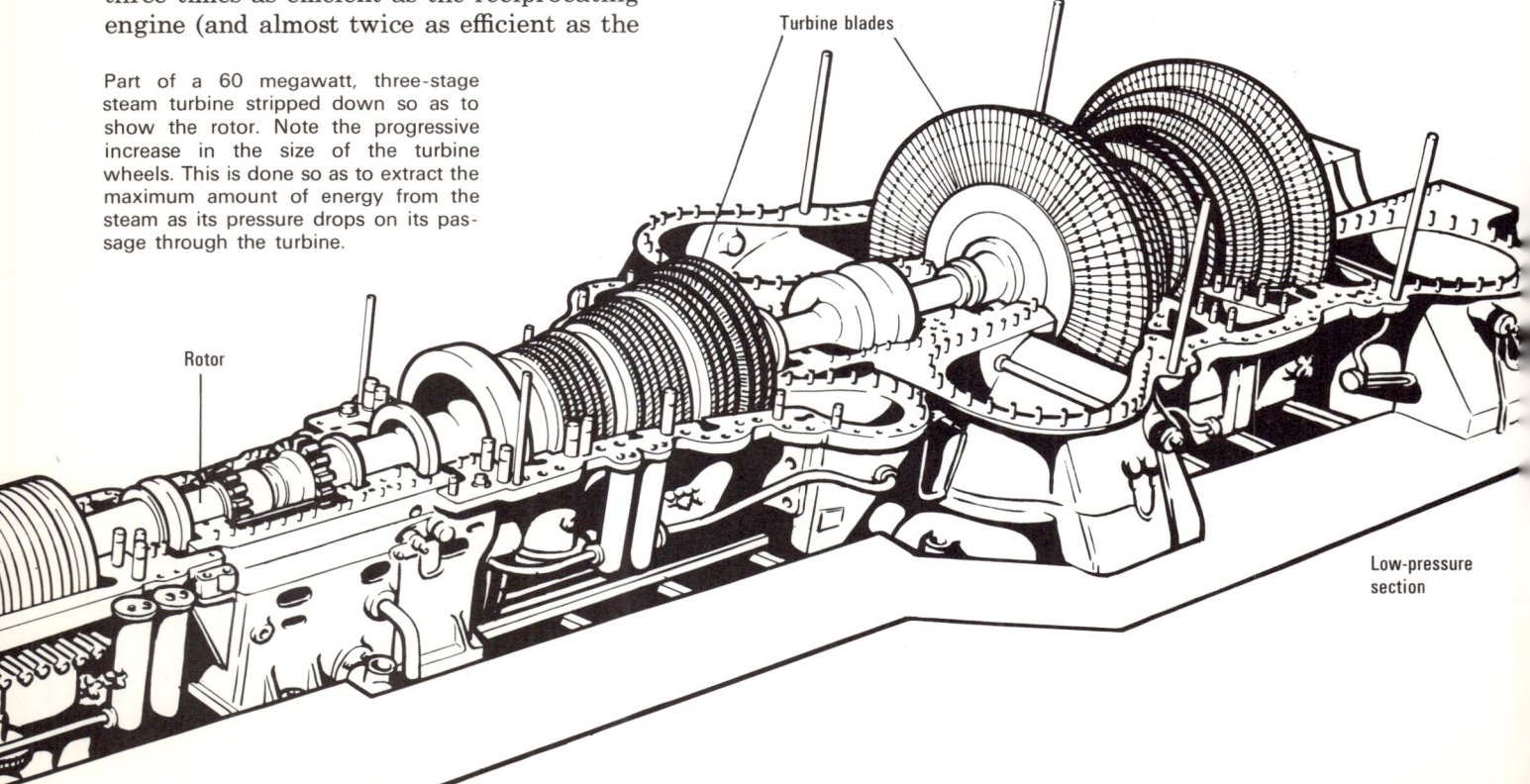

The water-tube boiler is the most efficient of the two and is the one used in conjunction with most high-powered turbines.

## The Steam Turbine

A steam turbine consists basically of a big shaft, or *rotor,* which rotates inside a cylindrical casing. Around the rotor are sets of wheels with curved blades. High-pressure steam passing through the turbine spins the blades and turns the rotor. Fixed on the side of the casing are rows of curved, fixed blades which are designed to direct the steam onto the rotor blades at the best possible angle. These fixed blades form the *stator.* The design of the turbine blades varies. In an *impulse* turbine the blades are well curved and are forced round by the pressure of the steam on them. In a *reaction* turbine the blades are much shallower and spin mainly by reaction. Most modern turbines use both kinds of blading.

This kind of turbine, in which steam passes along the rotor is called an *axial-flow* turbine. It is much more widely used than the *radial-flow* turbine in which steam is introduced at right-angles to the rotor.

Each set of fixed blades and moving blades is called a *stage.* Steam turbines are made up of many such stages. To use the expansion of the steam to full advantage, each stage is slightly larger in diameter than the one before. The pressure of the steam drops as it passes along the turbine, and the turbine is sometimes separated into individual sections: high-pressure, intermediate-pressure, and low-pressure. This arrangement gives greater efficiency. The steam leaving the low-pressure stages passes into a condenser cooled by cooling water. The partial vacuum produced improves the flow of steam through the turbine.

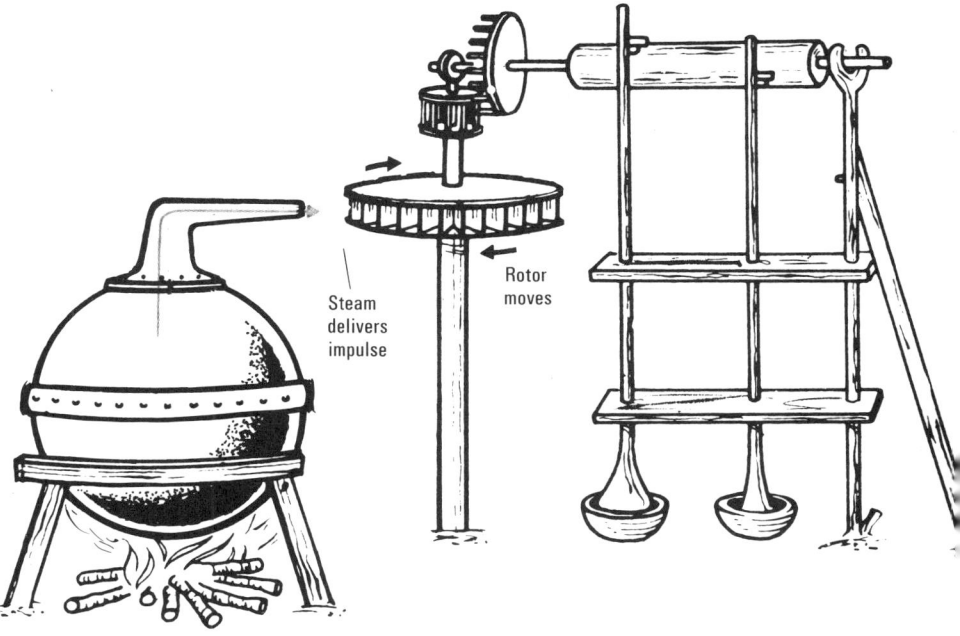

Although the steam turbine as we now know it was perfected by Charles Parsons in the 1890s, an interesting prototype was built in about 1629 by Italian architect Giovanni Branca. The picture above shows the kind of set-up Branca devised.

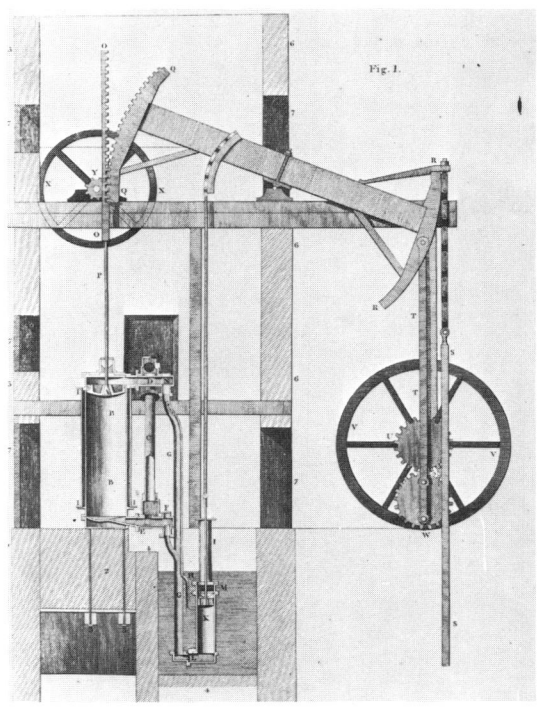

A diagram of one of Watt's steam engines. Unlike Newcomen's earlier inefficient engine, it has a separate condenser to condense the steam, allowing the main engine cylinder to remain hot. The up-and-down movement of the piston in the cylinder is communicated to the wheel on the right by means of a rocking beam and a sun-and-planet gearwheel arrangement.

### The Steam Engine Develops

Thomas Savery (1650-1715), who was a captain in the English army, built one of the first practical engines utilizing steam power. It was a simple pump, having no moving parts, only valves. Savery used the suction created by condensing steam to draw water from wells and mines.

But it was a Cornish ironmonger, Thomas Newcomen (1663-1729), who invented a practical engine, again a pump for mines. It incorporated a piston in a cylinder. A chain from the piston rod led over a rocking beam and was attached to the rod of the water pump. The rod was heavily weighted. This pulled the beam over as steam was introduced into the cylinder. The steam was condensed by a jet of cold water, producing a partial vacuum. The pressure of the atmosphere pushed down the piston to the bottom of the cylinder. Then more steam was introduced and the cycle was repeated.

Newcomen's engine was terribly inefficient. The great fault in design was that the cylinder was alternately heated and cooled. The incoming steam had to reheat the cylinder every time. The first to recognize this problem was the so-called 'father of the steam engine' James Watt (1736-1819). His solution was to condense the steam in a *separate condenser,* joined through a valve to the cylinder, which always remained hot. In this way

James Watt

Watt improved the efficiency of the steam engine out of all recognition.

Watt built his first full-size engine in 1769. Twelve years later he produced an engine which produced power on every stroke of the piston. Steam was introduced to above and below the piston in turn. This *double-action* principle is the one used in reciprocating steam engines today. At the same time Watt designed a 'sun-and-planet' arrangement of gear wheels so that the beam engine could produce rotary power. Later he invented the *governor,* which is an automatic device for keeping the engine running at constant speed. Watt's governor was an ingenious arrangement of rotating balls.

Watt was fortunate in teaming up with a successful industrialist from Birmingham, Matthew Boulton. Boulton had the facilities and skilled workmen Watt needed to build practical engines. By the 1780s the steam engines of Boulton and Watt, built at the famous Soho (Birmingham) works or under licence elsewhere, were pounding and hissing away in many countries. Strangely enough for such a brilliant engineer, Watt scorned the use of high pressures in steam engines. But in the early 1800s Richard Trevithick designed a successful high-pressure engine.

# Coal and Coal Gas

## Coal

For centuries coal has been one of the most important fuels, and it still is, although it is less important now than it once was. Other fuels, notably petroleum and natural gas, are replacing it. But coal is still widely used for domestic heating and in power stations for generating electricity. Enormous quantities of coal are made into gas, and into coke for use in metal-smelting furnaces, and elsewhere. And coal provides valuable raw materials for the chemical industry, too. But, here again, petroleum and natural gas are supplanting coal.

We mine coal from the ground, just as we mine minerals and ores (see page 128). The deposits may lie on or just below the surface or deep underground. It is in coal-mining that the greatest advances in mining methods have taken place. Continuous mining machines hew the coal from the coal face, where once the miners slaved. Lengthy conveyor systems convey away the hewn coal. Diesel and battery locomotives haul wagons along miles of underground tunnels.

We tend to call coal a mineral, too. But there is a great difference between coal and, say, iron ore. Iron ore is inorganic in origin, being formed as a result of inorganic, or non-living processes inside the Earth's crust. Coal, on the other hand, is organic in origin, being derived from plants which once lived on the surface. It consists mainly of carbon The actual proportion of carbon varies from one type of coal to another.

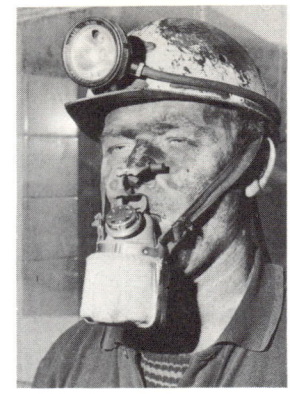

Coal miner wearing face mask, which enables him to breathe in dusty conditions underground. Note also the helmet and light. The cord leads to a battery strapped to his waist.

A great deal of the coal mined is extracted from surface, or opencast mines. Opencast working is possible where the coal seams lie within a few hundred feet of the surface. Huge dragline excavators, like that below, strip the overlying dirt from the seams, and then power shovels move in to scoop up the coal and load it into trucks.

Most of the coal in the world is derived from plants that grew in the Carboniferous Period of the Earth's history, which lasted from about 350 million to about 280 million years ago. During this period whole regions of the Earth were covered in swamps, in which lived gigantic ferns and trees. The plants died and fell into the swamps. Bacteria acted upon the dead vegetable matter and caused it to decay. As time went on, the decomposing vegetable matter became covered with mud and clay, then more plants grew, died, decayed, and so on. In places layer upon layer of decayed plant remains built up, sandwiched between mud and clay sediments. In time the Earth's crust folded this way and that; heat and pressure changed the sediments into rock and forced out most of the water from the decayed matter, which changed into coal. Thus were the coal measures formed.

Not all the coal was subjected to the same degrees of heat and pressure. Therefore there are different kinds of coals. *Peat* is not really a true coal, but it represents the first decomposition stage in the coal-forming process. *Lignite*, or *brown coal*, represents a further stage in the process. It contains a little less moisture, and somewhat more carbon than peat. Next there is *bituminous*, or *soft coal*, the most plentiful of the coals. It has a high carbon content and contains little moisture. The final product of the coal-making process is *anthracite*, or *hard coal*, which is over nine-tenths carbon and contains scarcely any moisture at all.

## Coal Gas

If coal is heated in a vessel out of contact with the air, it gives off a mixture of gases that burn readily. Also formed are vapours of other substances, together with tar and solid coke. These are the products of what is called the *destructive distillation* of coal. They are all extremely useful. In particular, coal gas is a valuable fuel for both domestic and industrial use.

Coal gas is a mixture of gases, and is made up mainly of hydrogen and methane, with a little carbon monoxide and ethylene. With these combustible gases there are small amounts of non-combustible gases such as carbon dioxide, nitrogen, oxygen, ammonia, and hydrogen sulphide.

The coal is distilled in huge fire-clay ovens, or *retorts*. The gas produced passes from these through a series of pipes to coolers, or *condensers*. Much of the tar given off during distillation and some of the ammonia separate from the gas in these pipes and in the condensers. The gas is then driven by a pump called the *exhauster* into *scrubbers*,

Above: Most coal is now won underground by means of powerful machines, such as this trepanner. A chain conveyor removes the cut coal.

Below: A control center on the surface continuously monitors the conditions in the underground coal mine or colliery. Nothing can be left to chance.

which remove any remaining ammonia in the gas by washing with water. The ammonia dissolves in the water. Next, *purifiers* remove hydrogen sulphide before the gas passes through meters into the gasholders, or *gasometers*. These gasometers are massive cylindrical tanks that store the gas above water. The more gas they contain, the higher they ride up above the water. They are the most prominent feature of any gasworks.

Usually *water gas,* or *blue gas* is added to coal gas to increase its heating value for domestic heating and lighting. This mixture is often known as *town gas*. Water gas is made by blowing steam through white-hot coke. It is made up mainly of carbon monoxide and hydrogen.

The by-products of gas-making are all of immense value to the chemical industry. Coke is used widely as a fuel and in all manner of processes, such as smelting iron ore. Coal tar yields a wide range of useful chemicals from which such products as dyes, drugs, insecticides, and plastics can be made. Ammonia is made into fertilizers.

# Petroleum and Natural Gas

## Petroleum

Petroleum has become our most important fuel. It is an invaluable raw material, too. Something like a thousand million tons of petroleum are produced every year from vast oil-fields in the United States, Russia, South America, the Middle East, and elsewhere. This is equivalent to about 8,000 million barrels, or over 30,000 million gallons! This seems an awful lot, but just think of the wide range of products we obtain from petroleum, or *oil,* as it is usually called. (The word 'petroleum' by the way, literally means 'rock-oil'.)

First and foremost there are the fuels that drive all kinds of engines: gasoline for automobile engines; diesel oils for the diesel engines of heavy trucks, locomotives, and ships; kerosene, or paraffin oil for the engines of jet aircraft and rockets, for industrial gas turbines, and for use in the home. Then there are the lubricants, the heavy oils and greases which literally keep the wheels of industry turning.

But oil is more than just a fuel. It is a reservoir of raw materials which the chemical industry converts into thousands of useful products, among them plastics and synthetic rubber, cleaning fluids and many other kinds of solvents, man-made fibers, and explosives.

Oil has replaced coal as a fuel for many applications because of several factors. It can be handled a great deal more easily and cleanly. It can be stored and transported more conveniently and occupies less space. Weight for weight, it has one and a half times the heating value of coal. And it produces little ash.

Crude oil is a dark brown or black (sometimes slightly greenish) treacly liquid when it comes out of the ground. It is made up almost entirely of various kinds of *hydrocarbons,* which are compounds that contain only hydrogen and carbon. The crude oil is of little use as it stands. First it must be separated into more useful parts, or *fractions.* This is done by the various processes of refining (see page 79).

## Natural Gas

Natural gas is a mixture of flammable gases which has also become an important fuel. It may be used by itself or mixed with gas made from coal or oil. A great deal of natural gas is used as a raw material for the chemical industry in much the same way as petroleum is. Natural gas is often found in association with petroleum in the great oil-fields of the world, although it is sometimes found alone.

Above: Oilmen set off an explosion and record the tremors resulting with a seismograph. Seismic surveying is a vital part of oil prospecting.

Below: An offshore drilling rig in position. Its legs are extended down into the seabed to make a stable drilling platform.

The United States and Russia are the world's biggest producers of natural gas. There are also vast gas-fields in Canada, North Africa, and the Middle East. In Europe, natural gas is produced from fields in Italy and France and under the North Sea, off the coasts of Britain and the Netherlands.

Like petroleum, natural gas is composed of hydrocarbons, which give off a lot of heat when they burn. The main hydrocarbons in natural gas are methane, ethane, propane, and butane. Propane and butane are usually removed and used separately as a fuel. They are obtained from natural gas by liquefaction. Under slight pressure they will liquefy, and they go by the name of liquid-petroleum gases, or LPG. Liquefied, they are used, separately or together, as a portable gas supply – for caravans, boats, isolated farms, and so on. Gas cigarette-lighters run on butane fuel, too.

## Prospecting

Oil and natural gas were formed millions of years ago (and presumably some still is being formed today). Tiny plants and animals which lived in shallow water died and sank to the bottom. There, certain bacteria caused

them to decay into a dark, slimy ooze. Eventually layers of mud and other sediment, shifted by rivers or the sea, piled on top of the ooze. In time the heat and pressure of the Earth's crust turned the sediments into rock, and the plant and animal ooze into oil and gas.

The gas and oil seeped through layers of porous rock until they came up against more solid rock and were trapped. They accumulated at the top of folds in the rock strata; or where a non-porous rock had slipped down across the oil-bearing, porous rock; or at the sides of salt domes which had thrust themselves up through the strata from deep underground.

Today, when prospectors are searching for oil, there is little evidence above ground to indicate whether oil lies below. But the oilmen do know from experience the kinds of rock formations in which oil is likely to be trapped – folds, fault-traps, salt domes, and so on. To a certain extent, the surface features give them some idea of what strata lie beneath. But they rely more on readings taken with certain instruments.

The *gravity meter* measures the strength of gravity, the 'pull' of the Earth. Variations in gravity provide a clue to the nature of the rock formations. The *magnetometer* records changes in the Earth's magnetism. Such readings again provide information about the rock strata. But the most useful of the prospector's instruments is the *seismograph*, a device that measures earth tremors. The prospector sets off explosions in the ground and records the shock waves on seismographs placed at various distances from the source of the explosion. The shock waves from the explosion travel down to the rock strata, which reflect them back again. The reflected shock waves are then recorded by the seismographs. From the seismograph traces, the shape, and to some extent the nature of the rock formations can be determined.

It is not only on land that the oilmen search for oil and natural gas. As oil-fields on land have become harder to find, the oilmen have extended their search to the seabed. And many profitable offshore fields are in production in many parts of the world.

**Drilling for Oil and Gas**

The crews who drill for oil and natural gas lead a tough and often dangerous life. One year they may be braving the way below zero temperatures of Alaska; the next they may be scorching in the relentless heat and wearying humidity of the Persian Gulf.

Their first job is to set up the drilling *derrick,* or *rig.* The rig is a rigid steel skeleton structure, looking something like an electricity pylon. Its function is to lift the long drilling pipes into and out of the borehole. It holds block-and-pulley systems for lifting. An engine of some kind (usually diesel) provides the power.

The most common form of drilling method is *rotary* drilling. A drilling *bit* at the end of a long pipe is rotated so that it bores into the ground. The bit, which is made of specially hardened steel, is nothing at all like the bit you use to drill holes in wood or metal. The oil-drilling bit is made up of a number of wheels with teeth, like gear wheels, which rotate when the pipe is turning and grind their way through hard rock. A different design is used for soft ground. The bit is attached to the drill stem by a joint called the *collar*.

The drill stem itself is made up of a number of hollow steel pipes, screwed together. The top pipe is attached to a pipe called the *kelly,* which has square or hexagonal sides. A rotating *turntable* on the floor of the drilling rig grips the kelly and turns it, thereby turning the drill stem and the drill bit. When the top of the kelly has sunk to table level, it is hauled out by means of the lifting tackle on the rig. Another section of drill pipe is inserted, the kelly is replaced, and drilling is continued. The drillers often line the hole they bore with a steel *casing* as they go along to prevent possible collapse of the sides.

A very liquid mud is forced down into the drill pipe during drilling. It flushes away the cut pieces of rock from the bit and back up the bored hole to the surface. It also provides cooling and lubrication for the bit.

If the drilling crew are lucky, they strike oil. They quickly bring up the drill pipe and seal off the casing. In some oil-wells, water below or gas above the oil deep underground provides sufficient pressure to force the oil to the surface. The oilmen then install a set of valves at the top of the well to control the oil and gas flow. This set of valves is often called a 'Christmas tree', to which it bears a vague resemblance. In other wells there is not enough pressure deep down to force the oil to the surface. Then the oil has to be pumped out.

Offshore drilling is naturally more difficult. The drilling crew, in addition to their usual problems, have to contend with wind, waves, storms, and sometimes hurricanes! One common way of drilling offshore is by means of a big platform that is a combination raft and drilling area. When in position, it lowers steel legs down to the sea bed. This is the kind of rig that is being used to drill for natural gas in the North Sea, which is not too deep.

The greater part of the world's oil reserves lie in the Middle East, often in desert regions. Here surveyors are mapping a route for a new oil pipeline.

Not a piece of modern sculpture but an oil refinery. In the tall towers distillation and separation of the various oil fractions take place.

Left: A drilling rig in action in the desert. The 100 foot high derrick is used for hoisting the drillpipes. New lengths of pipe must be added as the bit bites deeper into the ground. When the bit has to be changed, the whole pipe must be withdrawn and dismantled.

Right: A close-up of the floor of a drilling rig. The rig operators are withdrawing a length of pipe. The man in the foreground is controlling the hoisting machinery.

Below: An oil refinery at night, a strangely attractive sight. It is rather erie, though. Motors hum, equipment switches on and off, valves open and close, all apparently without human intervention. This is because oil refineries are almost fully automated.

## Transporting Oil and Gas

The oil and natural gas have to be carried from the oil- and gas-fields to processing plants and refineries which may be thousands of miles away. The method of transport used will depend largely on the location of the oil-field in relation to its market. But two main methods are used – pipelines and tankers.

The United States has the greatest underground pipeline system in the world, carrying oil and natural gas the length and breadth of the country. There are also long oil and natural gas pipelines in Russia and the Middle East. Oil from the Middle East is transported to Europe in specially constructed ships called *tankers* (see page 119). They are among the largest ships afloat. The biggest ones have a capacity of more than 2 million barrels.

## Oil Refining

In an oil refinery the thick, black crude oil is processed into a wide range of useful products, from gasoline to paraffin wax. As we mentioned earlier, oil is a mixture of quite a number of different hydrocarbons, which all have different boiling points. This makes it possible to separate them by *distillation*. If you heat up crude oil gradually, the low-boiling parts, or fractions of the oil, such as gasoline, vaporize first. As the temperature rises, the higher-boiling fractions, kerosene and diesel oils, vaporize in turn. The last to boil off are the heavy fuel oils and thick lubricants. In a refinery the crude oil is vaporized in a still, and the vapour goes to a great big tower called the *fractionating tower* which is designed to separate the fractions. The heavy oils separate at the lowest level of the tower, and gasoline at the highest. The various fractions obtained are not simple compounds, but still mixtures of hydrocarbons with only slightly different boiling points.

Gasoline is the most desired of the fractions, but any given crude oil produces only a certain amount of it. Further refinery processes aim at producing more gasoline and other lower-boiling fractions from less valuable fractions. Foremost among these processes is *cracking*. The heavy hydrocarbons from the tower *crack*, or break down into lighter ones when they are subjected to heat and pressure. This is called *thermal* cracking. In *catalytic* cracking, the heavy hydrocarbons are cracked in the presence of catalysts (substances which promote chemical reactions without being themselves chemically altered).

Cracking also yields gaseous products. A process called *polymerization* can be used to convert these gases into more valuable products, such as petrol. Polymerization is, in fact, the reverse of cracking. It involves building up small molecules into large ones (see page 38). And there are many more refinery processes, too, including reforming, hydroforming, and alkylation, all of which aim to produce more valuable products from less valuable crude oil fractions.

Above: Welding a length of natural gas pipeline. Pipelines are used extensively in gas and oil producing areas for transporting the raw materials to refineries or tanker jetties. They are also used for nationwide distribution in countries such as the United States which has abundant gas and oil resources.

Right: Burning off natural gas at an oilfield in Iraq. In some fields natural gas is not abundant enough to be handled economically and so it is burned at the well head after separation from the crude oil.

# Electricity and Magnetism

When you press an electric-light switch, the filament in the light bulb glows white-hot and produces light. Electricity has flowed through the filament and made it give off light and heat. And like heat and light, electricity is just another form of energy.

Electricity flows into our homes through a set of copper wires from the mains supply. They form part of an electrical *circuit*. We complete the circuit, and electricity flows when we press a switch. We break the circuit when we turn off the switch. The flow of electricity in a circuit constitutes an electric *current*. But what exactly is an electric current? To answer this we must look at the tiny atoms which make up all matter.

The atom consists of a central nucleus containing minute particles called protons and neutrons, around which circle even smaller particles called electrons (see page 26). These electrons have an electric charge which we term 'negative'. The protons have an equal, but opposite (positive) charge. Normally the nucleus has exactly the same number of electrons as protons, and is therefore electrically neutral. In certain circumstances an atom may gain an electron and become negatively charged, or lose an electron and become positively charged. Such charged atoms are called ions. A positive charge attracts a negative charge, but repels another positive charge. This is a basic law of electricity: unlike charges attract; like charges repel.

In certain materials, notably metals, the electrons are free to move about from atom to atom. And it is the flow of these free electrons which constitutes an electric current. Many materials, notably rubber, glass, and plastics, have scarcely any free electrons and are very bad conductors. They are known as insulators or dielectrics. Liquids and gases may conduct electricity, too. In this case it is not the electrons which move but the negative and positive ions.

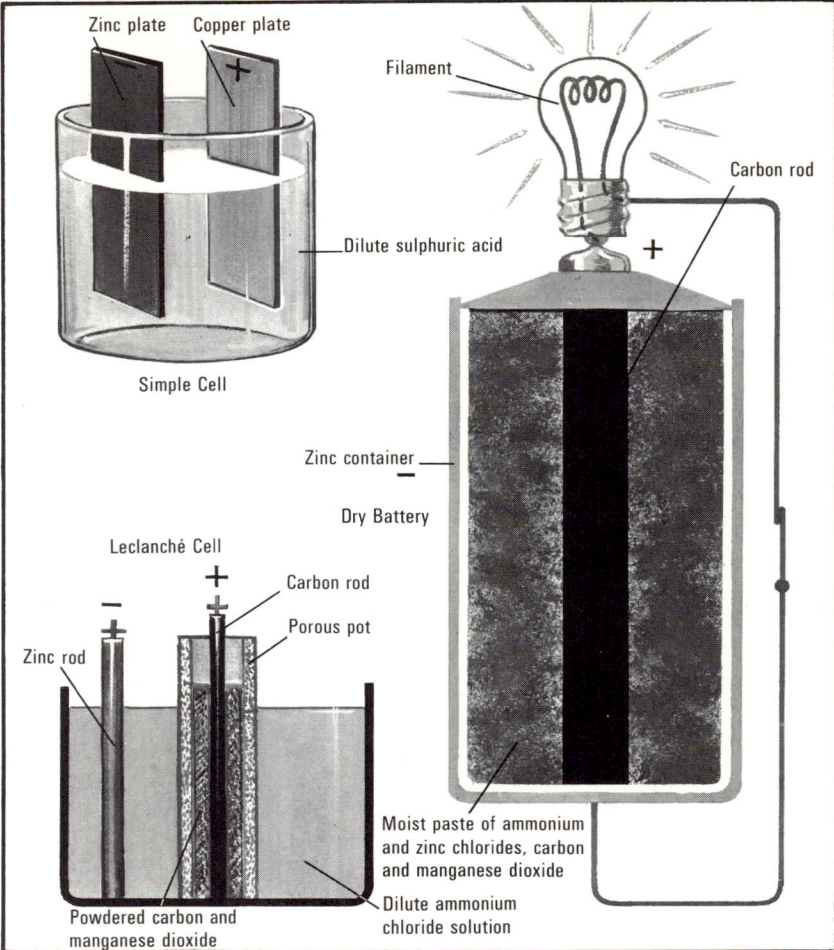

Three simple kinds of electricity producers. The simple cell was first devised by Italian professor Alessandro Volta. The simple cell suffers from polarization, or the accumulation of hydrogen gas on the positive electrode. Leclanché's cell and the dry battery equivalent therefore incorporate manganese dioxide to remove the hydrogen.

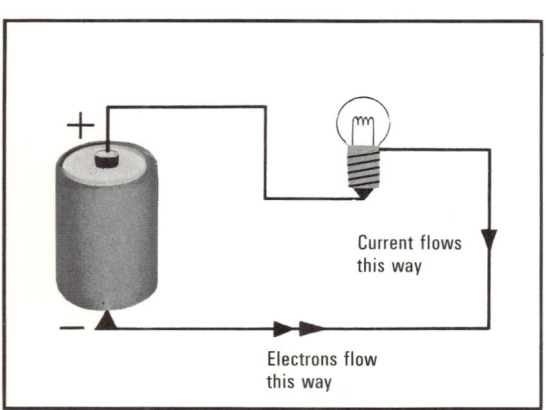

Left: By convention electric current is always considered to flow from positive to negative. In fact the flow of electrons constituting an electric current goes in the opposite direction.

## Voltage, Current, and Resistance

The free electrons in a conductor will not flow of their own free will to make an electric current. 'Pressure' must be applied to them from a source of electricity such as a battery or generator.

The unit of electrical pressure is called the *volt*, after Alessandro Volta (1745-1827) who built the first electric battery in 1799. The rate of flow of electricity in a circuit, that is, the current, is proportional to the

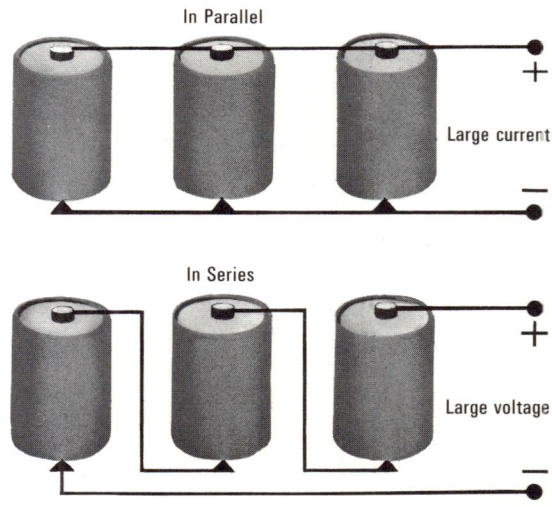

Right: Two ways of connecting batteries together. By connecting all the positive and all the the negative terminals together (in parallel) the voltage remains the same, but the current is trebled. When the positive terminal of one is linked to the negative terminal of another (in series), the voltage is trebled, but the current remains the same.

80

voltage applied. It is measured in *amperes*, or *amps*, a unit named after André Marie Ampère (1775-1836), who was one of the great pioneers of electricity. All materials show some resistance to the passage of electric current, and this resistance is measured in units called *ohms*, after George Simon Ohm (1787-1854), who summed up the relation between voltage, current and resistance in his famous law. Ohm's law states simply that current ($i$) equals voltage ($V$), divided by resistance ($R$).

**Electric Batteries**

The power line electricity we use to heat and light our homes is produced by generators (see page 84). But the electricity to power bicycle lamps, flashlights, and transistor radios comes from *batteries*. They work by changing chemical energy into electrical energy.

The simplest cell is the *voltaic cell,* named after Volta who developed it. It consists of two plates, or *electrodes* dipped into a conducting solution called an *electrolyte*. The electrodes are usually zinc and copper (or carbon), and the electrolyte is dilute sulphuric acid. Electric current flows when the two electrodes are connected by a wire. Zinc becomes the negative electrode and copper is the positive electrode.

The *Leclanché cell* is an improvement on the simple cell because it incorporates a *depolarizer* to remove hydrogen bubbles from the positive electrode as fast as they are formed. In the simple cell the accumulation of hydrogen leads to the cell quickly losing its voltage. The most familar form of cell, the dry cell, was developed from the Leclanché cell. It consists of a negative electrode of zinc, which forms the casing of the cell; a positive electrode of carbon; and a moist paste of ammonium and zinc chlorides, carbon, and manganese dioxide (the depolarizer) to act as the electrolyte.

Such single cells are called *primary* cells. Another important kind of cell is the *secondary,* or *storage* cell also called an *accumulator*. Such a cell must first be charged with electricity from an outside source. The passage of electricity through the cell causes chemical changes at the electrodes. Afterwards, when the cell is used as a source itself, the chemical changes that took place are reversed, and electricity is produced. The best-known example of such a cell is in the batteries used for automobiles. The automobile battery consists of six 2-volt cells in series. It is of the lead-acid type; the plates are lead, the electrolyte sulphuric acid.

**Electrolysis**

The charging of a lead-acid accumulator cell illustrates the chemical effect of an

A magnificent display of lightning in Tampa, Florida. Lightning is caused by the discharge of static electricity generated by movement of particles within storm clouds. The passage of lightning heats up the air suddenly, causing a shock wave we hear as thunder.

**Static Electricity**

Not all electricity is moving. There is static electricity, too. If you run a comb through your dry hair, you will often hear a crackling sound. If you then hold the comb just above your hair, you will find that your hair is attracted by it. Both effects are caused by *static* electricity, or *electrostatic* forces. As with current electricity, there are positive and negative electrostatic changes.

Static electricity is not as useful to man as current electricity. And electrostatic charges, as normally experienced, are relatively small. But large electrostatic forces can be built up. Lightning is the dramatic discharge of very high voltage electrostatic charges. It has been estimated that at the instant of a lightning flash the difference in potential between the thundercloud and the Earth is as high as a thousand million volts! Damage caused by lightning these days is comparatively light, thanks to Benjamin Franklin's invention of the lightning conductor in the 1750s.

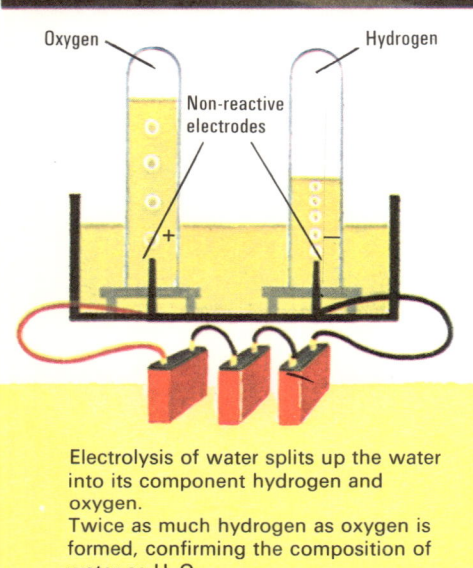

Electrolysis of water splits up the water into its component hydrogen and oxygen.
Twice as much hydrogen as oxygen is formed, confirming the composition of water as $H_2O$.

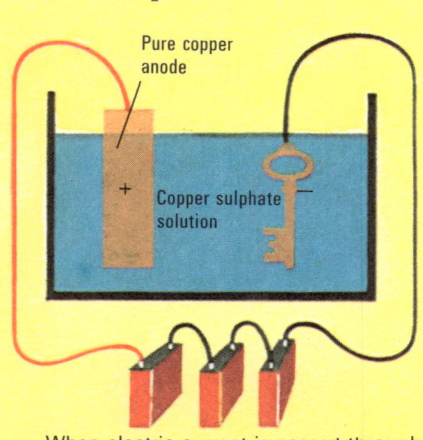

When electric current is passed through a solution of copper sulphate from a copper anode (positive electrode) to an iron key cathode (negative electrode), the key becomes plated with copper, this being an example of electroplating.

electric current. The passage of electricity caused chemical changes in the electrodes and the electrolyte. In this particular case, the chemical changes could be reversed and the cell made to produce the electricity again.

In general, however, the passage of electricity through an electrolyte produces irreversible changes. The process is known as electrolysis. It is a means whereby an electrolyte can be decomposed. For example, water can be split up into its component hydrogen and oxygen by electrolysis and sodium chloride (salt) into sodium and chlorine. The method of electroplating used in industry and the electrolytic refining of metals are extensions of the process of electrolysis.

## Magnetism and Electromagnetism

Magnetite is an iron oxide mineral which has the interesting property that it will pick up pieces of iron. And when it is suspended so that it is free to turn, it always aligns itself in a particular direction. Because of this it became known as *lodestone* ('guiding stone'). We call any substance with such properties a *magnet*. But for all intents and purposes there are only three magnetic substances: iron, cobalt and nickel.

The magnetism of a bar-shaped magnet is concentrated near the ends. When you allow the magnet to move freely, it always aligns itself approximately north-south. The English scientist William Gilbert (1540-1603) knew this and called the north-facing end the

Probably the biggest boon of electricity is that it 'lightens our darkness'. We are no longer dependent on the daily cycle of sunrise and sunset as our forefathers were. Human activity can continue if need be round the clock. Ordinary lighting utilizes the incandescent filament lamp, or light bulb. Electric discharge through gases at low pressure also produces light of a variety of colours. Discharge tubes are incorporated into this colourful sign advertising a motel in Tennessee.

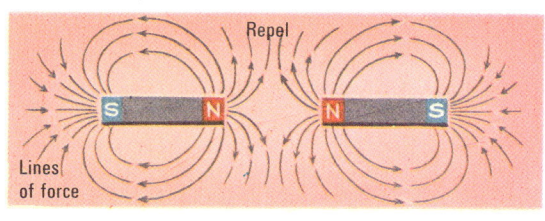
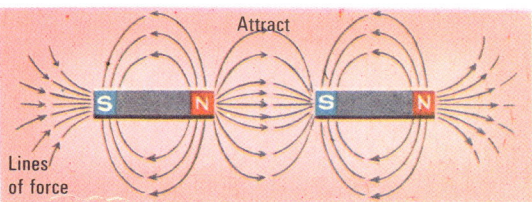

Left: Illustration of the basic laws of magnetism that like poles repel and unlike poles attract one another. Such lines of force can be "seen" by sprinkling iron filings over a card placed on top of the magnets.

Right: Diagram showing the magnetic field created round a coil carrying an electric current. The coil, termed a solenoid, behaves like a bar magnet when the current is passing, having a north and a south pole.

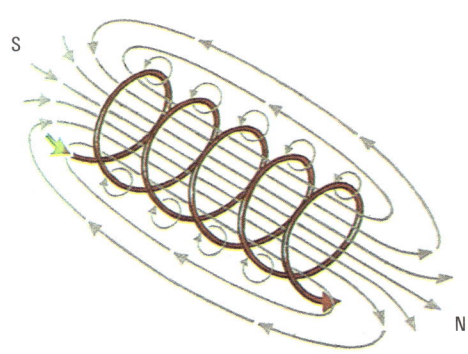

A solenoid

*north pole* and the south-facing end the *south pole*. He decided that the Earth itself must be a gigantic magnet in order to attract the ends of a magnet.

We know today that this is true. And, like our bar magnet, the magnetism of the Earth is concentrated at its magnetic poles. These poles do not correspond exactly with the geographical poles – they are in fact hundreds of miles away and continually changing their position. The directional property of a magnet is utilized in the compass, the instrument behind all navigation.

A simple magnetic compass

The basic rules of magnetism are very simple. If you hold the north (N) pole of one magnet near the south (S) pole of another, they will attract each other. But if you bring two S or two N poles together, they will push each other away. In other words: 'like poles repel, unlike poles attract.'

If you sprinkle iron filings onto a sheet of paper on top of a bar magnet and tap the paper lightly, you will see that the filings align themselves into curved lines. These lines show the lines of magnetic force around the magnet. The area around a magnet where the magnetic force exists is called the *magnetic field*. Thus we talk of the Earth's magnetic field.

### Electricity and Magnetism

Gilbert suspected that there is a connection between electricity and magnetism. But it was not until 1820 that anyone actually demonstrated it. In that year the Danish physicist Hans Christian Oersted (1771-1851) observed that a magnetic needle

The elements of an electricity generator producing alternating current. The magnetic field is provided by the two field coils (electromagnets). The armature is rotated, and its many windings cut through the magnetic lines of force. The current generated thereby is taken off through slip rings and brushes to power, in this case, a bulb. Direct current could be produced by replacing the slip rings with a commutator. (In the diagram the iron cores have been removed for clarity.)

is deflected when it is placed near a wire carrying an electric current, indicating the presence of a magnetic field. The broad field of study of the relationship between electricity and magnetism is known as *electromagnetism*. The great pioneers in this field, André Marie Ampère (1775-1836), Michael Faraday (1791-1867), and Joseph Henry (1797-1878), established the principles of electromagnetism on which most of our important electrical machines are based: electromagnets, electric motors, electric generators, and so on.

### Solenoids and Electromagnets

Concentric lines of magnetic force surround a wire carrying an electric current. When the wire is made into a coil and current is passed through it, it behaves like a bar magnet with a north pole and a south pole. As the current is increased, so the magnetic field becomes stronger. When the current is switched off, the magnetism disappears. Such a coil is called a *solenoid*. Solenoids are widely used in machines for activating switches and other equipment.

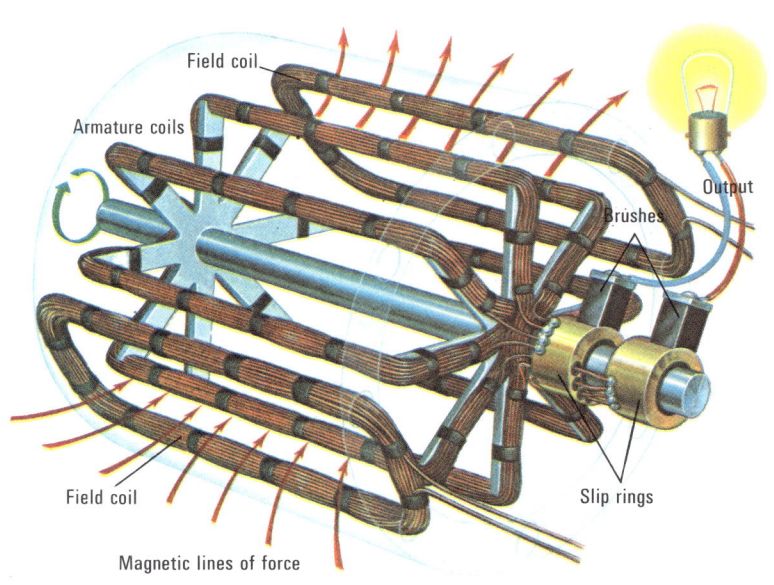

An even more useful extension of the same principle is used in the *electromagnet,* which is made by placing a piece of soft iron inside the coil of a solenoid. When the current is switched on, the iron itself becomes magnetized temporarily. This produces a much more intense magnetic field. Large electromagnets are used at junk yards and steelworks, for example, to pick up scrap iron. Small ones form the heart of the telephone and telegraph, electric bells, and so on.

### Electric Motors and Generators

When you pass an electric current through a conductor that is in a magnetic field, the conductor moves. This is because of the interaction of its own magnetic field with the external field. The electric motor works upon this principle.

In the electric motor the conductor consists of coils of many turns of wire, wound around a core. This constitutes the *armature,* which is free to rotate between the poles of a magnet (often an electromagnet). When current passes through a coil of the armature, one side of the coil will be forced down and the other side will be forced up because of the interaction of its magnetic field with that of the magnet. The coil will rotate, but only for about half a turn. To make it continue to rotate the current must be reversed. This is done by means of a *commutator* on the armature shaft.

There are three main types of direct-current motors – *series, shunt* and *compound* – described by the way in which the armature and field coils (of the electromagnet) are connected. Alternating-current motors of slightly different design are also widely used in which alternating current is supplied to armature and field coils. In the *induction* motor the armature coil receives its current by induction.

In an electric motor you pass current through a conductor in a magnetic field and make it move through that field. You can also do the reverse of this. You can move a conductor through a magnetic field and set up, or *induce* a current in it. This is an example of *electromagnetic induction.* The electric generator, or dynamo works on this principle.

The generator is constructed in much the same way as the electric motor. The current induced in the armature coils will naturally alternate in direction as the sides of the coil go alternately up and then down through the magnetic field. Thus, to produce one-way, or direct current a commutator must be fitted.

But generally no commutator is fitted. The current is taken from the armature by means of simple *slip rings.* Generators that produce alternating current are called *alternators.*

**Transformers and Induction Coils**

The principle of electromagnetic induction is also utilized in the construction of *transformers* and *induction coils*. In a transformer there are two coils wound round an iron core, one on top of the other – a *primary* coil and a *secondary* coil. When alternating current is applied to the primary coil, an alternating current is induced in the secondary coil. If the secondary coil has many more turns than the primary coil, the voltage of the induced current will be higher than the input voltage of the primary. Conversely, if the secondary coil has fewer windings the output voltage will be lower. Stepping voltage up and down in this way is vital to long-distance electricity transmission (see opposite). The induction coil used in the ignition system of automobiles works on similar lines (see page 208).

An electromagnet being used to lift scrap metal at a steelworks. The heavy iron mass is made magnetic by the passage of electric current through coils embedded within it.

# Power Stations

Electricity is our most useful form of energy. We use it for lighting, heating, cooking, and for powering, by means of electric motors, all kinds of machines, from vacuum cleaners and hair driers to milking machines and locomotives.

Where does all this electricity come from? The only way we have at present of producing electricity in the large amounts we require is by means of electricity generators. Batteries produce electricity, but they are no use for large-scale generation because they are nowhere near powerful enough.

Generators convert mechanical energy into electrical energy (see opposite). Some kind of engine, or prime mover is needed to drive them round. The majority of the huge generators are driven by turbines, often called *turbo-generators,* or *turbo-alternators,* because they produce alternating electric current (A.C.). In two-thirds of the world's electricity generating stations, or power stations, steam turbines are used to drive the generators. In most of the others, water turbines are used, providing what is called *hydroelectric* power.

In isolated areas which are not served by main's electricity, small generators powered by petrol or diesel engines may be used. Hospitals usually have a diesel generator for emergency use. Some outlying farms have generators driven by wind turbines. They are a modern version of the windmill.

## Conventional and Nuclear Stations

In a conventional power station the high-pressure steam needed to drive the huge turbines is raised by burning fuel in a furnace beneath a water-filled boiler (see page 72). The largest steam-turbine generators have an output of 500 megawatts or more. This is enough electricity to light 5 million electric-light bulbs! A power station will have a number of such generators. All kinds of fuels are used in the furnace. Practically all of the early power stations burned coal, and many stations still do. They do not burn the coal in big lumps as they once did, though. Now they burn coal that has been ground up into a very fine powder. A great many stations, however, have now gone over to oil as fuel, and even natural gas, where this is in plentiful supply.

When demand for electricity is high, for example, in mid-winter, power stations may have to supplement their normal power supply. They do this by bringing in extra generators, which may be turned by diesel engines or by gas turbines. These kinds of

Electricity is distributed from the power stations to the consumers via overhead cables carried on pylons. The cables are suspended from the steel towers by ceramic or glass insulators. Here engineers are replacing some of the insulators.

Some of the world's largest power stations are hydroelectric. The energy of falling water is harnessed to spin turbogenerators. Here engineers are examining an impulse-type turbogenerator.

engines are particularly valuable for coping with temporary high loads because they can be started up and be in operation in a short while. Ordinary steam plants require a long time to 'warm up' before they are ready for operation.

In nuclear power stations the heat to turn water into steam is provided by a nuclear reactor. Thereafter the steam is used to drive a conventional steam turbine in the usual way. A nuclear reactor is a means of releasing the heat of nuclear fission – the

splitting of atoms – in a controlled way. The nuclear power scheme comprises the reactor itself, a biological shield, heat exchangers, and the power plant.

The *reactor* contains a 'core' of fuel, usually uranium, in a pressure vessel. A gas or liquid *coolant* extracts heat from the core. The *biological shield* protects the reactor operators from the dangerous radiation given out by the core. It is usually made of reinforced concrete about 10 feet thick. The reactor coolant gives up its heat to circulating water in one or more *heat exchangers*. And the heat turns the water into steam, which drives the turbines in an adjacent *power plant*. You can find out more about how nuclear reactors work in the section on atomic energy (see page 34).

**Hydroelectric Power Stations**

In hydroelectric power schemes man harnesses the energy of falling water and uses it to spin water turbines, which in turn drive the electricity generators. The world's largest power plants are hydroelectric. The largest ones have outputs of 5,000 megawatts and over, almost twice those of the largest steam generating plants.

An engineer working inside the low-pressure section of a stripped-down steam turbine. Multi-stage turbines are used in power stations to extract the maximum energy from the expanding steam. The high-pressure stages consist of smaller diameter turbines.

Hydroelectric plants are usually situated in mountainous regions where the engineers can utilize the natural fall of the land to provide the 'drop' for the water. Often they dam a river to increase the *head,* or the distance the water has to fall to the turbine. Pipes called *penstocks* carry the water from a high-head site to the turbines.

There are two basic kinds of water turbines: impulse and reaction. In the *impulse* turbine, such as the Pelton wheel, the water is directed through nozzles onto a series of buckets arranged round the circumference of the wheel. Impulse turbines are suited to high-head sites. *Reaction* turbines spin by reaction to the water. They are fully immersed in the water and are more suited to lower-head sites.

Reaction turbines may be of the *Francis* type, with spiral vanes, or the *Kaplan*, or feathering-propeller type, which looks like a ship's propeller. The angle of the blades of the Kaplan type can be altered to suit the flow through it.

An unusual hydroelectric station is the *tidal power station* at the Rance estuary in Brittany. The scheme utilizes the rise and fall of the tide to generate power. Movement of the water back and forth spins turbines set in the walls of a barrage across the estuary.

**Electricity Transmission**

The power station generators produce electricity at a certain voltage, say 10,000 volts. This is not high enough to transmit over long distances. There would be too much power lost during transmission. Power losses are lower when the electricity is transmitted at high voltage (say 400,000 volts). Most generating stations produce alternating current, which can easily be 'stepped up' to a higher voltage by using a transformer (see page 84).

The high-voltage electricity is transmitted across country by *transmission lines* carried high above the ground on pylons. At electricity substations the high voltage is 'stepped down' again and reduced to a useful level: about 230 volts in the case of the ordinary domestic user.

In most countries there is an interconnecting network, or *grid* of transmission lines between the power stations. When the demand for power in one station's sector is more than it can cope with, power flows into this sector from another station.

Above: A Swiss hydroelectric installation. Penstocks lead from a distant reservoir to an underground turbine hall. In the foreground is the distributing station. Right: The control room of an electricity generating station. Below: Circuit-breakers and switch-gear at a power station where the electricity generated is fed into the national grid.

# Pottery

The bits and pieces of pottery we handle every day – cups, saucers, plates, jugs, and so on – started life as shapeless lumps of dirty clay taken from the ground. The skill of the potter has transformed clay into useful and attractive objects. The soft clay has been set into shape and hardened by baking. This kind of pottery is one of the most familiar forms of what are called *ceramics*. The word comes from the Greek and means 'potter's clay'. Bricks are other familiar kinds of ceramic products. And there are many more.

## Types of Clay

Clay of some kind or another is found, generally beneath the top-soil, practically everywhere in the world. Therefore it is cheap and plentiful. It is formed when rock breaks down under the action of chemical processes and the weather. There is a wide variety of different clays, but they all have, to a greater or lesser extent, two basic properties that make them useful to us.

First, they become *plastic* when mixed with water. This means that they can be shaped and moulded without falling apart. Second, they become extremely hard and strong when they are *fired,* or exposed to great heat. What actually happens is that a kind of glass is produced which acts as a binding agent between harder particles.

Clay to be baked into bricks is used more or less as it is taken from the ground. But few natural clays by themselves are ideal for making pottery. They are therefore blended together, often with other materials, to give a clay with suitable properties. Such a manufactured clay is called a *clay body*. Three different basic clay bodies are used to make the main kinds of pottery in use today, which are earthenware, stoneware, and porcelain.

Pottery was one of the earliest crafts practised by man. The fine Greek vase above was made many centuries ago. The decoration is characteristic.

Deft hands can quickly transform a shapeless lump of wet clay into an attractive pot.

## Types of Pottery

*Earthenware* is one of the cheapest and commonest kinds of pottery, and it is the one most amateur potters are familiar with. It is easy to work and is fired at a relatively low temperature, but it is dull, porous and therefore absorbent. It must be *glazed,* or given a glassy coating if it is to be used to hold liquids. The cheap, so-called 'china' used for everyday tableware is white-glazed earthenware. On the surface it looks like real china, but when it chips or cracks you can see the dull, porous body showing through.

*Stoneware* is an extremely hard and strong kind of pottery that is used for jugs, heavy dishes, storage jars, and so on. Unlike earthenware, it has a *semi-vitreous,* or glass-like body and can therefore hold liquids without having to be glazed. But for many uses it is glazed with salt. Sewer and drainage pipes are familiar examples of salt-glazed stoneware. Stoneware usually fires at a higher temperature than earthenware to a dull grey or brown.

*Porcelain* is the ultimate in pottery. It is brilliantly white, not only on the surface like other white ware, but the whole way through. It is also *translucent,* which means that if you hold it up to a strong light, the light will shine through. Porcelain is often called *china* or *chinaware* because the secret of porcelain-making was discovered long ago by the Chinese.

The material from which porcelain is made is called the *paste*. True porcelain is known as *hard-paste* porcelain. It is harder and requires a much higher firing temperature than any other pottery. This high temperature makes it completely vitreous. The main ingredients are pure white *kaolin* and *china stone,* also called *Cornish stone* and *petuntze.* China stone, which is a kind of granite rich in feldspar, melts and forms the glassy bond which holds the kaolin together. It is what is known as a *flux.* Ball clay may be added to the paste to make it easier to work. Pure quartz, that is, silica, is often included, too.

An artificial kind of porcelain, called *soft-paste* porcelain, is made from a mixture of a white clay and a powdered mass of glass, sand, or broken china. It fires at a much lower temperature than true porcelain. When broken, it reveals a grainy, chalky texture.

*Bone china* is an artificial kind of porcelain that is used for high-quality tableware. It is translucent like true porcelain but is not quite as white. Bone china is so-called because its main ingredient is bone ash, which acts as a flux. The other ingredients are kaolin and china stone.

### Shaping Pottery

The shaping of a piece of clay into a beautiful piece of pottery is the most fascinating part of the potter's craft. For the simplest shapes he may use his hands alone.

To make rounded, symmetrical shapes, however, the potter generally makes use of the *potter's wheel.* This is a simple device for rotating the piece of clay while the potter shapes it with his hands. The action of shaping a piece of wet clay on the wheel is known as *throwing.* To begin the throwing operation, the potter throws a piece of wet clay into the center of the wheel-head and starts it turning. With wet hands he alternately squeezes and flattens the clay until it is of the same consistency throughout.

Then he begins to shape it. He plunges his thumb into the center of the spinning clay while his fingers guide the outside. By moving his thumb towards his fingers he makes a bowl-shape. The bottom of the 'bowl' is the base of the pot to be made. Next he builds up the walls of the pot by squeezing the sides of the bowl gently between thumb and fingers and drawing his hand gradually upwards. This forces the clay into a thinner, higher wall. He uses the fingers of both hands to produce the final shape.

### Mechanical Shaping

The method of hand throwing described would not be quick enough to produce the vast quantities of tableware needed in the home. Commercially, therefore, potteries employ a mechanical throwing process, in which one surface of the piece is shaped in a mould, and the other by a special cutting tool. Every piece shaped on the same machine is identical to every other piece.

The traditional European pottery kiln has the appearance of a bottle. The "bottle" part covers the chamber being heated and serves as a chimney. It is an updraught kiln in that the hot gases are drawn from the furnace upwards through the pottery chamber. Modern pottery kilns are of the continuous, tunnel type.

Various stages in the 'throwing' of a pot. The potter centers a piece of well-kneaded clay on the wheel and then quickly and deftly 'draws up' the sides of the pot with his fingers. Then he shapes the neck and carries out any finishing touches, such as making a lip or attaching a handle or spout.

Flatware, such as saucers and plates, is made on what is called a *jigger*. The jigger operator flattens out a ball of clay into a kind of pancake, called a *bat*. He slaps down the bat on top of a plaster mould, which has the desired shape of the inside of the saucer. It is mounted on a rotating wheel-head. He presses the clay firmly against the rotating mould to shape the inner surface of the saucer. He shapes the outer surface by gradually lowering a shaped blade onto it, which trims the excess clay and leaves a smooth surface of the correct shape.

Cups and other hollow ware are made in a slightly different way on a machine called a *jolley*. On this machine the clay is pressed *inside* a plaster mould, which shapes the outer surface of the cup. A shaped knife blade trims off the surplus clay from inside to shape the inner surface of the cup. The handle is made separately, usually by moulding, and joined to the cup body with a little wet clay when the body is leatherhard.

**Firing Pottery**

The whole object of firing pottery in a furnace, or *kiln*, is to change some portion of the constituents in the clay into a kind of glass. When the glass cools, it binds together the other materials in the clay into a rigid mass. The forming of this glassy, or vitreous bond is called *vitrification* (from the Latin word *vitrum*, meaning glass). The amount of vitrification varies from product to product, according to the composition of the clay and the temperature of firing.

Earthenware, which is fired at quite a low

To mass-produce hollow ware like these pots a simple device called a jolley is used. The damp clay is placed in a mould and a knife blade is inserted to shape the inner surface.

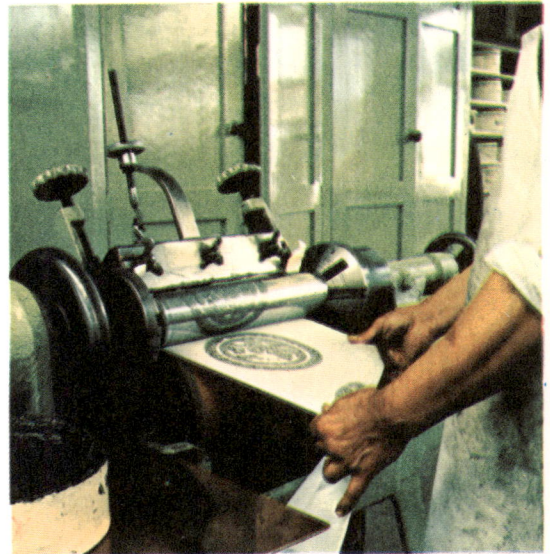

An engraved design in potter's colors being removed from the engraving roller. The design will now be transferred to the surface of a plate and then fired to provide permanent decoration.

Left: A Lebanese potter practising his age-old craft. He is 'throwing' his pots on a simple potter's wheel, operated by a foot treadle. Modern wheels are driven electrically.

Right: Clay figures being applied to a vase for decoration. They are moulded separately, and will become white after firing in the kiln.

temperature (about 1000°C), has little vitreous bonding and is quite soft and porous. Stoneware and porcelain, which are fired between about 1200°C and 1400°C, are both hard and vitreous. Porcelain, in addition, is translucent.

A piece of pottery cannot be fired while it is still wet from the potter's wheel, or even at the leather-hard stage. The steam produced from the water it contains would split the clay apart. First, therefore, the newly shaped pottery must be allowed to dry thoroughly. Drying must be very carefully controlled. Clay shrinks quite a lot when it dries. If the drying process is too fast, thin parts of the piece of pottery will dry out and shrink before the thicker parts. And cracks will result.

## Glazing Pottery

Glazing a piece of pottery involves coating the surface with a thin layer of a clay-like liquid mixture called *glaze* and then firing it. At a fairly low temperature (800 – 1000°C) the glaze melts into glass and flows evenly over the piece to form a glossy coating. A variety of different glazes can be made by adding certain compounds to the basic clay-like mixture. Simple transparent glazes may be tinted various colors by adding certain metal oxides. Adding tin oxide or zinc oxide makes the glaze an opaque white. One of the oldest glazes, used widely for stoneware, is *salt glaze*. It is applied by throwing salt into the kiln just before the end of firing. The salt vapour produced settles on the pottery to form the glaze.

Sometimes the glaze is applied to air-dried pottery, and firing of the clay and of the glaze takes place at the same time. But often, especially for high-quality ware, the glaze is fired separately. The first firing of the clayware before glazing is known as the *biscuit*, or *bisque* firing. The second firing for the glaze is called the *glost* firing.

## Decorating Pottery

Painting is probably the most popular means of decorating pottery. It may be done on clayware before it has been glazed, and is then called *underglaze*. Special pigments must, of course, be used in such paints to withstand the heat of firing. *Overglaze* decoration is applied after the piece has been glazed. The so-called *enamel* colors must be fired briefly after application to fuse them with the glaze. Commercially, *transfer-printing* is widely practised. Engraved designs in potter's colors are printed on paper from copper engravings. Then the design is transferred to the surface of the pottery.

Above: Part of an attractive coffee set in earthenware. Earthenware by itself is porous and hence of little use for holding liquids. Consequently it has to be glazed to make it watertight. Decoration may be applied under the glaze or over it. Or the glaze itself may provide the decoration.

Painting is a favorite method of decorating pottery. The picture right shows a colorful selection of painted ware outside a Spanish pottery.

91

# Industrial Ceramics

## Bricks

Bricks are among the commonest of our building materials. Joined together with mortar or cement, they form strong walls which resist the action of the weather. Like pottery, they are made from certain kinds of clay. The weak clay is shaped into blocks and then fired into a hard, compact mass.

Clay for brick-making is taken from the ground by huge power shovels. The raw clay is crushed fine and *screened,* or sieved, to make sure that there are no lumps in it. Then it is mixed with water and kneaded into a doughy mass, which goes to the brick-making machine. The cheapest bricks, called *wire-cut* bricks, are made by extrusion. In the brick-making machine, a large screw called an *auger* forces the clay out through a rectangular hole in a long ribbon. This works in much the same was as a kitchen grinder.

Left: Newly cut 'green' bricks leaving the cutting machine. They need to be cut slightly bigger than the finished bricks to allow for shrinkage during drying and firing. Drying is very necessary prior to firing to prevent cracking occurring.

Below: Sun-dried bricks, or adobe are still used for building houses in hot countries such as Mexico.

Next, the extruded column is sliced into individual blocks by the parallel wires of a cutting machine, which looks like a giant egg-slicer. The blocks thus formed are cut slightly larger than the finished bricks to allow for shrinkage during firing.

Better-quality and more expensive bricks are made by a pressing process. The clay used is only slightly damp and very stiff. The bricks are formed one by one by forcing the clay under high pressure into moulds. These bricks are known as *pressed* bricks or, in stronger qualities, *engineering* bricks.

Before firing can begin, the bricks have to be dried in hot air. If they were fired while they were still hot, the steam produced inside the bricks would weaken and probably crack them. Firing of the bricks takes place in the brick kiln at a temperature between 800° and 1100°C, depending on the kind of clay used. The bricks may be fired in batches or in a continuous process.

The continuous process is much more efficient than the batch, or periodic one, and requires less man-power. The dried bricks are stacked on flat wagons and drawn through a tunnel, which may be hundreds of feet long. In the middle of the tunnel is the firing zone, where the bricks are fired. The dried bricks are pre-heated as they move along the tunnel towards the firing zone. When they leave this zone, they gradually cool off. At the end of the tunnel, they are cold enough to be handled.

## Insulators

Ceramics play as important a role in industry as they do in the home. Industrial porcelain, for example, which is made from more or less the same ingredients as pottery porcelain, is invaluable in the electrical industry. It is a first-class insulator which resists even the strongest electric currents. The big insulators of electricity pylons of the national grid, which support high-voltage power lines, are generally made of porcelain. Nearer at home it may be used in electric plugs, sockets, and fuse boxes.

## Refractories

But it is the great resistance to heat of many ceramic materials which makes them so valuable to industry. Substances that can withstand high temperatures are called *refractories*. Among the commonest are the *firebricks* used to line the inside of industrial furnaces, such as metal-smelting furnaces. Fire clay is used to make the firebricks used in domestic grates. But it is not resistant enough for furnaces such as the open-hearth steel-making furnace, which operates at about 1500°C. For such a furnace, bricks rich in silica (sand), dolomite, or magnesite must be used. To resist still higher temperatures, the bricks must contain a very high proportion of alumina (aluminum oxide).

Alumina, in fact, is a common refractory material. It not only has great resistance to heat (melting point 2050°C), but is also very hard and an excellent electrical insulator. These properties make it ideal for its most familiar use – the white insulating body of the spark plugs of automobile engines. It is also a good material for making cutting tools and an excellent abrasive, or grinding agent. This is because it remains hard despite the great heat generated by cutting or grinding. This is true of all good refractories. Some of the best cutting and grinding agents are synthetic carbides (combinations with carbon) of such metals as tungsten.

One of the most spectacular uses of ceramics today is in the heat shields of spacecraft, which have to withstand terrific heat when they re-enter the Earth's atmosphere. (See page 19.)

Porcelain is not only used for making fine china. It has many industrial uses as well. It is widely used for example in the electrical industry for it is an excellent electrical insulator. Porcelain insulators are used for suspending high-voltage cables from electricity pylons.

To make synthetic rubies alumina powder, mixed with a coloring agent, is fed into an oxyhydrogen flame and fuses in the 2,000°C temperature into tiny globules. The globules collect as a so-called boule, which cools to form synthetic ruby. Synthetic sapphires can be made in a similar way.

### Synthetic Gems

Diamond is the hardest substance known and is the finest and the rarest of our gems. Diamonds were formed deep down inside the Earth's crust by the action of great heat and pressure on carbon. Scientists have succeeded in imitating the diamond-making process, and quite a quantity of synthetic diamonds is now being made. Synthetic rubies can be produced in a similar way from powdered alumina. Synthetic gems have a number of industrial uses. For example, synthetic diamonds are used in drill bits.

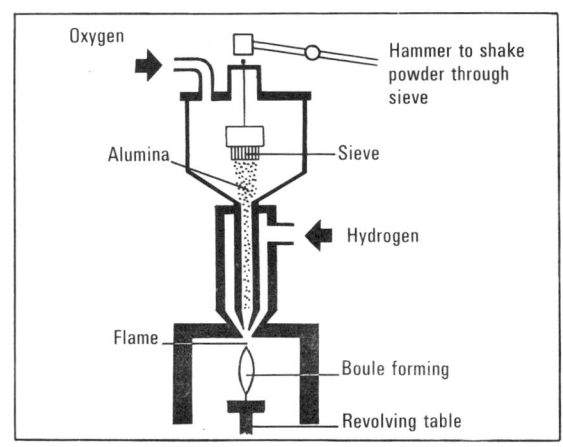

# Glass

## Making Glass

Glass has become an essential part of our way of life today, as a quick glance around you will show. That glass should be so popular is not surprising when you consider its properties. It is transparent and can be shaped easily. It is unaffected by practically everything so that it will not rot, rust, or taint foodstuffs. And it can be cleaned very easily. Last, but by no means least, it is extraordinarily cheap.

The ideal substance for making glass is silica. When heated very strongly, silica melts. It forms glass when it cools. The glass is not crystalline, as you might think, but is what is known as a *supercooled liquid*.

Silica by itself, however, is of no use for making ordinary glass. It melts at far too high a temperature to be economical. Substances called *fluxes* are therefore added to it to make it melt and form glass at lower temperatures, which is much more economical. The common fluxes used to make ordinary glass for bottles, windows, electric-light bulbs, and so on are soda ash (sodium oxide) and lime (calcium oxide). That is why this glass is known as *soda-lime* glass.

Ordinary glass is made by charging sand, soda ash, and limestone in the correct proportions into the glass-making furnace. The sand provides the silica; the soda ash, the soda; and the limestone, the lime. At white heat these ingredients melt. Waste glass, called *cullet,* is also added to help the mix melt more easily. Most glass-making furnaces operate continuously. The ingredients are fed into one end, and the molten glass is removed continuously from the other.

Soda-lime is the commonest but by no means the only kind of glass. Other kinds are made by varying the 'recipe' of manufacture. In other words, different fluxes are used with the silica.

## Special Glass

*Lead glass,* also called *flint glass,* contains a high proportion of red lead, a lead oxide. This glass has an extremely fine luster and brilliance. When expertly cut, it gleams and sparkles like diamond. It is widely used for high-quality tableware and also for optical purposes. Glass with a very high lead content is used in the atomic-energy industry and elsewhere because it absorbs harmful radiations. Scientists can safely watch their experiments with radioactive materials through windows made from this glass.

Another important kind of glass is the heat-resistant glassware used in the kitchen. Pour boiling water into a cold bottle made from ordinary glass, and the bottle will almost certainly crack and fall apart. The hot water makes the inside of the glass expand so quickly that the stresses set up cause the glass to crack. Pour boiling water into glass ovenware, for example, and nothing happens. The glass is made so that it expands scarcely at all when heated. It contains mainly silica and boron oxide, and is called *borosilicate glass*. Borosilicate glass (such as Pyrex) is used not only for cooking ware but also for laboratory equipment, industrial pipes, and so on. It is also used as an electric insulator.

*Silica glass* represents the extreme in heat-resistant glass. It can be heated red hot and then plunged into ice-cold water without even cracking. It is very nearly pure silica in composition. It is made by heat-treating a borosilicate glass. The use of silica glass is limited, however, because it is very expensive to make.

Opposite: Molten glass from the furnace is fed between rollers and emerges in the form of a sheet. Patterned rollers are employed to produce a textured surface.

Glass-blowing is an ancient craft still widely practised. The glass-blowers dip the end of an iron blowpipe into the molten glass and blow a 'bubble' in the gob of glass that sticks to it. By deft manipulation of the molten glass they can fashion all kinds of intricate objects.

Below: The amazing property of a glass ceramic, which can withstand extreme temperature changes without shattering.

A fine example of Venetian glass showing superb craftsmanship.

A much purer form of silica, *fused silica,* is used for very-high temperature wares in laboratories and in industry. It is made by fusing pure quartz crystals in an electric furnace. A transparent form, called *fused quartz,* is of great value in precision optical and other instruments because it is transparent to a wide range of radiation, including ultra-violet and infra-red (heat) as well as visible light.

Spectacle lenses have to be made of a very pure kind of glass which contains no flaws or impurities to distort the light passing through it. Such glass is called *optical glass.* It is used not only in spectacles but in the lenses of all kinds of optical instruments, such as cameras, microscopes, telescopes and projectors. It is made from only the finest and purest raw materials and contains a variety of ingredients besides the basic ones to give it the required optical properties. But it is the care taken during the manufacturing process which largely determines the optical purity of the glass being made.

Optical glass absorbs a lot of the light passing through it. Camera and binocular lenses are therefore coated, or 'bloomed' to cut down this absorption. Objects seen through bloomed lenses are much clearer, especially when the light is poor.

Colored, black, or opaque glasses can be produced by adding to the basic 'recipe' of glass manufacture compounds of certain metals, including copper, chromium, nickel, cobalt, manganese, and iron.

## Shaping Glass

Glass leaves the glass-making furnace as a red-hot, syrupy liquid, and it is in this state that it is shaped. This may be done in a variety of ways. The most important method is by *blowing.*

Traditionally the blowing is done with the mouth by skilled craftsmen. But these days high-speed, automatic machines have taken over for all except the finest and most intricate pieces. Bottles, jars, and containers of all kinds account for much of glass production. They are made by blowing air into molten glass in a mould of the desired shape.

In a bottle-making machine, for example, lumps, or *gobs* of red-hot glass are dropped

### Safety Glass

Automobile windshields and aircraft windshields must withstand severe impacts without shattering. The kind of special glass used for such purposes is called *safety glass*. It is far stronger than ordinary glass and does not shatter into sharp, dangerous splinters. Two main kinds of safety glass are produced – toughened and laminated glass.

*Toughened* glass is made by heating glass sheet in a furnace and then rapidly chilling the sides of the sheet with blasts of cold, compressed air. This process is called *tempering.* When toughened glass breaks, it shatters into hundreds of blunt, rounded particles which cannot cause bad cuts. *Laminated* glass is a kind of glass and plastic 'sandwich'. Two sheets of flat glass with a layer of plastic in between are heated and pressed together to form a strong bond. Laminated glass is safe because when it breaks the pieces of glass stick to the sandwiched layer of plastic.

Aircraft windshields must be especially strong to withstand the impact of birds, for example. They are therefore made of laminated, toughened glass and combine the qualities of both kinds. Several glass and plastic layers are built up to give a thickness of an inch and a half, or more.

Toughened glass shatters into a myriad of rounded pieces when it breaks, instead of dangerous sharp splinters.

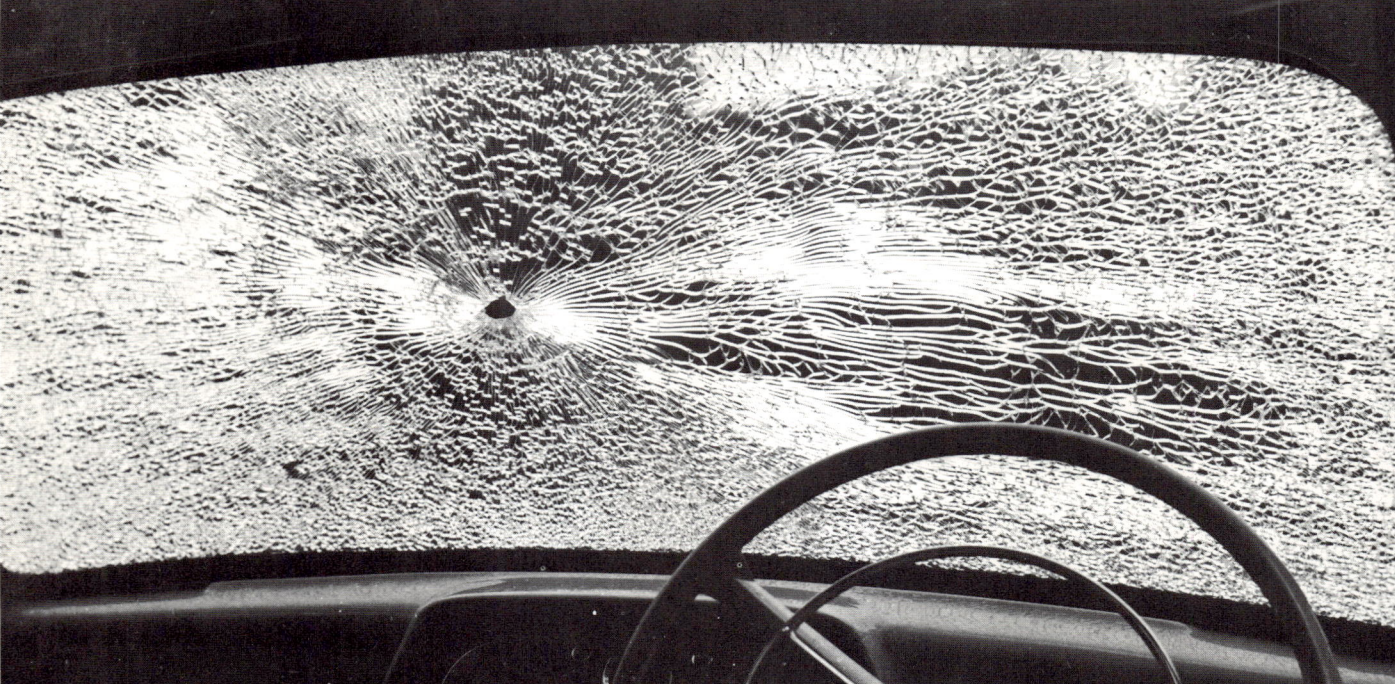

into waiting moulds. Compressed air is blown into these preliminary, or *parison* moulds to form the neck and a hollow shape. The partly shaped glass pieces are then transferred to finishing moulds where they are blown into their final shape. For wide-necked containers such as jam jars, the gob of molten glass is *pressed* into the parison mould by a plunger before being finally blown.

*Pressing* into a mould, with or without subsequent blowing, is in fact quite a common method of shaping glass. The articles to be formed must, of course, be of such a shape that the plunger can readily be withdrawn. Dishes, blocks, lenses, and headlamp glasses are typical pressed products. Blocks and lenses can also be made by *casting*. The molten glass is simply poured into moulds. It must be somewhat hotter than the glass used in other methods of shaping so that it forms more easily into the shape of the mould.

Surprisingly enough it was the production of flat glass for windows which used to cause glass-makers their biggest headache. The earliest way of doing it was to blow a big glass bubble, burst it, and then twirl it round rapidly so that it flowed into a more or less flat sheet. At the center of the sheet, where the neck of the bottle was, the glass was naturally much thicker. This produced the familiar 'bulls eye' glass which you can still sometimes see in old houses.

Today, flat glass is made in one of three ways. Ordinary plain window glass, called *sheet* glass, is made by drawing a wide ribbon of molten glass vertically upwards from the glass-making furnace. The faster the ribbon is withdrawn, the thinner it is. The glass has a natural, brilliant, or 'fire-polished' finish, but it tends to be uneven and distorted because of the way it is drawn from the furnace.

The better-quality glass used for shop fronts, windshields, glass doors, and so on, called *plate* glass, is made to be perfectly flat. A ribbon of glass is drawn horizontally from the furnace between rollers. This prevents any distortion occurring. However, the surface of the glass becomes slightly uneven because of the action of the rollers on it. It is therefore ground and polished accurately afterwards by a series of high-speed rotating discs. Grinding and polishing, however, takes away the natural fire finish of the glass, and, what is more important, greatly adds to the cost of production.

Plate glass is now rapidly being superseded by *float glass,* which is better and cheaper. The basis of the float process is that a ribbon of molten glass from the furnace is floated on a bath of molten tin. The surface of the tin is perfectly flat, which means that the glass layer floating on top of it will be

Production of continuous-filament fiberglass. The fine strands of glass emerging from the bottom of the furnace are collected and sized to keep them together and wound on a high-speed winder.

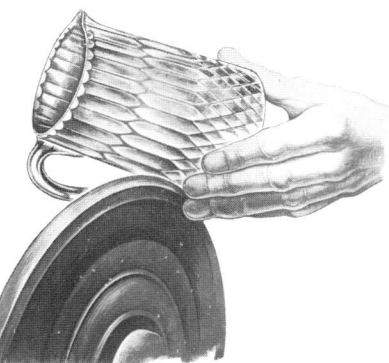

'Cutting' glass for decorative effect is usually done by grinding. Lead glass is particularly beautiful when expertly cut, flashing brilliantly like diamond.

Delicate designs are applied to glass by engraving with small abrasive copper wheels.

97

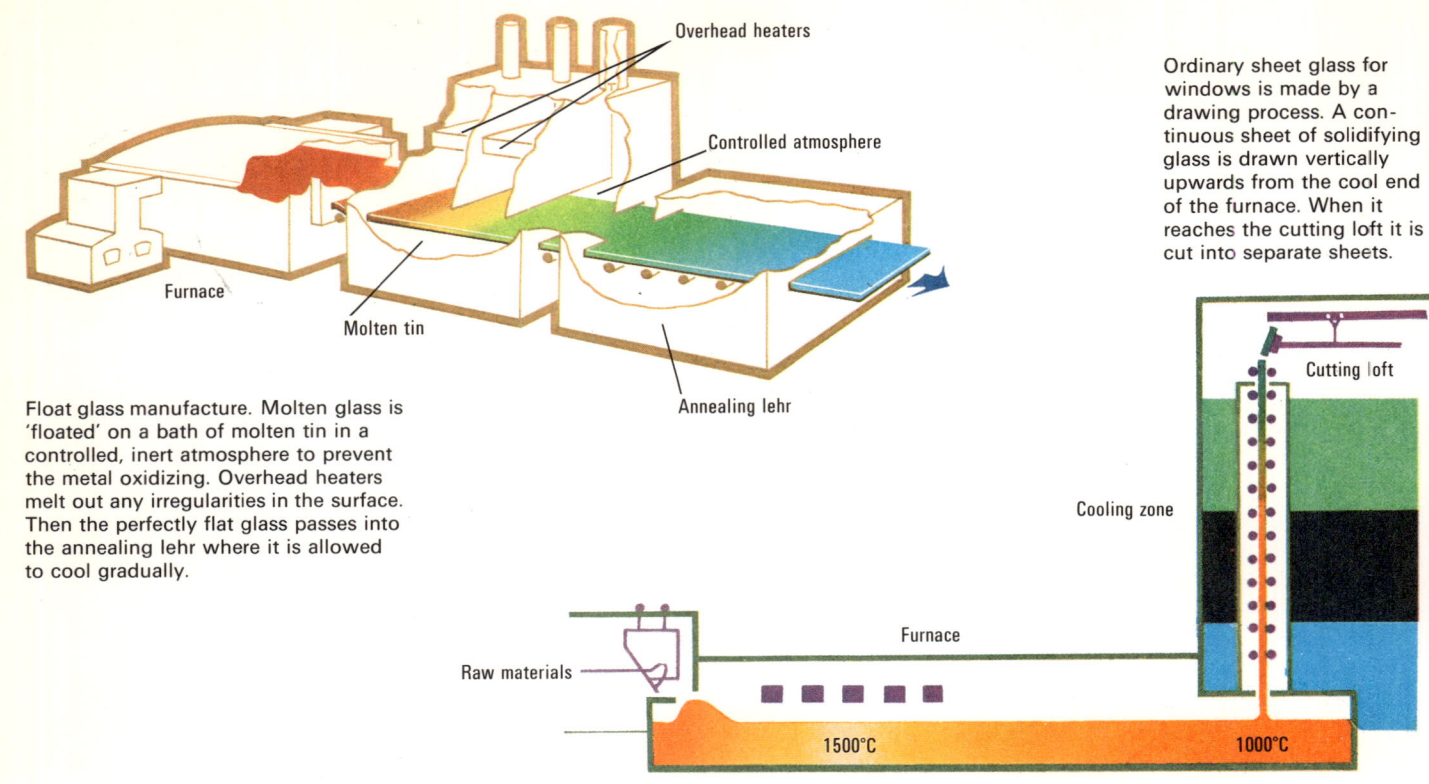

Float glass manufacture. Molten glass is 'floated' on a bath of molten tin in a controlled, inert atmosphere to prevent the metal oxidizing. Overhead heaters melt out any irregularities in the surface. Then the perfectly flat glass passes into the annealing lehr where it is allowed to cool gradually.

Ordinary sheet glass for windows is made by a drawing process. A continuous sheet of solidifying glass is drawn vertically upwards from the cool end of the furnace. When it reaches the cutting loft it is cut into separate sheets.

perfectly flat, too. Also, the molten tin does not mark the glass, which thus retains its original fire finish.

A method of drawing is also used to make glass tubing. A ribbon of molten glass is wound around the inside of a rotating, hollow tube. Compressed air forced through the tube forms the glass into tubing. As soon as it is formed, the tubing is drawn away. The faster it is drawn, the smaller is its diameter and the thinner its walls.

After the glass has been shaped, no matter what method has been used, it must undergo a process called *annealing*. Annealing involves reheating the glass in an oven called a *lehr* and allowing it to cool slowly at a definite rate. It eliminates stresses and strains that are set up in glass which is cooled down too quickly.

**Decorating Glass**

Most glassware is left plain, or undecorated. Any beauty and elegance it might possess lies wholly in its dimensions and proportions and in the skill of the craftsman who fashioned it. Some glassware, however, is decorated to give it additional beauty. This can be done in a variety of ways, including cutting, engraving, etching, and sandblasting.

*Cutting* is one of the commonest methods. It involves cutting into the glass to make a number of faces, which catch and reflect the light better. Expertly cut lead crystal glass flashes and sparkles with all the colors of the rainbow from every angle. Glass is 'cut' by grinding away the surface with a grinding wheel of carborundum or some other abrasive. In *engraving* craftsmen use fine, rapidly spinning copper wheels to 'grind' away delicate designs on the glass.

*Etching* is a method in which a design is etched, or 'eaten' away by a very strong acid. The part of the glass that is not affected by the design is painted with an acid-resistant chemical. Then the article is dipped in hydrofluoric acid, which attacks and eats away the unprotected design. In *sandblasting* coarse sand is blown against the glass to produce a rough, translucent surface. A design can be worked out by blasting through a stencil.

For special effects the glass may be decorated with colored enamels. These enamels fuse to the surface of the glass when heated, or fired in a kiln.

**Fiberglass**

One of the most interesting and unusual uses of glass is as fine fibers. Fiberglass may be produced in continuous strands, or *filaments*, or in short, staple form. It is a very special kind of synthetic fiber. A coarse, short-fibered form is called *glass wool*.

Fiberglass can be successfully woven into cloth to make such things as curtains and tablecloths. These fabrics have excellent properties. They resist staining, creasing, stretching, rotting, and attack by moths, and they are completely fireproof. They are also easy to wash and need no ironing. They can be printed in bright colors, which do not fade easily.

It may seem strange that cloth can be

## Stained Glass

The most spectacular use of stained glass is in the windows of churches. A stained-glass window is made up of a large number of differently shaped pieces of colored glass set in a lead framework. The art of making stained-glass windows was well developed as long ago as the Middle Ages. Some of the finest examples of medieval stained-glass can be seen at Châtres Cathedral, France.

Designing is the first stage in making a stained-glass window. The artist draws first a small colored sketch and then a full-size one, called a *cartoon*, which shows the exact shape, size, and color of the various pieces of colored glass. Craftsmen then cut glass to fit the pattern. The glass they use is specially made so that it is uneven in thickness and contains minute bubbles and impurities. The varying thickness gives an interesting variation in color. The impurities present scatter the light and make the glass 'glow'.

The various pieces of cut glass are assembled like a jigsaw puzzle on a piece of flat glass. The artist then paints in details of faces, lines, drapery, and so on, on the glass with colored enamel paints. The pieces of painted glass are fired slowly and the enamel fuses with the glass. Then the pieces are fitted together in lead strips to form the completed design.

Above: The huge Baptistry window at the rebuilt Coventry Cathedral is one of the finest modern examples of the use of stained glass in Britain.

Below: Examples of Lebanese colored glassware. The imperfections in the glass add to rather than detract from its charm.

made from glass, which, in its familiar form, is extremely brittle and shatters into splinters readily. Glass fibers, however, are so extremely fine that they are flexible enough for textile spinning and weaving. They are stronger than any other natural or synthetic fiber.

Bundles of glass fiber rods have the peculiar property of 'bending' light. They have been used in medicine and industry for peering into awkward places, for example, inside the throat.

The most familiar use of bulk fiberglass, or *glass wool* is as a material for insulating the roofs of houses against heat loss. It is supplied for this purpose in the form of thick, 'fluffy', rolls. It makes an excellent heat insulator because of the air trapped among the fibers. Glass wool has similar uses in industry, too.

Often these days you come across fiberglass automobile bodies, boat hulls, suitcases, fishing rods, vaulting poles, and so on. These items are not made of glass fibers alone, however. They are made of plastics (polyesters) which have simply been strengthened, or *reinforced* with glass fibers. The reinforced plastic may be built up in various ways. Often the plastic is poured over and worked into a coarsely woven fiberglass mat laid over a mould. In another method, alternate layers of plastic and glass fibers are laid down on one another until the desired thickness is built up. Spraying a mixture of plastic and glass wool onto a mould is a third method.

*Continuous-filament* fiber-glass is made by melting down glass marbles of known quality in a furnace. The molten glass leaves the furnace through a series of fine holes to form delicate filaments, which are gathered into a single strand and wound onto a high-speed reel. Staple fiberglass is produced by breaking up the emergent filaments by a blast of high-pressure steam. Glass wool is usually made in a different way. Molten glass is introduced into a rapidly rotating metal dish with fine holes in the sides. Fine fibers form as the glass is flung out through the holes.

# Telegraphy

Telegraphy was the first means of sending messages by electricity. It is the simplest because the information to be transmitted is described in pulses of electricity over, usually, a single cable. Devices called *teleprinters* may be installed at both ends of the cable to convert the message into electrical pulses and then, at the receiving end, to convert the pulses back into typewritten letters. A *telegram,* or *cablegram* sent through post offices or telegraphic networks may be typed out on a strip of gummed paper that is torn off the machine and stuck onto a telegraph form for delivery. A network of land-wire circuits and undersea cables links the world to provide telegraphic communication between most countries. Some *radio-telegraphy* is transmitted by means of orbiting satellites.

One of the earliest telegraph systems used five wires in parallel. Letters of the alphabet were identified by combinations of positive or negative pulses of electricity in the five wires. Such a system was slow and could be used only where the demand on it was light, such as for sending messages between railway stations.

One wire can be used with a pattern of pulses in sequence to identify individual letters. The *Morse code* provides a suitable pattern, with combinations of dots and dashes to represent all the letters of the alphabet. A disadvantage of the Morse code, however, is that patterns for different letters take different lengths of time to transmit. This makes it difficult for a machine to operate automatically to type out a transmitted message.

A more suitable code for use with printing machines is the *five-unit code,* invented in France in 1874 by Emil Bandot. In this code, each signal contains five equal units, each unit having two possible conditions, say, 'positive' and 'negative' or 'current' and 'no current'. The regularity with which these signals can be transmitted by machine means that it is possible to send messages from several machines simultaneously along the same cable, the messages jigsawing in between one another in sequence.

A teleprinter can be used to send a five-unit code message. Early teleprinters produced a perforated paper tape that was then run through an automatic transmitter. Today, a message typed out on a transmitting machine is almost simultaneously typed out on a remote receiving instrument.

## The Teleprinter

The teleprinter is powered, like an electric typewriter, by an electric motor. Each five-unit code combination is preceded by a start signal, which sets the receiving teleprinter in operation in step with the transmitter. A stop signal at the end of each combination brings the receiver to rest until the next signal arrives. The electric motors are provided with accurate speed governors so that they keep precisely in step.

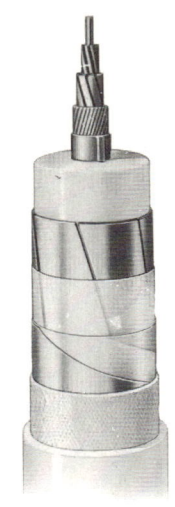

An armored, reinforced and insulated electric cable suitable for submarine applications.

Samuel Finley Breeze Morse was born in 1791 and died in 1872. He studied to be an artist and achieved some success in this field. He did not become interested in the electric telegraph until he was about 40, and did not achieve success until 1844, when he demonstrated his telegraph before the American Congress.

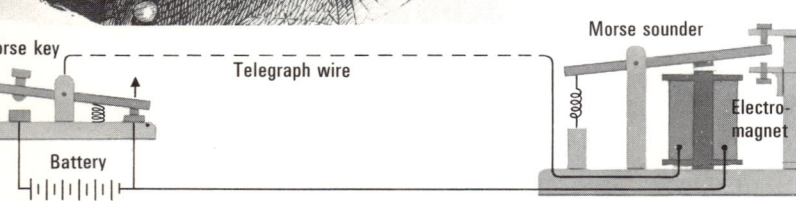

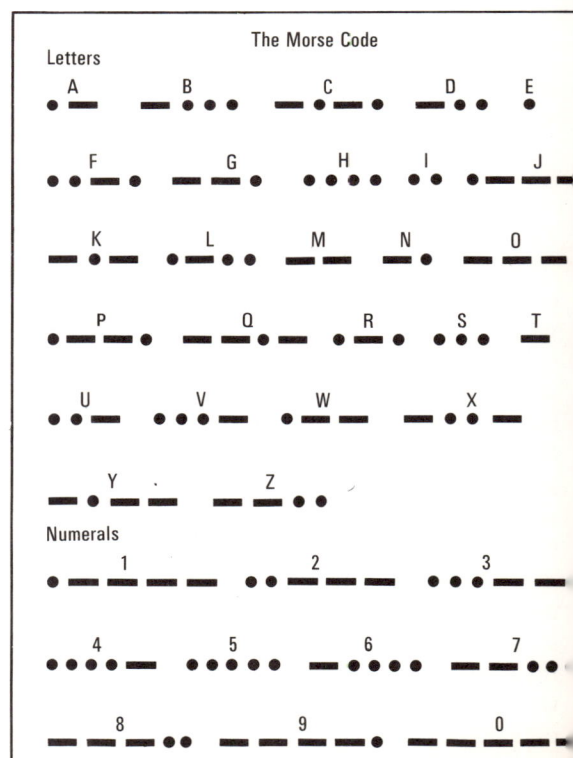

The teleprinter has a keyboard similar to that of a typewriter, but the code does not provide sufficient combinations for both capital and small letters as well as figures and punctuation marks. And so, teleprinter messages are printed in all capitals or all small letters. When a key is depressed, its bar falls on the serrated edges of five code bars that run at right angles to the key bars. This slides the code bars to the appropriate position for the given letter. The machine transmits a signal governed by the positions of each of the five bars, and then releases the key so that another may be depressed. The transmitting machine usually provides its own printed record of the message sent, so that any errors can be spotted and corrected.

## Telex (or TWX)

Regular users of the telegraph system are connected to automatic telegraph exchanges, which are similar to automatic telephone exchanges. They enable any subscriber with a teleprinter to get a quick connection to any other subscriber. Business companies can thus send messages and statements of all kinds by telegraph direct to anyone equipped with a teleprinter. The system can also be used for transmitting data to a computer.

The *Telex* system is established between many countries. To make a call, the Telex subscriber first dials a number that identifies the country required, and then another that connects him to the machine he is calling. Some long-distance services are provided by short-wave radio.

### Telegraph Cables

The world has a network of telegraph cables around it. Cables are usually understood to be those laid underwater to carry telegraph and telephone messages across oceans. These cables can transmit eight telegraph messages totalling 400 words a minute. They are laid by special cable-laying ships that pay out the cable over their stern or bows as they cross the ocean. Submerged electronic *repeaters*, enclosed in watertight steel capsules, are installed at intervals along the cable to amplify the signals being transmitted.

A light cable, insulated with gutta-percha, was laid across the English Channel in 1850, but quickly failed. Another was successfully laid the next year wrapped in hemp yarn and reinforced with an armouring of iron wires. The first successful transatlantic cable was laid in 1866.

Above: A cable-laying ship at work. Note the characteristic sheaves at the bows, through which the cable is paid out. Cable-layers have a special hold which can accommodate hundreds of miles of cable.

Below: A Telex operator "typing" out a message on a teleprinter. To connect with another subscriber she first dials a number; then she types out the message, which is reproduced by the receiving machine.

# Telephony

Since Alexander Graham Bell first developed a practical telephone in 1876, its use has grown and spread until there are over 255 million telephones throughout the world. Nearly half of these are in the United States, Japan is second with more than 15 million, and Britain is a little way behind with about 14 million.

Modern telephone instruments normally have a separate 'mouthpiece' and 'earpiece', or transmitter and receiver. The *transmitter* consists of a carbon-grain microphone. When you speak into the mouthpiece, your voice vibrates a diaphragm. Carbon grains packed behind the diaphragm are compressed according to the nature of the sound hitting the diaphragm. A property of the carbon grains is that their electrical resistance varies with the pressure on them. And so an electric current passing through them gets stronger or weaker as the pressure on them gets greater or smaller. In other words, the resulting current 'copies' the pattern of the sound waves. This current is passed along the telephone wire to the receiver of another telephone to which the first instrument is linked.

The telephone *receiver* also has a diaphragm. This diaphragm is vibrated by current passing through an electromagnet, just like a miniature loudspeaker. In this way the current passing along the telephone wire is changed back into recognisable sound.

Before a conversation can be held between two telephones, the two instruments must, of course, be linked electrically. A modern national telephone network has millions of miles of lines connecting millions of telephones. Most telephone lines consist of copper wire bound together in cables, some cables containing more than 4,000 wires. These may be carried overhead on poles or be buried underground or even run under the sea. Many long-distance connections involve radio links, possibly by means of an orbiting satellite (see page 13).

More and more telephone exchanges are becoming automatic in operation, but operators are still required to deal with person-to-person calls, calls to manual exchanges, problems, suspected faults, and so on.

### Bell

Alexander Graham Bell (1847-1922) was born in Scotland, but spent most of his life in the United States. His father was a teacher of the deaf, and Graham Bell became greatly interested in this problem, too. He thought that a primitive telephone equipment could be used for teaching deaf-mutes to speak. Bell carried out a series of experiments to determine how vowel sounds are produced, and the use of electrically driven tuning forks to produce the sounds gave him the idea of 'telegraphing' speech.

He first tried to develop a 'harmonic telegraph', which would transmit several telegraph messages simultaneously over the same wire. He failed, but decided that it might be possible to transmit the sounds made by a human voice over a telegraph wire. His transmitter consisted of an electromagnetic coil near to an iron diaphragm, with a similar device for a receiver. One day, the diaphragm got stuck to the electromagnet. Bell's assistant tried to pluck it free, and Bell, in another room, heard the twang on his receiver. With some modifications, he found he could transmit recognisable voice sounds. The first intelligible sentence was carried by wire in March 1876.

The Bell Telephone Company was formed the following year, and Bell received recognition in many countries. He continued his life of invention, and produced a variety of other devices, particularly in the field of recording and in the teaching of speech to the deaf. He spent the latter part of his life mostly on his estate on Cape Breton Island, Nova Scotia.

Alexander Graham Bell produced the second great communications device—the telephone—about 30 years after Morse had revolutionized communications with his electric telegraph. The Bell Telephone Company which he founded in 1877 is still a world leader in the communications field.

# The Telephone Exchange

A pair of wires from each telephone runs to switching equipment in an exchange. Some older or smaller exchanges are still operated manually. An operator sits in front of a *switchboard* carrying plugs and sockets, to one pair of which each phone is connected. When a phone is lifted to make a call, a light goes on at the switchboard. The operator plugs a personal headphone into the caller's socket and enquires what number he wants. He can then put the caller's plug into the socket of the required number to establish contact between the two phones and operate a switch to ring a bell in the telephone of the called person.

Most exchanges now are partially or completely automatic. On an automatic exchange, a lifted telephone sends a signal to the switching equipment, which operates a *dialling tone* to show that it is ready to receive a call. The caller can then indicate the number he wants, usually by turning a numbered dial on his instrument. This sends appropriate pulses of electricity to the exchange, one pulse for the figure 1, two pulses of electricity for the figure 2, and so on.

A series of selector switches in the exchange are energized by the pulses to complete a circuit, which applies a *ringing tone* to the line and operates a bell in the telephone being called. If this telephone is already occupied by another connection, the equipment informs the caller by applying an *engaged tone* to the caller's line.

The switching equipment in telephone exchanges is being continually improved. Electromagnetic switching equipment is being gradually replaced by electronic switching, which uses far fewer moving parts and so is much faster, as well as being much more compact.

# DDD

DDD, or Direct Distance Dialing, has become possible with the great increase in the number and 'quality' of long-distance telephone circuits. To make a dialled trunk call, several extra figures must be dialled before the required number. These figures identify the route to the necessary local exchange. They are fed to a 'register-translator', which determines the correct charging rate for the call and passes on instructions to set up the required chain of connections. The British register-translator is commonly known as *Grace,* or 'group routing and charging equipment'.

Gradually, automatic dialling facilities are being extended until eventually it should be possible to dial directly to another telephone almost anywhere in the world.

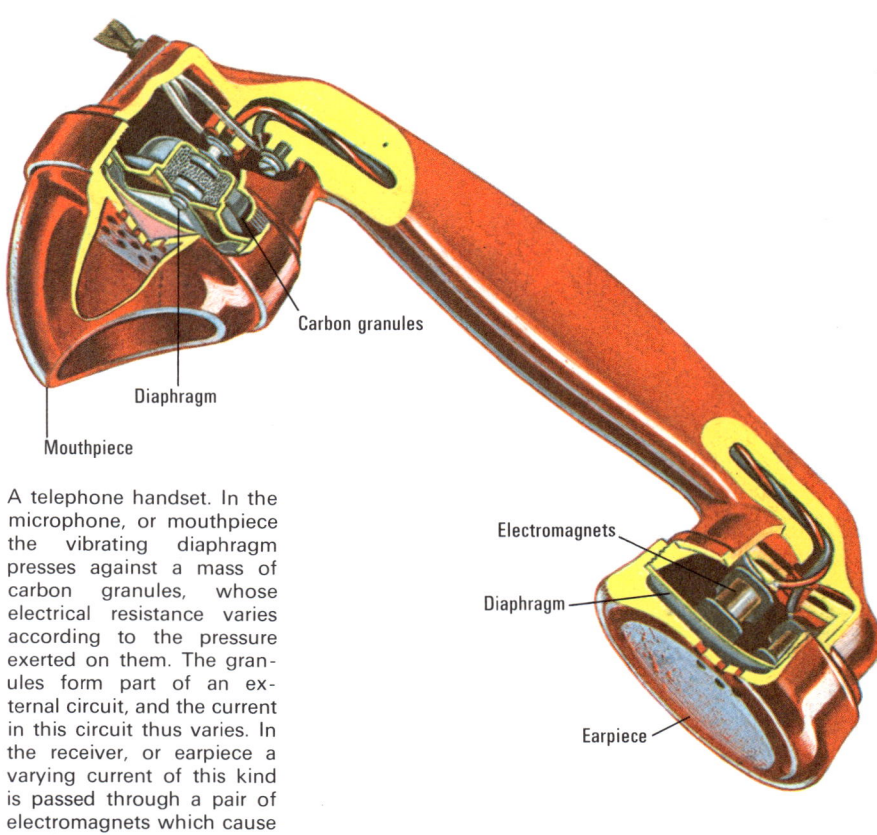

A telephone handset. In the microphone, or mouthpiece the vibrating diaphragm presses against a mass of carbon granules, whose electrical resistance varies according to the pressure exerted on them. The granules form part of an external circuit, and the current in this circuit thus varies. In the receiver, or earpiece a varying current of this kind is passed through a pair of electromagnets which cause a diaphragm to vibrate, emitting sound.

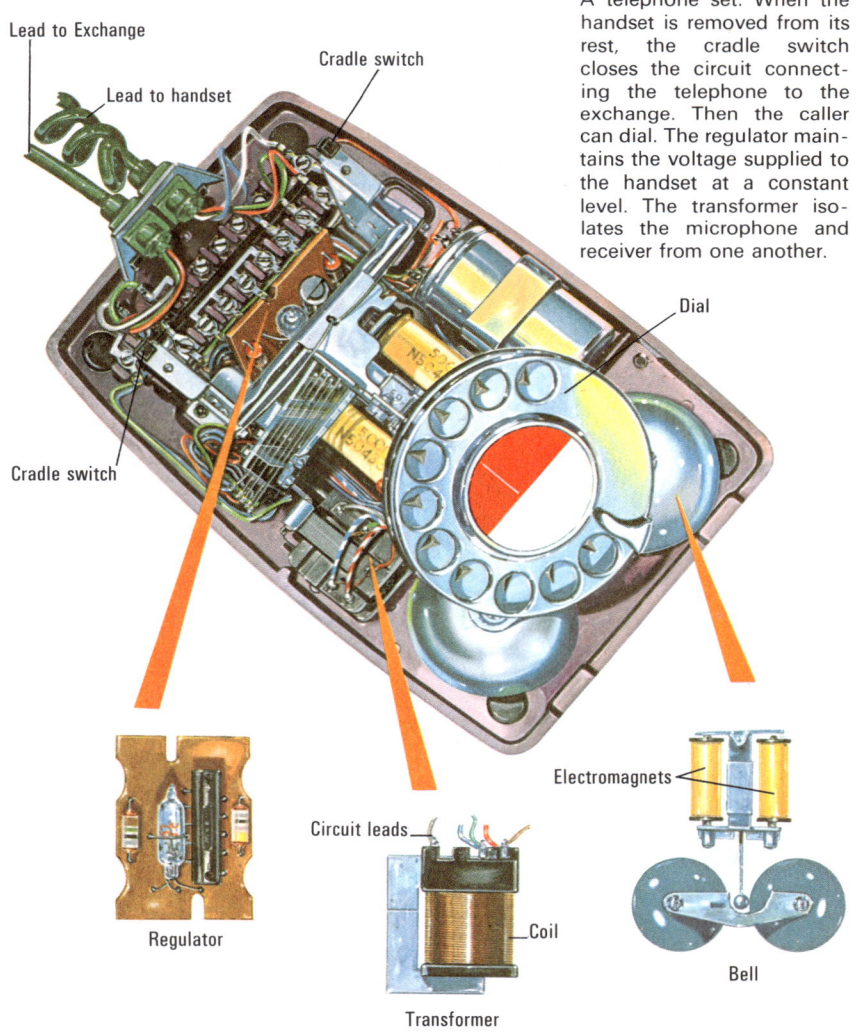

A telephone set. When the handset is removed from its rest, the cradle switch closes the circuit connecting the telephone to the exchange. Then the caller can dial. The regulator maintains the voltage supplied to the handset at a constant level. The transformer isolates the microphone and receiver from one another.

# Sound Recording

**Phonograph Recording**

In 1877, before a gathering of his friends, Thomas Edison recited 'Mary had a little lamb' into a primitive microphone. The microphone was connected to a blunt pin, which rested on a rotating cylinder covered in tinfoil. He replaced the microphone with an equally primitive loudspeaker, and rotated the cylinder. A distorted, but recognisable imitation of his voice came out of the speaker, astonishing everyone present.

Edison's machine was the first phonograph, or *gramophone*. A modern phonograph works broadly on the same principles. Sound waves are used to generate movement in a *cutting stylus* (by means of various electronic apparatus). This stylus makes a groove in an appropriate material. When another, *playback stylus* is led along the groove, it is moved in a similar way to recreate the original sound. Early recordings were made on wax cylinders. Modern records are plastic discs.

The first step in the production of a modern record is to make a *master disc*. (We are speaking now of the industrial process. The real first step, of course, is the rehearsal of artists, a choir, or an orchestra in a studio). A flat blank aluminium disc, coated with a shiny lacquer, is placed on a recording lathe, which looks rather like an elaborate, and very heavy, phonograph turntable and pickup.

The disc press, showing the stampers for each side of the disc in position and ready to come together. The piece of plastic to be stamped is seen resting on the lower stamper.

A recording studio during a recording session. The sound-proof control booth can be seen in the background. Note the positioning of the various microphones to provide suitable balance.

The sound signal from the recording microphone is fed as a variable electric current to a coil in the cutting head. The variations in current cause a chisel-shaped cutting stylus of sapphire or diamond to vibrate. The stylus is electronically heated to soften the lacquer, and so cuts a wavy groove into the rotating disc.

This master disc is sprayed with a silver solution to make it electrically conducting, and electroplated with a thin layer of nickel. The nickel shell, when peeled off, is a replica of the original, but with ridges instead of grooves. It is used, in turn, to produce a stronger version of the master disc, known as the *mother*.

From the mother disc are made several *stampers*, which are like the earlier nickel shell in having ridges corresponding to the grooves of the original recording. These stampers are fitted into large presses to produce the final records that will be sold in the shops. *Extended-playing* (E.P. – 45 revolutions per minute) and *long-playing* (L.P. – $33\frac{1}{3}$ rpm) records are now usually made of vinyl plastic.

**Stereo Recording**

Two kinds of records are available now: mono and stereo. In *mono* the sound is reduced to a single 'component' for recording purposes. But the sound for a *stereo* recording has two components. These come, in the simplest case, from two microphones set at different angles to the source of the sound. If the two components are kept separate in the recording and played back

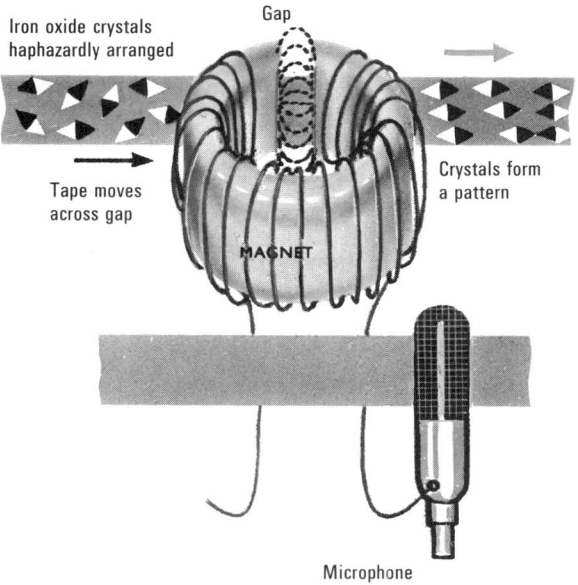

through two loudspeakers an appropriate distance apart, the effect will be 'three-dimensional', almost as if the listener were actually in the room with the people making the recording.

The groove in a mono record is cut laterally. That is, the stylus moves from side to side, while the depth of the groove remains the same. Stereo records are cut differently. The stereo recording cutter is modulated (varied) in two directions at the same time, one for each component, or channel of sound, as it cuts a groove. These directions are at right-angles to each other, and at 45° to the vertical, so that, in effect, one channel of sound is recorded on one wall of the groove, and the other channel on the other wall. The width and depth of the groove being cut thus change continuously.

**Tape Recording**

The sound signal that is fed to the master disc comes usually from a tape recording. This enables the producer to control more closely the final product. He can cut, or edit the tape; he can record part of it again; he can vary volume and tone.

A tape recorder uses a $\frac{1}{4}$-inch ribbon of strong plastic (usually PVC or 'Mylar') tape, coated on one side with a thin, smooth layer of ferrous oxide, which is magnetic. In a tape recorder, the tape is passed across the face of a *recording head,* which consists of a small electromagnet. The electrical signal from the microphone is passed to the coil of this electromagnet, causing a varying magnetic field. The alignment of the particles of ferrous oxide on the passing tape alters to match the magnetic field.

To hear the recording, the recorded tape is drawn past a *playback head,* similar to the recording head. (One head may sometimes serve both functions.) The magnetic 'pattern'

Above right: A modern tape recorder with microphone connected ready for recording. Above left: The principle of tape recording. A pattern of sound waves entering the microphone is converted to a variable electric current. Passage of this current through the coil of the electromagnet sets up a corresponding varying magnetic field in the gap. This field rearranges the iron-oxide crystals in the tape into a characteristic pattern.

in the particles of ferrous oxide on the tape induces an electrical signal in the coil of the playback head. The signal is amplified and fed to a loudspeaker, which reproduces the original sound.

A third, *erase* head is usually provided on a tape recorder before the recording head to clean a previous recording off the tape. It generates a magnetic field that neutralizes any existing magnetic pattern on the tape.

**Film**

Another method of recording sound is used in conjunction with motion pictures. Here the sound pattern is printed onto movie film alongside the frames of the pictures as variations in the density of the film or as variations in area.

The principle of variable-area sound recording on film. The variable electric current produced in the microphone is amplified and then fed to the coil of a mirror galvanometer, causing the mirror to be deflected. The mirror reflects a triangular patch of light onto a slit over the film. Depending on how much the mirror is deflected, a narrow or broad section of the triangular patch falls on the slit, producing a variable-area sound track. On playback this variable-area track is converted back into sound via a photocell and loudspeaker.

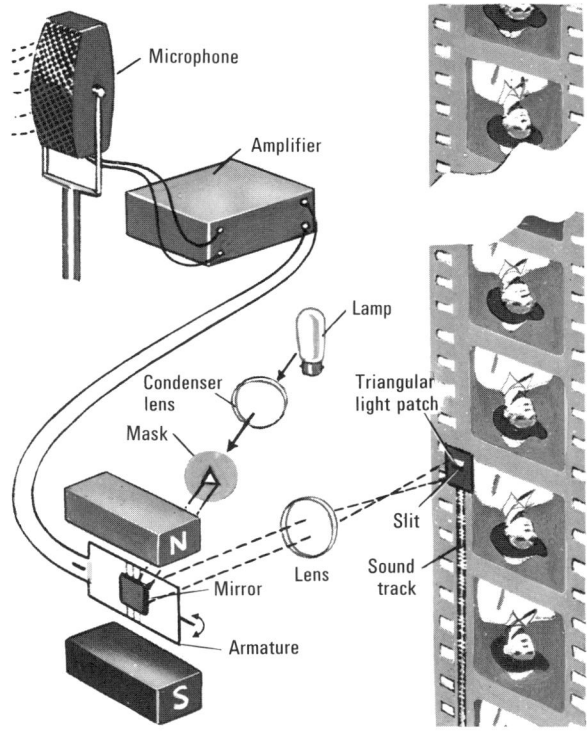

# Electronics

In this modern world electronics plays a vital role. Radio, television, radar, and computers are just a few of the things that rely on electronic devices for their operation. It is well straight away to distinguish between 'electrics' and 'electronics'. Both relate to the flow of those tiny, negatively charged particles of matter called electrons (see page 26). But 'electrics' involves the flow of electrons – that is, electric current – in conductors, such as copper wires. And 'electronics' involves the flow of electrons through valves, tubes, transistors, photoelectric cells, and similar devices.

Many electronic devices are designed to control and vary the strength and direction of electric currents. In radios, for example, one kind of valve is used to convert the alternating current from the mains electricity supply into direct current to operate the various components in the set. This is called *rectification*. Another kind of valve takes the weak signal that comes from a radio transmitter many miles away and strengthens, or *amplifies* it into a useful electric current. These valves work by generating and then controlling electrons in a certain way. And this, broadly speaking, is basic to most electronic devices.

## Valves

Radio and television sets, record players and tape recorders are just a few of the examples of the use of valves. A valve is a glass bulb in which there are two or more plates or wires called *electrodes* to carry electric current. The electrodes are connected to pins in the valve base which fit into sockets to make contact with the electrical circuit. Most valves are evacuated, but some may be gas-filled like an electric-light bulb is.

The simplest valve is the one used to rectify alternating current into direct current. It has just two electrodes and is known as a *diode*. The negative electrode, or *cathode* gives out a stream of electrons when it is heated. (This effect is called *thermionic emission*, and valves are generally termed *thermionic* valves.) It may consist simply of a wire filament similar to the filament of an electric-light bulb, or it may be a coated tube which gives off electrons when it is heated by a coil. The positive electrode, or *anode* is a simple metal plate which attracts and 'collects' the electrons that are emitted from the cathode.

The flow of electrons from cathode to anode means that electric current flows through the valve. Of course, the electrons will flow only if the anode is positive. If it is given a negative charge, it will repel the electrons and no current will flow. Consider, then, what happens when an alternating current, which is alternately positive and negative, is applied to the anode. The valve will let current pass only for the positive part of the alternating current. Thus, although an alternating current was applied to the valve, only the positive part is allowed through, constituting a unidirectional, or *direct* current.

A valve with three electrodes is called a *triode*. It is a valve that amplifies signals sent to it. It has a similar kind of cathode and anode to the diode, but in addition it has a third electrode, or *grid* between them. The grid, which is a fine coil of wire, controls the flow of electrons from the cathode to the anode.

The grid achieves control by means of its charge. When it has a strong negative charge, it will allow fewer electrons through than when it has a weak negative charge. And it is very sensitive in that only a slight variation in its charge produces a large variation in electron flow. Thus, when the weak signals from, say, a radio aerial are made to vary the charge on the grid, much bigger variations are produced in the electron flow through the valve. This is how the triode amplifies weak signals.

The most common amplifying valves, however, have four or five electrodes (called *tetrodes* and *pentodes*, respectively), the extra electrodes being grids.

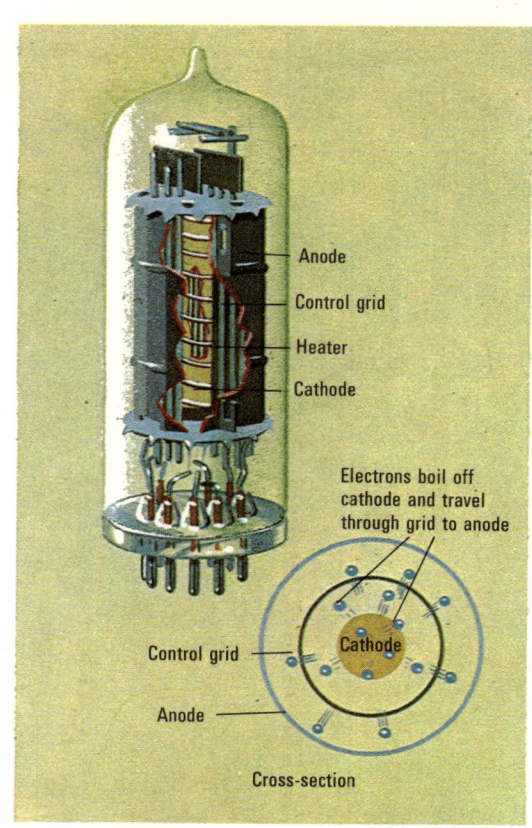

A typical miniature thermionic valve. It is a simple triode, with three electrodes. A heated cathode emits electrons, which are drawn towards the anode. Their passage is controlled by altering the potential, or voltage on the control grid. A slight alteration in the voltage on the control grid gives rise to a large alteration in anode current, thus affording a means of amplification.

Below: Standard symbols for common valves and semiconductors used in electronic equipment.

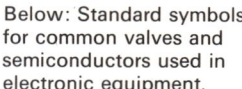
Thermionic diode

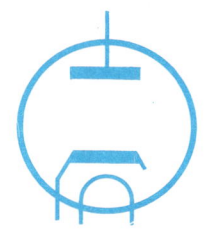
Thermionic pentode

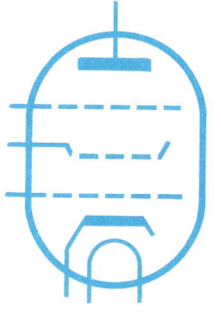
Transistor

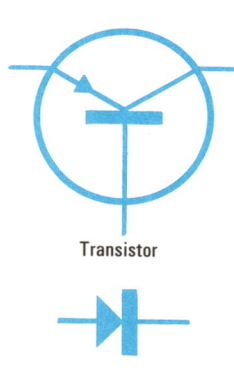
Semiconductor diode

## Tubes

The most common type of electron-beam tube is the *cathode-ray tube* used, for example, in a television set. The part that generates and controls the electron beam is known as the *electron gun*. This gun has a cathode which is heated to generate electrons. The anode has a hole in its centre through which the electrons can pass. In between there is a control grid which controls the electron intensity. A focusing electrode focuses the electron beam. A similar device is used to produce the beam in an electron microscope.

The set-up described so far, would produce a central spot on the fluorescent surface of the cathode-ray tube. But we can deflect the beam (which you remember is negative) by applying a magnetic or electric field to it. This is done by means of deflector electrodes, or deflector coils – one set for up-and-down deflection, another for side-to-side deflection. Suitable signals fed to the deflectors can make the beam trace patterns on the screen. In an *oscilloscope* the pattern may be only a simple wave form. In a television receiving tube it is a composite picture. (See page 107.)

Another kind of tube is used to generate and amplify the ultra-high frequency radio waves called *microwaves*, which are used in radar apparatus (see page 234). *Klystrons* and *magnetrons* are two types of microwave tubes.

## Transistors

Transistors are rapidly replacing valves in most electronic apparatus. They can do the same thing as valves but they have many advantages. Not least is their size. A transistor the size of an aspirin tablet can perform the same function as a 4 inch-high valve, hence, their use in pocket radios and spacecraft equipment. Transistors are also much tougher than valves. They require much less power and work immediately without the need to 'warm up'. And they do not get overheated.

Transistors are made up of crystals of a *semi-conductor* such as germanium and silicon. This is a material in between a true conductor like a metal and a true non-conductor like glass. Semi-conductors have peculiar electrical properties because of the presence of minute traces of certain impurities. With one kind of impurity a few free electrons are produced in the semi-conductor which can constitute an electric current. This produces an *n*-type semi-conductor. With another kind of impurity a few electrons are removed from the semi-conductor (a *p*-type), leaving what are called 'holes'. The movement of these 'holes' also constitutes an electric current.

Certain combinations of *n*-type and *p*-type semi-conductors give transistors the rectifying or amplifying properties of valves. They are made up of three layers: *n-p-n* (a *p*-type sandwiched between *n*-types) or *p-n-p*.

Microminiaturization of electronic equipment has been made possible by using *integrated circuits* – complete electronic circuits in tiny chips of semi-conductor.

The electron microscope has brought about a revolution in microscopy. Since it uses a very short wavelength electron beam instead of relatively long wavelength light rays, much greater detail can be resolved. Magnifications of up to nearly 2 million times have been claimed for the most advanced electron microscopes!

Lifesize transistors. It was the introduction of devices like these which revolutionized electronics in the 1950's.

A typical cathode-ray tube. The heated cathode emits a stream of electrons which is accelerated down the tube by the anodes. Its intensity is controlled by the control grid. The first, low-voltage anode acts as a focusing electrode to produce a fine beam. Voltages are applied to the deflection coils to make the electron beam trace a characteristic path.

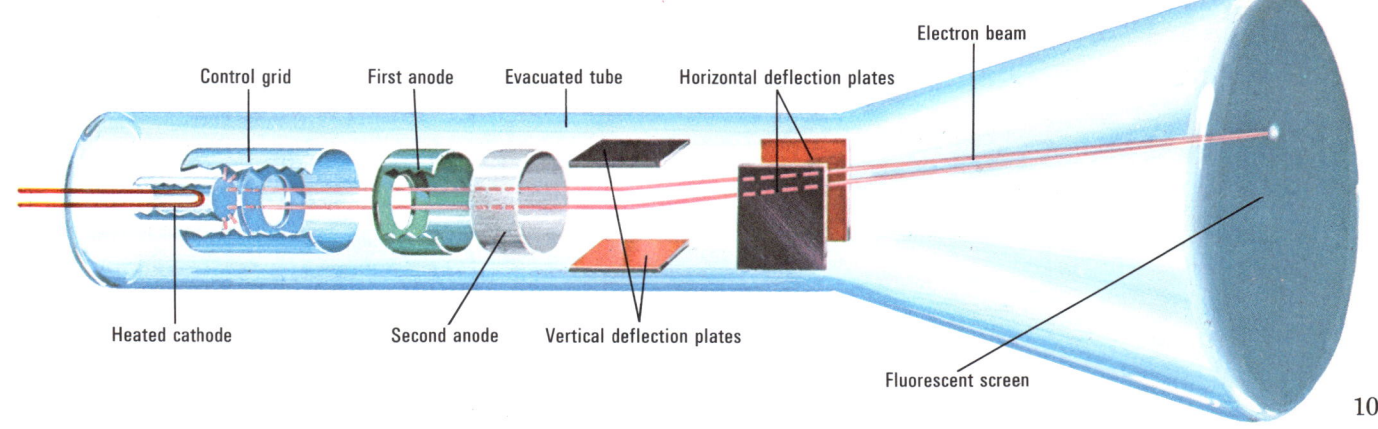

107

# Radio and Television

## Radio

Radio was born when it was discovered that electrical signals could be transmitted from place to place without any need for wires to carry them. Hence the name 'wireless'; a name that has seemed rather inappropriate to anyone who has examined the workings of one of the old radio sets!

Radio signals are carried by electromagnetic waves, which travel through space at the speed of light. In a broadcasting studio, sound waves are converted by the transmitter into radio waves, which travel through the air from aerial to aerial to be converted back to sound waves by the receiving set.

## Radio Transmission

In the studio, or at the outside broadcasting unit, sound enters a microphone. The microphone contains some form of mechanical device, such as a metal ribbon or coil or a crystal, which vibrates to convert the sound waves into an electric current. The current varies at the same rate, or *frequency* as the original sound.

This electric current is said to be at an *audio-frequency*. It is fed through an audio-frequency amplifier, which increases its strength, and then combined with a so-called *carrier* signal that is at a higher, *radio-frequency*. The carrier signal is altered, or *modulated* by the audio-frequency signal.

The modulated carrier signal is fed to the transmitting aerial. There electric currents cause radio waves to be emitted in all directions. The waves travel with the speed of light (that is, about 300 million metres or 186,000 miles a second). When they hit a receiving aerial, they set up oscillating waves in it, similar to those in the transmitting aerial.

## Radio Reception

These received waves may be very weak, and so, when the radio receiving set is 'tuned' to receive them, they are fed into a radio-frequency amplifier to be strengthened. Then they are passed to a *detector,* which separates the audio-frequency signal from the carrier wave.

There now remain patterns of currents identical to those that left the microphone in the broadcasting studio. They are amplified as necessary by an audio-frequency amplifier and fed to a loudspeaker. The loudspeaker acts like a microphone in reverse. The electric currents flow in an electromagnet to make a diaphragm move back-

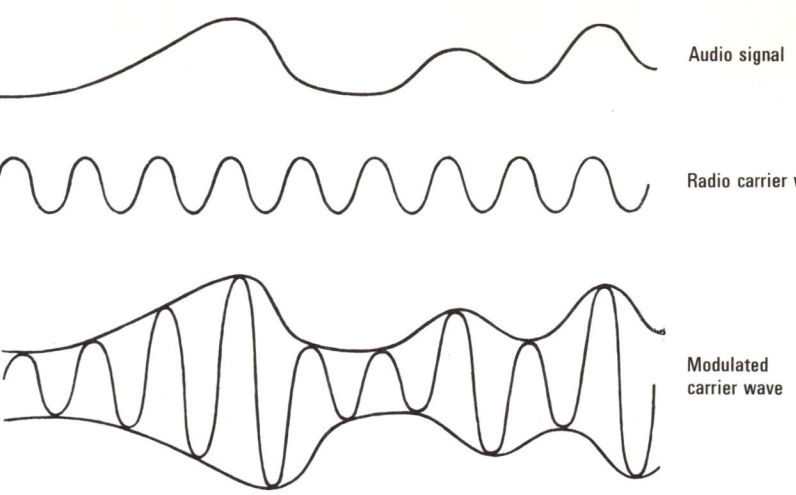

Illustration of the principle of modulation. The sound, or audio signal from the microphone (top) is combined with the higher frequency radio carrier wave (middle) to produce an altered or modulated carrier wave (bottom) that is broadcast. In the radio receiver the audio signal is removed from its carrier wave, a process called detection.

Guglielmo Marconi in 1902 with his radio transmitter (right) and receiver. The previous year he had transmitted radio signals across the Atlantic with similar apparatus.

## Marconi

Guglielmo Marconi (1874-1937) produced the first practical wireless telegraph when he was twenty-one. He was living at Bologna with his Italian father and Irish mother when he learned of the discovery by Hertz of electromagnetic waves. He realised that these waves might be used to carry messages. Though with little scientific training, he set to work to reproduce Hertz's experiment. He succeeded in making a spark appear across a gap in a metal ring by applying an electric voltage across another gap in apparatus a few feet away.

Marconi then tried to make the wave do something more useful. His first success was when he rang an electric bell in a room two floors below. He then connected a Morse key to the transmitter until he could send messages to his brother on the other side of a hill. His brother acknowledged receipt of the signal by firing a gun.

Marconi offered his invention to the Italian government, but they were not interested and so he came to Britain where he demonstrated his apparatus, first on the roof of the GPO London and then over 4 miles on Salisbury Plain. The Admiralty was particularly interested, and Marconi went on to improve his equipment to send messages over sea. In 1899 he sent the first wireless telegraph across the English Channel. On December 12, 1901, he transmitted the first long-distance wireless signal across the Atlantic from Poldhu in Cornwall to St. John's, Newfoundland.

wards and forwards to set up sound waves in the air. These sound waves are as near as possible the same as those waves that originally struck the microphone.

## AM and FM

The carrier wave used in the transmission of radio signals is modulated in one of two ways. Either the power of the carrier wave is varied to match the pattern of the signals (called *amplitude modulation*, or AM) or the frequency of the carrier wave is varied (called *frequency modulation*, or FM).

In the past most radio transmission has been by amplitude modulation. An AM wave can travel a long distance because the long, medium and short waves used are reflected by the ionosphere to follow the curvature of the Earth. Unfortunately much of the electrical interference is also AM, which sometimes makes reception rather difficult.

Frequency-modulated waves are not reflected around the curve of the Earth, and can be received only as far as the horizon. But they are not so easily interfered with by outside sources, and so produce more satisfactory reception. The quality of FM reception is also higher because these waves can carry higher frequencies more accurately.

## The Ionosphere

Radio waves, like light waves, travel in a straight line. So, unless they were reflected, they would travel straight out into space from the transmitting aerial. Listeners out of sight of the aerial would not receive its radio signals. Fortunately, most radio waves are reflected from electrically charged 'layers' between 50 and 300 miles up in the Earth's atmosphere. This region is called the *ionosphere* because it consists mainly of ions, or electrically charged atoms. Short waves are reflected back more strongly than longer waves, and so short-wave radio is used for long-distance transmission.

## Television

Television depends, like radio, on the production of a suitable electromagnetic signal that can be transmitted, received, and converted back into its original pattern. In sound transmission, the chain of events begins with a microphone, which changes sound waves into electrical signals, and ends with a loudspeaker, which converts the electrical pattern back into sound waves that resemble the original sounds.

In picture transmission, the key unit is the *camera,* which looks at a scene and changes the image into electrical impulses. At the

Closed-circuit television (CCTV) in action. CCTV is being increasingly used in industry, hospitals, stores, schools and many other places for information, training and educational purposes.

The principle of the image orthicon, or TV camera. The mosaic emits electrons according to the amount of light falling on it. These electrons travel to a target plate where they set up a pattern of electric charges. An electron beam 'scans' the target and is reflected back from each point depleted in intensity to an extent determined by the charge there. A variable electron flow, or electric current thus results.

other end, the equivalent of the loudspeaker is the cathode-ray tube, which converts the pattern of electrical impulses into a visible image.

## Television Transmission

Inside a television camera is a light-sensitive *plate*. An image of the outside scene is formed on this plate by means of a lens or lenses. The material of the plate is such that electrons are released wherever light falls on it. (This is called the *photoelectric effect*.) Brightly illuminated areas produce most electrons, and darker areas produce proportionately less. The electrons travel to another plate, called the *target,* where they produce a pattern of positive electrical charges equivalent to the original pattern of light.

A steady electron beam from an electron 'gun' scans the target, moving rhythmically across from left to right in narrow, horizontal bands. At the end of each band it flicks rapidly back to the left side and shifts down a fraction of an inch to start the next

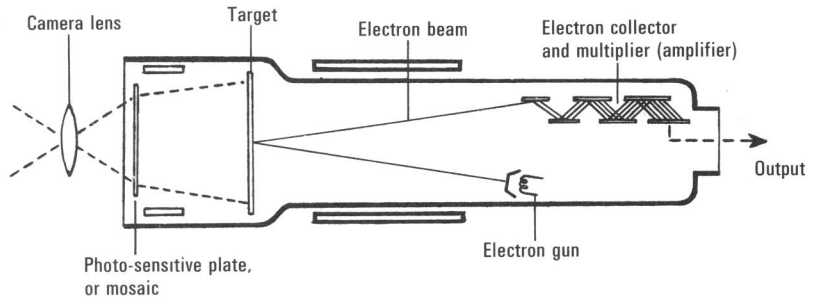

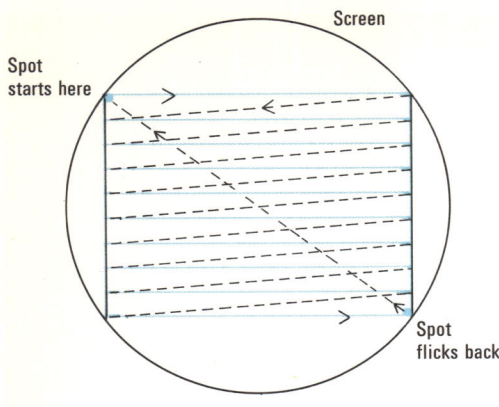

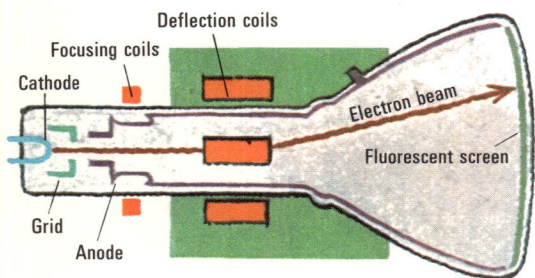

Top: A simple scanning arrangement. The electron beam in the cathode-ray television tube is made to scan back and forth across the screen in a regular manner in sympathy with a similar beam in the camera tube. Variations in beam intensity in the camera tube are paralleled by variations in intensity in the receiving tube. In this way the image seen by the camera tube is reproduced on the screen. In the television receiver the grid controls brightness, while the deflection coils control scanning (bottom).

### Video–Tape

We can record sound on tape recorders. The sound is converted by a microphone into variable electrical signals. These signals are made to 'imprint' a variable 'pattern' on a magnetic tape. When the tape is played back through the tape recorder, the variable magnetic 'pattern' on the tape is converted back to electrical signals and thence to sound (see page 105).

The electrical signals from the television camera can also be recorded on magnetic tape in a similar way, together with the appropriate signals from the sound microphone. Such a tape is called a *video-tape*. When the tape is played back, the original electrical signals are reproduced, and they can be transmitted in the usual way as a television picture. The video-tape has to travel very much faster than that of a normal tape recorder in order to accommodate all the signals from the scanning mechanism. The signals are often recorded as a zig-zag pattern on a specially wide magnetic tape.

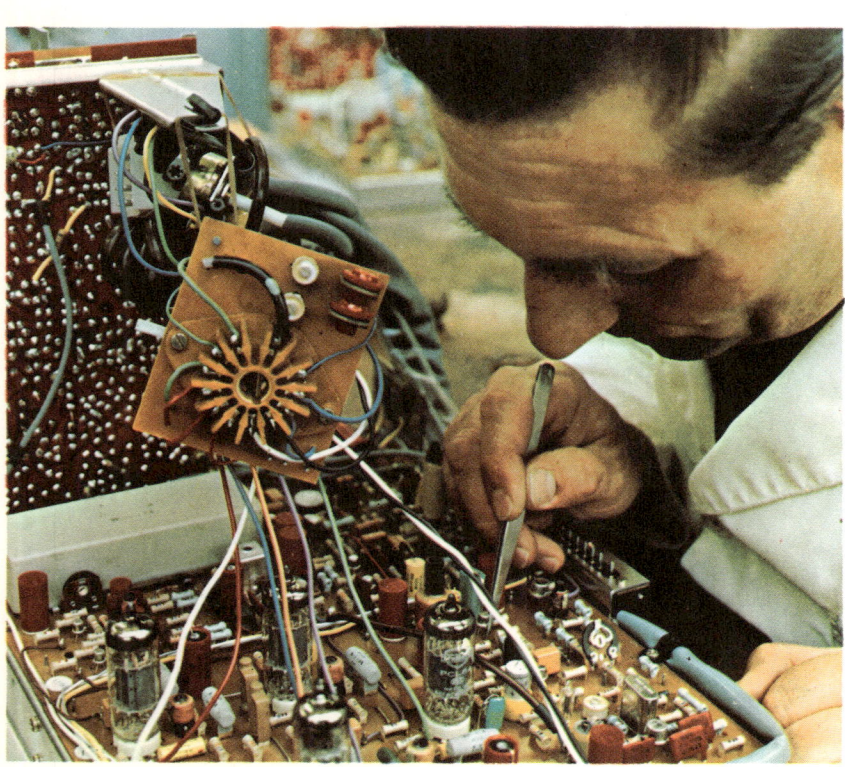

An electronics engineer tests a colour television receiver before it leaves the assembly line. Note the complexity of the circuitry.

band. When it reaches the bottom right-hand corner it flicks back up to near the top left and scans a second set of bands between those of the first sequence. This whole scanning process takes only 1/25th of a second, and is repeated continuously. There may be a total of 405 or 625 horizontal lines in the double sequence in Britain. In other countries different line standards may be used. In the United States, for example, 525 lines is used; in France, 819 lines.

Where the electrons in the scanning beam fall on an area of the target containing a large positive charge, they are attracted and held. In other areas, a larger number of electrons are reflected back to a collector. As a result, the current flow into the collector is proportional to the charge on the target and so to the brightness of the original image.

This current is amplified and fed to the transmitter, where *synchronizing pulses* are added that will lock the scanning beam of the receiver in step with that of the camera. The 'sound' signal is added before the final output is fed to the transmitting aerial. For transmission very-high (VHF) and ultra-high (UHF) radio frequencies are used.

### Television Reception

A television receiver picks up a signal in its aerial, in the same way as a radio does. The *video*, or vision signal is separated from the *audio*, or sound signal and fed to the cathode-ray tube as a series of electrical impulses, similar to those that left the camera at the television studio. In the tube is an electron gun that operates like the gun in the camera to scan across a fluorescent screen.

The strength of the electron beam varies with the strength of the signal. As the electrons strike the surface of the fluorescent screen, they make it light up with a pattern of brightness that also depends on the strength of the signal. The electron beam is made to scan the screen of the cathode-ray tube at the same rate as, and in step with, the beam in the television camera. And a picture is created on the screen of the television receiver which corresponds to the scene focused on the plate in the camera.

### Color Television

The system described above gives only patterns of brightness and darkness, and hence a *monochrome* picture. A color picture may be transmitted and seen by using various quantities of the three primary colors— red, blue, and green. These three colors, taken in the appropriate combinations and strengths, will provide nearly all the colors seen in Nature.

Hence, color television uses three separate signals, one for each of these primary colors. The lens system of the color camera splits the image by means of filters into three. These are passed into three separate camera tubes that produce output signals corresponding to their colors.

In the color receiver, the cathode-ray tube has three electron guns. One gun receives the signals for red, one those for blue, and one those for green. The fluorescent screen is coated with thousands of tiny spots of phosphors in groups of three. In each group, one phosphor produces red light, one blue and one green. A *shadow-mask* behind the screen has holes so lined up that the beam from each electron gun hits only its appropriate color phosphor. The result is, or should be a true color representation of the original scene.

Above: Scene in a television studio during the shooting of a historic drama.

Below: The elements of a color camera and receiver. Note the three camera tubes and the three electron guns in the receiver, each of which deals with light of one particular color.

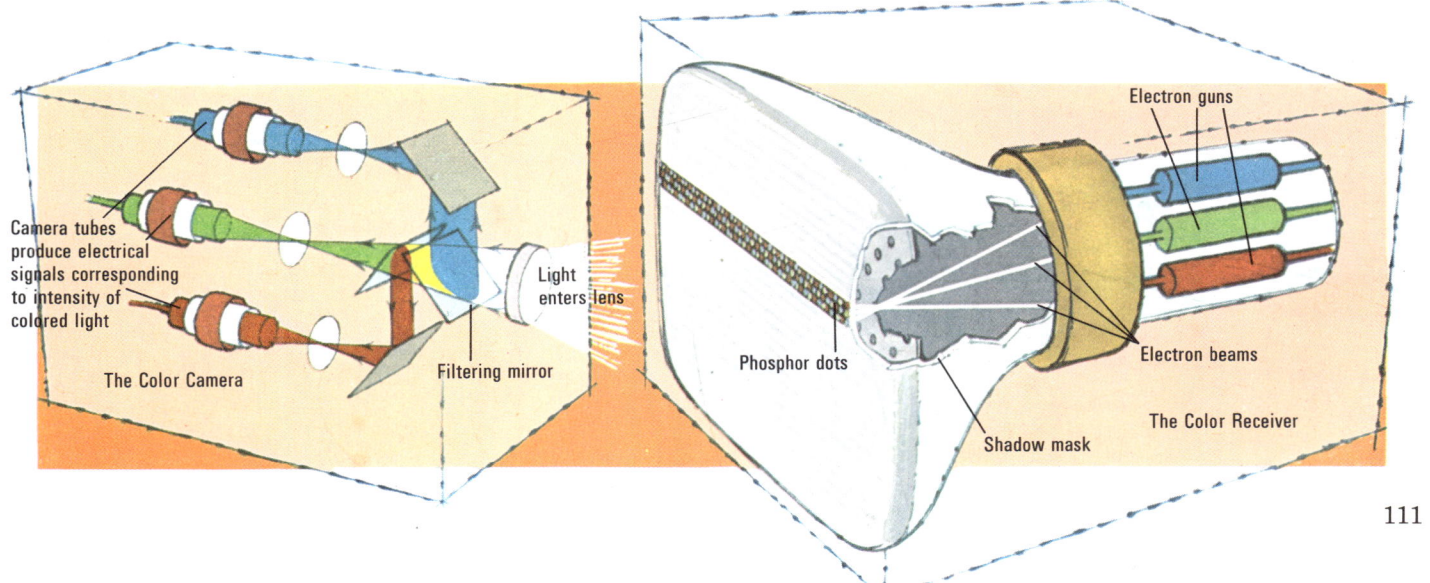

# Computers

When machines were first introduced into factories, some people opposed them and tried to smash them because they thought that machines should not displace human effort. When Charles Babbage designed a computing machine in 1822 that would calculate mathematical tables, this aroused a similar reaction, and the lack of public sympathy was such that the machine was never completed.

Various forms of calculating machines were constructed in subsequent years, but the first true computer was built in America in 1944. Now, of course, computers are taking over many of the routine jobs in industry and commerce. A computer is unable to do anything that a man cannot; its value lies in its ability to do simple jobs, like adding numbers, much more quickly than a man. As a result, tasks can now be undertaken with the aid of computers that were quite impracticable before.

## Types of Computer

The two principal kinds of computer are digital computers and analogue computers. A *digital* computer is used for making most numerical calculations. A number may be represented in a machine in a variety of ways. The milometer on an automobile or a bicycle represents numbers by teeth on a gear wheel. A bicycle cyclometer is a simple computer. The front wheel tells the cyclometer every time it has completed a revolution by knocking round a little star wheel. When this star wheel has rotated an appropriate distance it in turn moves on a dial marked in tenths of a mile. Every complete revolution of this dial moves on another dial marked in miles and so on.

Many desk calculating machines work on this principle. Sets of gear wheels are used to add numbers together. Mechanical machines, though, are severely limited in their speed, and modern high speed calculators work electronically.

An *analogue* computer does its calculations by measuring a physical quantity that may be quite different from that concerned in the particular problem. A simple example is a thermometer. The heat of a furnace, say, causes the liquid in a tube to expand. The expansion of the liquid is measured, and its length is described in terms of temperature. Most analogue computers are used by scientists and are designed to do a specific job.

The components of a computer are known as its *hardware*. The information and instructions that are fed into it are known as *software*.

## Hardware

The heart, or perhaps more appropriately, the brain of a computer is the *central processing unit*. This is the essential operating part which makes decisions, performs arithmetic calculations and controls input and output units. Input units provide the central processor with information, or *data*. Output units handle data that have been processed. To process information supplied to it, the central processor is divided into three parts—a storage, or memory unit, a control unit, and an arithmetic unit.

The *memory unit* stores data until they are required. It also stores a set of instructions, called a *program*, which tells the computer what to do with the data. Data and instructions that are needed immediately over and over again are stored *internally* in the memory unit. Internal storage depends on ring-shaped magnetic cores, which can be magnetized in either direction. Large amounts of other information which are required only occasionally are stored *externally* on magnetic tapes, drums, discs, and cards.

The *control unit* selects data and instructions from the memory unit, interprets them, and controls the carrying-out of the instructions in the *arithmetic unit*. The arithmetic unit adds, subtracts, and compares (less than, equal to, or greater than) data supplied to it, as instructed by the control unit.

The arithmetic unit does the calculations at very high speed. It uses *binary* numbers for this. Binary arithmetic uses only two digits, 1 and 0, so that a computer can work

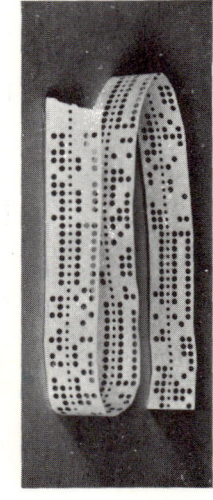

Punched paper tape of the kind that is often used to feed information into computers.

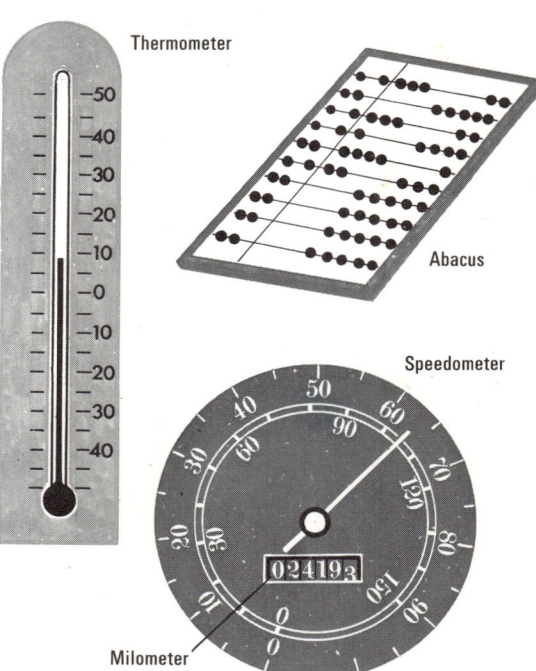

Some simple computing devices. The thermometer is an analog computing device since it indicates one quantity (temperature) in terms of another (length of mercury in the column). Similarly, the speedometer is an analog device, with movement of a pointer indicating speed. The abacus and milometer on the other hand deal directly in numbers and are therefore digital devices.

Left: Input into many computers is by means of punched cards. Here an operator is loading punched cards into the card reader.

Right: The great thing about the computer is its speed of operation. This is readily apparent in the print-out of results. This line-printer, for example, prints out at 900 lines a minute!

on a simple 'on' or 'off' principle. A circuit that is 'on' (or electrically positive) might represent one digit, and a circuit that is 'off' (or electrically neutral) might represent the other. Binary digits are often called *bits*.

Data may be fed into the computer in a variety of ways. Punched cards and tape are common input media, and magnetic tape, discs, and drums are also widely used. For some applications the instructions may be written on the face of a cathode-ray tube (like a television screen). Each input medium naturally requires its own input device to 'translate' the information and instructions supplied to it into a form that the computer can utilize, that is, so-called computer 'language' (binary).

Similar media are used to provide the output from the computer after computation is completed. The processed data may also be printed out to give a visual record.

### Software

The person who prepares a program of instructions for the computer is the *programmer*. Before he can begin the program, he must be provided with a description in plain language of the job to be done which clearly defines what problems have to be solved and details the operations step by step. From this *job description* the programmer plans a *systems flow chart*, in which he uses symbols to represent an overall view of the operations that will have to be performed on the computer. Next he draws up an *operations flow chart* which goes into the operations in great detail. The programmer uses this chart as a guide when he constructs the actual program.

The computer, as we have seen, can operate only on data supplied to it in the form of binary numbers. Thus the program must be presented to the machine in binary language. Early programs were laboriously written directly in binary machine language. But today the programmer writes his program in what is called a *computer*, or *source language*. The computer itself, when suitably programmed, will translate this source language into binary machine language. Commonly used computer languages include Algol, Fortran, and Cobol.

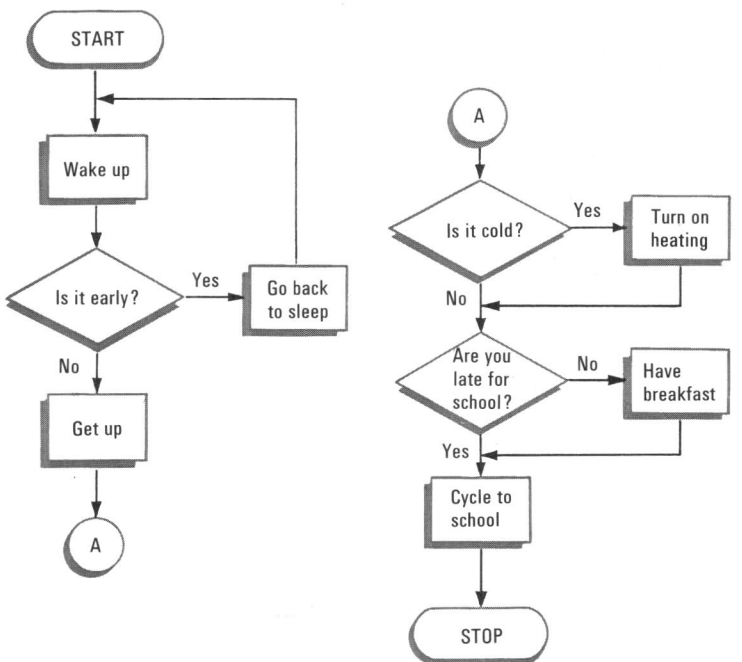

A trivial example to illustrate the procedure of flowcharting. The simple action of going to school in the morning has been broken down into a number of consecutive stages. The computer programmer flowcharts his problem in a similar way before he begins to write his program.

# Sound

We all have our favorite sounds: church bells ringing in a distant steeple, the song of the skylark, the murmur of a mountain stream, the crash of the dinner gong. And we have our dislikes, too: the ear-splitting roar of a pneumatic drill, the squeal of rats, the screech of chalk on the blackboard. How are these sounds – pleasant and unpleasant – produced?

If you could climb into the belfry of the steeple where the bell was ringing and feel the bell, you would feel it vibrating. And vibration is the clue to the production of sound. A vibrating object passes on its vibrations to the surrounding air. It *compresses* the air as it moves outwards. Then it moves back, and the air *expands* back to occupy the space which the object has left. The air in the immediate vicinity of the vibrating object itself passes on the vibrations to air adjacent to it, and in this way the vibrations spread out from the source in all directions. These vibrations which travel through the air are what we call *sound waves*. We hear them as sound when they strike our eardrums and cause them to vibrate. The vibrating eardrums send impulses to the brain, which interprets them as sound.

There are basically two kinds of sounds. There is *musical sound* – sound made by a vibrating body that sends out regular vibrations at regular intervals. And there is *noise* – sound made by a body vibrating irregularly and at irregular intervals. A vibrating tuning fork and a violin string are examples of bodies which send out very regular vibrations, producing a musical sound. A badly fitting window frame rattling in the wind is a typical example of a body producing irregular vibrations and therefore noise.

Sound waves travel in other things besides air – in water, for example. Submarines and naval ships use underwater sound detectors called *hydrophones* to listen for the propellers of enemy shipping. Sound, surprisingly enough travels faster in water than in air. In air sound travels at about 1,100 feet per second (750 miles an hour) but more than four times as much (4,700 feet a second) in water. In fact, the denser the medium sound travels in, the higher its speed. It does not travel in a vacuum, of course, because there is no medium present to pass on the vibrations. The *speed of sound* is very important in aeroplane flight, because strange things start to happen when the plane nears sonic speed. But many aircraft these days fly at supersonic speed (see page 225). Sound travels much slower than light, which gives rise to interesting effects. It means that you hear a clap of thunder several seconds after you see the lightning.

Pitch, intensity and quality are three properties of sound that enable us to distinguish one sound from another.

## Pitch

A whistle is a sound with a high pitch, a growl is a sound with a low pitch. The factor that determines the pitch of a sound is the rate, or *frequency* of vibration of the object generating the sound. The frequency is generally measured in complete vibrations, or cycles per second (cps). On average a

Moving in either direction (above), a vibrating string compresses the air in front of it and creates a partial vacuum behind. Waves of compression are transmitted through the air away from the vibrating string (below).

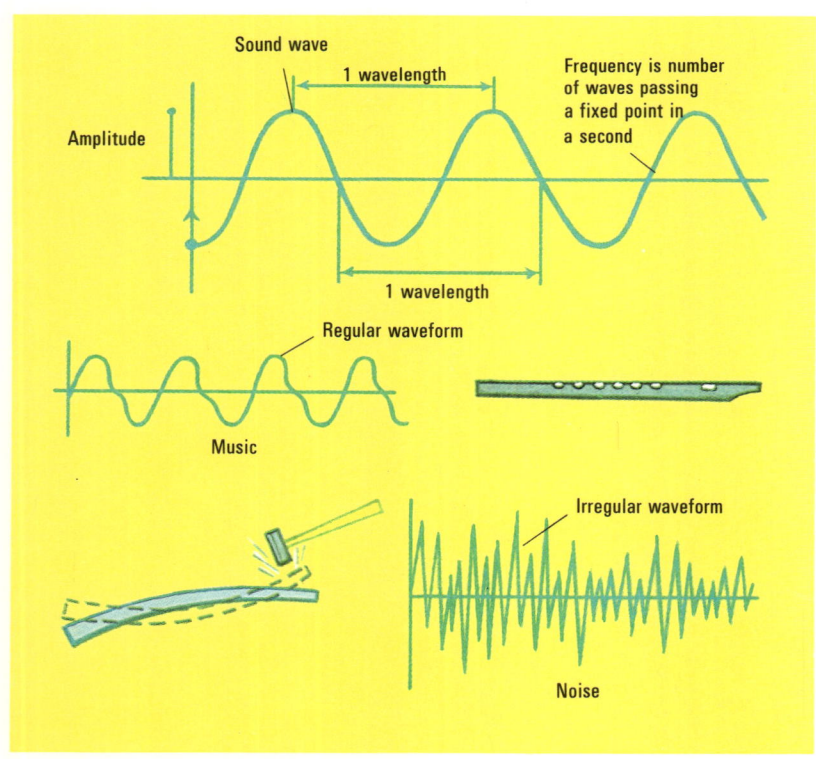

Illustration of some of the features of sound waves, and demonstrating the difference between a musical sound and noise.

person can hear sounds with frequencies between 20 and 20,000 cps. Other animals can detect sounds beyond human ears – dogs and bats, for example, can hear much higher frequencies. We call sounds beyond the audibility of human ears *ultra-sonic*. Sound waves with frequencies of millions of cycles per second have interesting properties. They can be used, for example, for drilling and for detecting flaws in metal.

Sometimes the pitch of a certain sound seems to alter. The pitch of a police siren seems much higher when the police car is racing towards you than when it is disappearing into the distance. This apparent change of pitch with moving objects is known as the *Doppler effect*. It can easily be explained. Since the object is coming towards you, the sound waves are 'crowded' closer together, and more of them reach you in a given time. So it appears that the frequency of sound is higher. The reverse happens when the object moves away. Less waves hit your ear because the sound waves are effectively 'stretched' farther apart. And the frequency appears lower.

### Intensity

The intensity of a sound is a measure of the energy of the vibrations which cause it. A large vibration will set up intense sound waves, which will have a large *amplitude* (movement to and fro). The more intense the sound, the louder it is when it reaches us. *Loudness* is the apparent intensity of the sound as we hear it. The farther we are from the source of the sound, the fainter the sound will be.

We measure comparative sound levels with a special meter in units called *decibels*. A sound of 0 decibels can hardly be detected by the human ear. A whisper is about 20 decibels. A jet engine about 40 yards away is almost 130 decibels, very near the limit when noise becomes painful to the ear. The modern world has become disturbingly noisy, noise being another aspect of pollution.

### Quality

If you tune, say, a guitar string and a violin string to the same pitch and pluck them, you might expect the sounds produced to be the same, (assuming they are of the same intensity). But they are not. Each string produces a note with a different quality from the other, even though the fundamental pitch is the same. It is this quality which gives the sound from every instrument its individuality, and makes it possible for us to distinguish between, say, the violin and the guitar or the trumpet and the French horn.

What causes this difference in quality? When an object such as our violin string

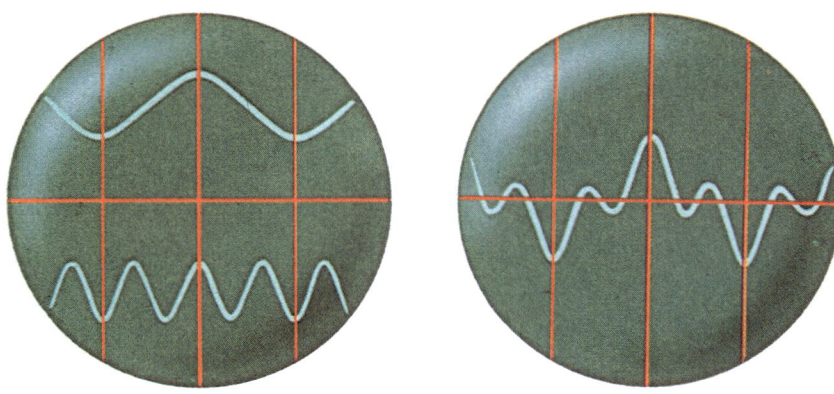

Sound vibrations can be pictured on an oscilloscope. The tracings shown here demonstrate how the fundamental vibration of a musical instrument is modified by a harmonic. The left-hand scan shows the fundamental with the overtone beneath it. The right-hand scan shows them combined.

The vibration of air in organ pipes. The wave form in the pipe shows the extent the air vibrates. The air is still at a node and vibrates most at an antinode. An organ pipe always has an antinode at its open ends. The left-hand diagram shows the fundamental vibration of an open pipe. The next one shows a harmonic of that pipe. It has many more too. The right-hand diagram shows the fundamental of a closed organ pipe. The closed pipe has only half as many harmonics as the open pipe and therefore produces a rather different sound.

vibrates, it sends out a main vibration called the *fundamental*. But it sends out other, though weaker vibrations, too. For example, the violin string vibrates as a whole to produce the fundamental vibration. But each half, third, and quarter of the string may also vibrate separately. These additional vibrations are known as *overtones*, or *harmonics* of the fundamental frequency. It is the superimposition of the overtones on the fundamental which gives every musical instrument its particular tonal quality.

### Resonance

Some sopranos have been known to shatter glass when they hit a high note. They have also been known to make a gas flame flicker. It is not magic but a phenomenon called *resonance*, or sometimes *sympathetic vibration*. Some bodies have a certain natural frequency at which they would vibrate if they were struck. When vibrations of that frequency fall on them, they will begin to vibrate themselves. If the stimulating vibration is prolonged, the induced or sympathetic vibrations may become bigger and bigger. In the case of the soprano, the vibrations in the glass became so violent that they have shattered it.

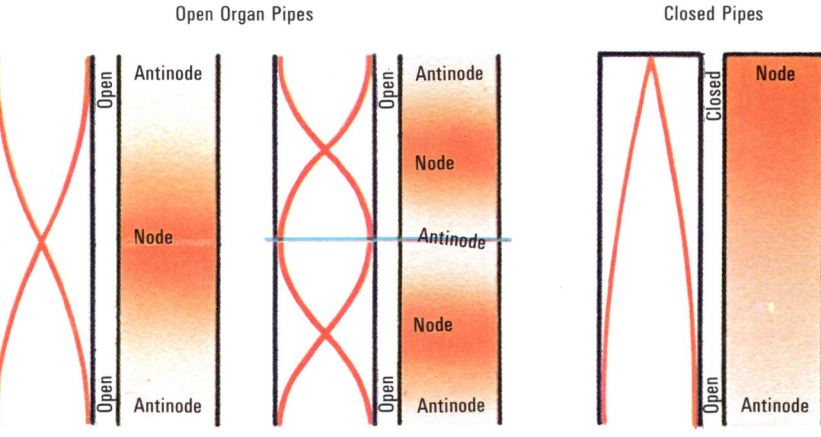

# Hydrostatics

When you place a piece of wood in water, it floats. If you push it under the surface and then release it, it will rise again. This indicates that the water exerts an upward force on it. This effect is known as *buoyancy*. Place a heavy object in the water and it will sink. If you suspend the object from a balance, you will find that it weighs less in the water than out of it. The loss in weight is due to the upward force of the water on it, called the *upthrust*. The object will naturally push aside, or displace some of the water when it is immersed. And if you weigh this displaced water, you will find that its weight will be equal to the apparent loss in weight of the object due to the upthrust. You will then have demonstrated the truth of the principle of Archimedes. The principle applies not merely to water but to all fluids, liquids and gases alike. It forms one of the fundamental laws of hydrostatics, a science concerned with the study of fluids at rest.

When the object is lighter than water, it will float. Then we have a special case of Archimedes' principle which we call the *law of flotation;* when an object floats in a fluid, the weight of the object is equal to the weight of the fluid displaced.

The most important example of the application of Archimedes' principle is in ships (see page 118). Ships are made of steel and weigh thousands of tons. But they can float because they displace a great amount of water. A ship sinks into the water until its total weight is just balanced by the upthrust of the water it displaces. Another important application of Archimedes' principle is in the *hydrometer* which is an instrument designed to measure the densities of liquids.

## Liquid Pressure

When you fill a bucket with water, its weight increases, that is, the total force of the water pressing downwards on the bottom of the bucket increases. *Pressure* is the force acting on a unit area of surface, say a square inch. As you fill up the bucket, the weight of water increases, but the area of the bucket bottom remains the same. Therefore the pressure on it increases. In other words pressure increases with the depth, or *head* of the water. Pressure acts equally in all directions, too. If you puncture your bucket with holes at intervals down the side, the water spurts out sideways.

We make use of or take account of liquid pressure in many ways. For example, we utilize the pressure of a large head of water in hydroelectric power schemes. The head used may be a natural head, or it may be created by damming a river gorge. The dams, of course, must be designed so as to allow for the increase in pressure with depth (see page 156). Deep-sea divers have to take great account of water pressure.

When a body is immersed in a fluid, it experiences an upthrust which is equal to the weight of fluid displaced. In hot-air ballooning the balloon is filled with light, hot air. When the weight of the air displaced and thus the upthrust is greater than the weight of the balloon, the balloon will be forced to rise.

Below left: Illustration of the principle of Archimedes. The loss in weight of the 1000 gm weight when immersed in water is found to be the same as the weight of water it displaces. Below right: The principle of the hydraulic jack. A force of 10 lb acting over 1 square inch creates a pressure of 10 lb a square inch that is transmitted throughout the liquid.

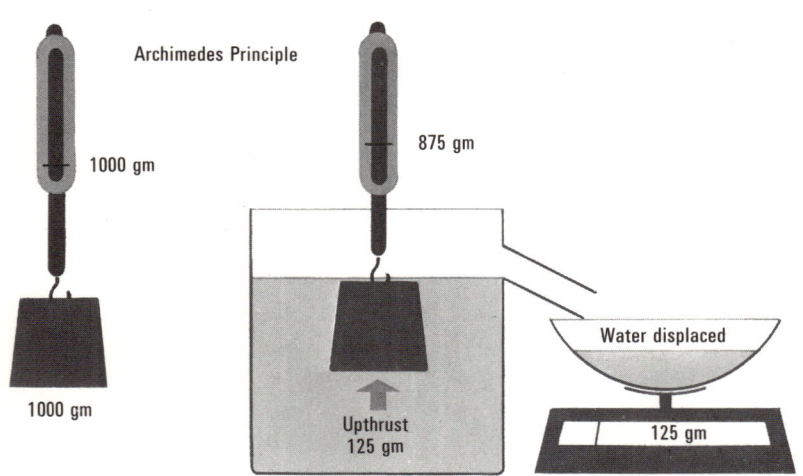

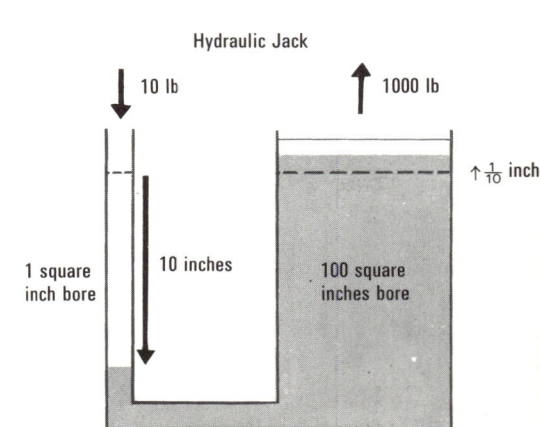

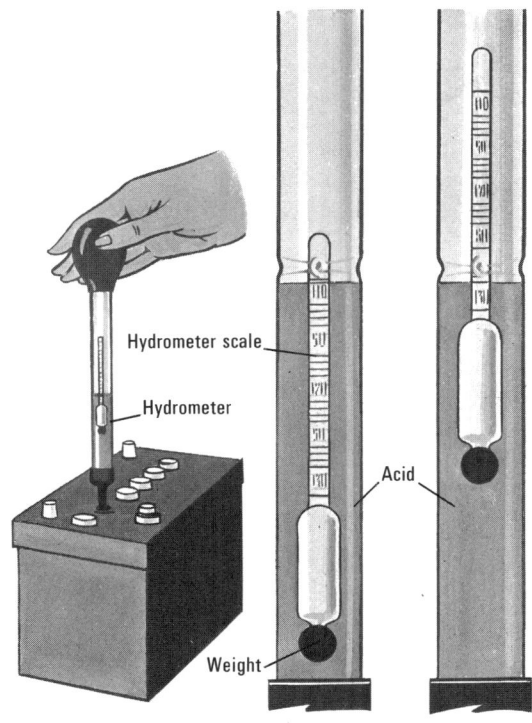

Left: A hydrometer being used to test the specific gravity of the acid in an automobile battery. The hydrometer is contained inside the sampling tube, through which the acid is withdrawn. The level at which it floats is determined by the specific gravity of the acid. A low specific gravity indicates that the battery is not charging properly.

Right: An aneroid barometer. Variation in air pressure causes the corrugated lid of a partially exhausted box to move in or out. This movement is magnified and indicated by a pointer. The aneroid barometer must first be calibrated against a standard instrument.

Aneroid Barometer

Another fundamental law of hydrostatics is that pressure can be transmitted through a fluid in all directions. The first effect the pressure has, though, is to compress the fluid. Gases compress easily, liquids hardly at all. And for all intents and purposes we can consider liquids as being incompressible. We use many machines based on the principle that liquids transmit pressure. Foremost among them is the *hydraulic press*. In essence the press consists of a narrow and a wide cylinder fitted with pistons and connected and filled with liquid. Suppose that the narrow cylinder has an area of 1 square inch and the wide one an area of 100 square inches. If you push down on the small piston with a force of 10 lb, a pressure of 10 lb per square inch will be transmitted through the liquid. When it acts on the large piston (area 100 square inches), it exerts a total force of $10 \times 100 = 1,000$ lb. Thus the original force has been greatly magnified. Giant presses exerting 50,000 tons pressure or more are used for shaping metals (see page 137). Hydraulic devices working on a similar principle are used in all kinds of machines. The main foot brakes of automobiles work hydraulically. So do the undercarriage of an aircraft, the tipping mechanism of a truck, and so on.

## Air Pressure

We are not really aware of it, but the atmosphere presses down on us with a force of 14·7 lb per square inch. This is what we call atmospheric pressure. It results from the weight of atmosphere above us. As we go higher, the depth of atmosphere above us becomes less. Therefore the atmospheric pressure falls.

We measure atmospheric pressure with a *barometer*. The most accurate is the mercury barometer. If a long glass tube is filled with mercury and inverted in a bowl of mercury, the mercury level in the tube will fall to a certain level. This column of mercury is being supported by the pressure of the atmosphere, and if the pressure varies, so will the height of the column. At sea level the normal air pressure will support a column of mercury about 30 inches or 760 millimetres high. A less accurate barometer, called the *aneroid* barometer utilizes the movement of the top of a partially exhausted box to indicate change in pressure.

## Gas Laws

Gases, as we mentioned earlier, are highly compressible, and air is no exception. Robert Boyle (1627–1691) investigated how the volume of a given mass of gas varied as the pressure on it is varied. He established what is known as *Boyle's law*. He found that pressure ($P$) is inversely proportional to the volume ($V$) if the temperature remains constant. In other words if you double the pressure, you halve the volume. Jacques Charles (1746–1823), a hundred years after Boyle, discovered that the volume of gas is directly proportional to its absolute temperature ($T$, the temperature measured from absolute zero) if the pressure remains constant. Taken together Boyle's and Charles's laws make up the fundamental gas laws that can be expressed as $PV = RT$, where $R$ is the so-called gas constant.

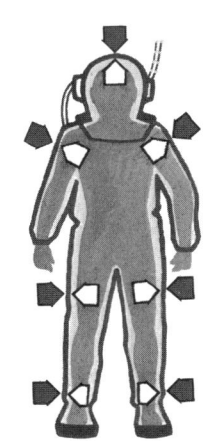

In a liquid the pressure increases regularly with increase in depth, becoming about 4½ lb a square inch greater for every extra 10 feet. At 300 feet depth the pressure is about 10 times that at the surface. Deep sea divers wear suits that are supplied with compressed air to help resist the tremendous pressures.

# Ships

Ships of all shapes and sizes ply to and fro across the oceans of the world carrying passengers, mail, and an astonishing variety of cargo. And there are sturdy, manoeuvrable *tugs* to guide large ships into port; thick-hulled *icebreakers* to keep northern shipping lanes open in winter; *cable ships* to lay submarine telephone cables; *dredgers* to keep navigation channels deep and navigable; *lightships* to warn shipping of danger; *whalers* to catch whales; and so on.

**Passenger Liners**

The largest liners operate on the transatlantic routes between Europe and the United States. The largest and longest of them is the *France*. It is 1,035 feet long, and can carry more than two thousand passengers. The crew alone number over a thousand. It has a cruising speed of 30 knots (35 miles an hour). Rivalling the *France* on the Atlantic crossing is the 963 feet long *Queen Elizabeth 2*, Cunard's £25 million successor to the other two famous *Queens* – *Mary* and *Elizabeth* – which were taken out of service in the 1960s. These huge passenger liners are nothing less than floating luxury hotels.

A passenger liner can be distinguished from other vessels by the extent of its superstructure rising above the hull. This may house five or more decks containing the public rooms and cabins, navigation bridge, sun decks, and so on. In modern ships the superstructure is made of aluminum.

To increase the comfort and safety of the passengers most large passenger liners are fitted with *stabilizers* to reduce rolling in rough weather. Stabilizers are fins projecting from the sides of the hull under water. When the ship begins to roll, they move automatically to try to make the ship roll in the opposite direction. This reduces the overall rolling motion. Liners also have the latest navigational aids. The *Queen Elizabeth 2*, for example, has a satellite navigation system.

**Ferries**

Some of the best-known ferries operate between French and English ports across the

One of the train ferries which operate between the Channel ports. Railway lines on the lower deck match those on the quayside so that coaches can be rolled on and off.

The *France*, the world's longest ever liner, which has a length of 1,035 feet. Launched in 1961 for the Atlantic crossing, she is 4 feet longer than the *Queen Elizabeth*, which at 83,774 gross tons was the largest passenger liner ever built.

English Channel: Dover-Calais, Newhaven-Dieppe, and so on. Cross-channel passenger ferries carry large numbers of passengers and mail, with only a little cargo, and connect with rail services at both ends of the passage. They are normally known as *cross-channel packets*. They have powerful engines so that they can maintain their scheduled services in all weathers. Most are fitted with stabilizers to improve passenger comfort when the sea is rough.

Operating over the same routes as the passenger ferries are the *car ferries* and *train ferries*. Car ferries carry both cars and passengers. Practically all of the below-deck space is taken up by the cars. Train ferries are designed mainly for transporting sleeping cars, but they may also take passengers, cars, and trucks. They have tracks on the decks which connect with the tracks at the terminal ports.

## Cargo Ships

Many thousands of cargo ships, or *freighters*, ply to and fro across the seas. Some of them carry a few passengers as well. *Cargo liners* are fast cargo ships that operate on a fixed schedule. *Tramps* are freighters that have no fixed ports of call, but travel whenever and wherever there is trade.

General cargo ships carry all kinds of assorted goods for different shippers. The cargo is stowed in holds below deck. On deck are a number of short, stocky masts, to which are attached long poles, or *derricks* with a pulley block on top. These are used to swing cargo in and out of the holds with the help of winches.

*Bulk carriers* are specially designed freighters used for transporting bulk cargoes, such as coal, oil, sugar, grain, and ore. Bulk oil carriers, or *tankers* are a special case and are dealt with separately. Bulk carriers are generally long and low, with their engines well aft.

## Tankers

Oil tankers are among the most distinctive of ships. They are very long and ride low in the water when laden. They have scarcely any superstructure. In many of the largest tankers, some of which are more than 1,100 feet long, the engines, accommodation, and bridge are all situated in the stern. Other tankers are designed with the bridge, together with more accommodation, in the bows or amidships.

The main cargo space in the tanker is split into several different compartments by partitions, or bulkheads. This prevents dangerous surging of the oil in rough weather which could well cause the vessel to capsize.

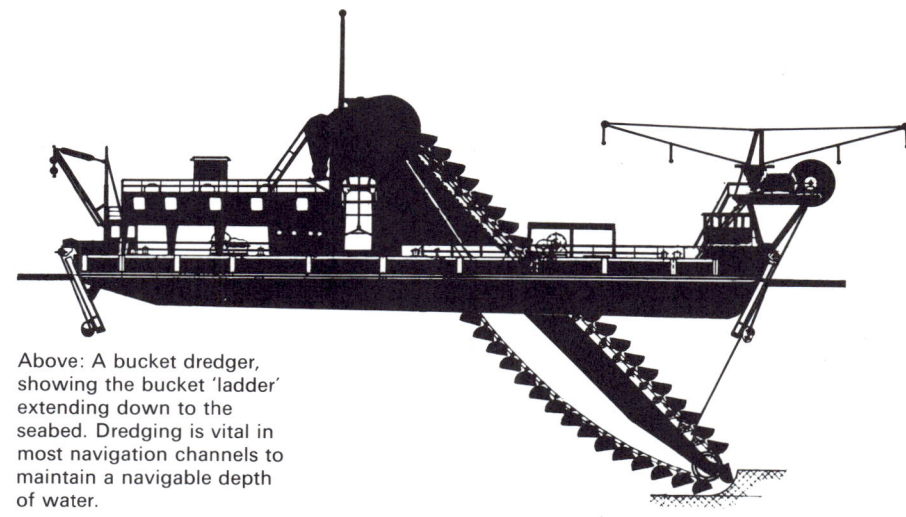

Above: A bucket dredger, showing the bucket 'ladder' extending down to the seabed. Dredging is vital in most navigation channels to maintain a navigable depth of water.

Right: Lightening a 200,000 deadweight tanker in the North Sea to reduce its draught so that it can enter the port of Rotterdam-Europoort, the world's busiest port. This kind of operation will become increasingly necessary as more and more large tankers come into service.

Below: Ships in port. The ship in the dock is a passenger liner. Note its extensive superstructure. With it are the tugs which are needed to manoeuvre the ship in the port area. The far ship is a freighter with some passenger accommodation. Note the derricks fore and aft.

# Ship Construction and Propulsion

As we have seen, ships vary a great deal in size, shape, and so on, which is not surprising because they are all designed for different purposes and for travelling at different speeds. It is not only the above-water shape that differs but also the below-water shape. In general a ship's hull is designed to be as streamlined as possible so that the water resistance, or *drag* is at the minimum. The ideal streamlined shape is that of an elongated tear-drop. And, broadly speaking, most fast vessels have this sort of shape. For vessels such as tankers and heavy freighters the emphasis is not on speed but on carrying capacity. And the hull is then almost box-shaped, with only slight curving at the bow.

The person in charge of ship design is the *naval architect*. When he has prepared what he thinks is a suitable design, he makes extensive tests with scale models in a testing tank, where he can simulate most kinds of conditions that the full-size ship will encounter at sea. When he is fully satisfied he draws up detailed plans from which the vessel will be constructed.

The hull and in fact most of a ship is made by joining thousands of steel plates together. In the past the main method of doing this was by *riveting*: red-hot rivets were hammered home through holes in overlapping plates. But today practically all ships are made by *welding* the steel plates together (see page 140). This not only gives a stronger joint, but also reduces weight because no overlap is required. The plates are first cut to shape from standard-sized steel sheets. This is done by oxyacetylene cutting torches operated either manually or automatically by machines controlled electronically. Any plates that have to be curved – in the bow and stern sections, for instance – are bent in massive hydraulic presses.

Most ships these days are built by the method of *prefabrication*. Whole sections of the hull are built up, under cover, in an assembly shop. And then they are assembled in the building berth. This is much quicker than building up the whole hull, plate by plate, out in the open.

As the hull takes shape, it is enveloped in scaffolding, from which the welders and later the painters work. Much of the heavy machinery is installed below decks before construction of the hull is completed. Then the ship is launched down a greased wooden slipway and is towed to the fitting-out yard to be finished. There the remainder of the machinery is installed, the superstructure is added, and the cabins furnished and decorated. Then comes the sea trials during which the ship is 'put through its paces' to make sure that it reaches the performance demanded by the prospective owners.

## Propulsion

Deep down inside the hull of the ship in the engine room lie the powerful engines which turn the ship's propellers. A propeller has three or move curved blades. When it rotates,

Practically all ships built these days are made up of steel plates welded together (right). Prefabrication techniques are invariably employed in which whole sections are assembled and then welded in position (left).

it 'sucks' water from in front and thrusts it out behind, just as an electric fan sucks in air. The thrust forces the propellers and hence the ship, forwards. In effect the propeller 'screws' itself through the water.

Most ships are powered by steam engines, which include reciprocating steam engines and steam turbines (see page 72). The majority have steam turbines. High-pressure steam from the ship's boilers spins round the blades of the turbine rotor, which is connected through gearing to the propeller.

The steam is provided by burning fuel, usually oil, in a furnace beneath a boiler. There are two basic kinds of boilers. In the *fire-tube,* or *Scotch* boiler, hot gases from the furnace pass through tubes in the boiler water. In the *water-tube* boiler, the furnace gases pass over tubes containing the boiler water. This kind of boiler is more efficient and is preferred for most ships.

The latest submarines and a few surface ships have a nuclear reactor to provide heat to raise the steam for their turbines (see page 34). Heat produced from the *fission,* or splitting of uranium atoms in the reactor core passes through a heat exchanger, where it turns water into steam. Reciprocating steam engines are used in some of the older ships, but they are much less efficient than steam turbines. In these engines, the steam is used to drive pistons back and forth in closed cylinders. The reciprocating (back-and-forth) motion is changed into rotary motion to spin the propeller shaft. The same kind of engine is used to power steam locomotives (see page 72).

Above: The hull of a ship taking shape in its building berth. Left: A typical ship's propeller, often called a screw. Phosphor bronze is an ideal material for making propellers since it is easy to fabricate and has excellent corrosion resistance. Below: With hull complete and painted, a ship is launched down the slipway. It goes next to the fitting-out yards, where it is completely finished and made ready for its sea trials.

# Submarines

Beneath the surface of the ocean lurk some of the most deadly warships ever built – submarines. The latest nuclear submarines are armed with guided missiles with nuclear warheads (such as *Polaris*) that can be fired from beneath the water at targets thousands of miles away. They can travel submerged at speeds up to 45 knots (52 mph), and stay submerged for months without needing to resurface for refuelling or for air.

Most submarines have a long, cigar-shaped body, with a structure called a *conning tower* rising above it, which houses the bridge for surface navigation and usually the navigation room as well. Some submarines, however, have a tear-shaped body, which is a more perfect streamlined shape for fast underwater movement. Like other ships, submarines are propelled by propellers and steered by a rudder at the stern. Near the bow and usually at the stern as well are movable side fins called *diving hydroplanes*. They are operated to control the angle of descent or ascent when the submarine is diving or surfacing.

The principle of submarining is very simple. Submarines are made with a double hull. In the gap between the outer and inner hull are a number of tanks which can be filled with water and emptied at will. Filling these *ballast* tanks with water makes the submarine sink. Emptying the tanks of water makes it rise.

On the surface the submarine is handled just like any other ship. To dive, air is released from the *main* ballast tanks. Water rushes in, and the submarine sinks until it is just below the surface. It is said to be in a condition of *neutral buoyancy*, so that it neither rises nor sinks. The submarine begins to dive when the *auxiliary* ballast tanks are flooded.

When the vessel has reached the depth required, the auxiliary ballast tanks are 'blown'. Compressed air forces the water out. With the auxiliary ballast tanks empty once more, the submarine returns to a state of neutral buoyancy and remains at the depth it has reached. Surfacing is simply the reverse of diving.

A nuclear submarine is powered by a steam turbine. Heat to change water into steam to drive the turbines comes from a nuclear reactor. The great advantage of the nuclear power plant is that it needs no air to operate. It therefore works as well under water as on the surface. Nuclear submarines can therefore remain submerged for very long periods.

A conventional submarine has two kinds of propulsion. On the surface it uses a diesel engine, which needs air to operate. Under water it uses electric motors driven by batteries, which need no air to operate. But the batteries run down and they must be recharged with electricity regularly. Recharging is done on the surface using an electricity generator driven by the diesel engine, which, of course, needs air. During these periods the submarine is very vulnerable to attack.

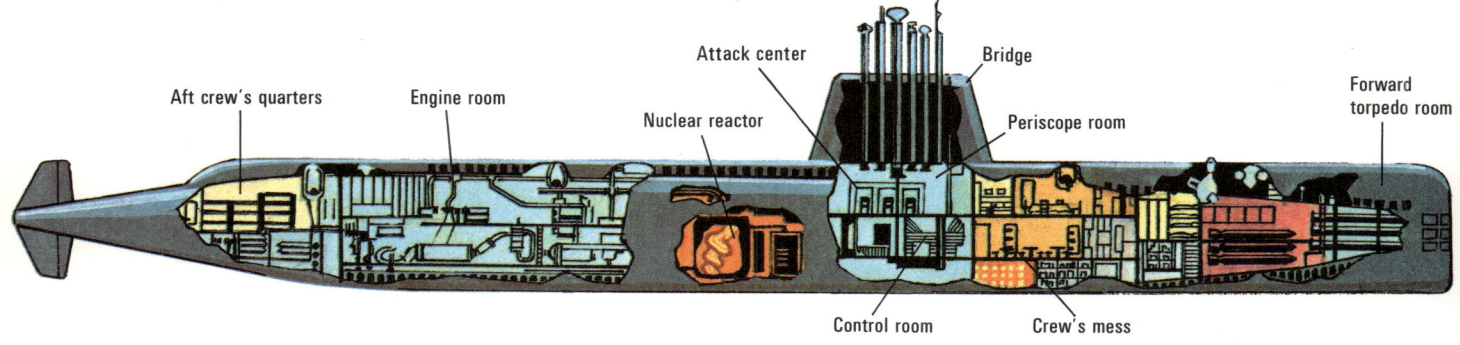

The layout of the US Navy's first nuclear-powered submarine *Nautilus*. The US Navy now has more than 40 nuclear submarines in operation armed with nuclear-warhead rockets that can be fired from underwater.

One of Britain's nuclear submarines at Portsmouth Dockyard. Note the bulbous bow section, which gives excellent underwater performance, and the diving hydroplanes on each side. In operation the hydroplanes function in a horizontal position.

# Hydrofoils and Hovercraft

The cruising speed of the great transatlantic passenger liners such as the *France* and the *Queen Elizabeth 2* is about 30 knots (35 mph). That of most other passenger and cargo ships is generally not much above 20 knots (25 mph). How slow this is compared with land and air transport.

The trouble with water transport is that a boat uses most of its power just overcoming the resistance of the water, or *drag* on its hull. The less hull there is in the water, the less will be the drag and the greater will be the speed. You notice that speed-boats are designed so that the hull gradually lifts out of the water with increasing speed. At full speed, only the stern, where the propellers are, remains in the water. This works for small, light boats, but is no good for heavier vessels. They cannot get up enough speed to lift themselves out of the water in this way.

But they can be fitted with what are virtually underwater 'wings', called *hydrofoils*. The hydrofoils are fitted on struts beneath the hull. As the boat gathers speed, the flow of the water past the hydrofoils generates lift in much the same way that an aircraft wing generates lift when it travels through the air. As the boat's speed increases, the hydrofoils lift it further and further out of the water until the hull is right out of the water. Only the struts and the hydrofoils themselves are now causing drag.

You can therefore imagine how much faster these hydrofoil boats can travel through the water. Already there are quite a number of large hydrofoil boats in regular service with cruising speeds of more than 50 knots (57 mph). The fastest hydrofoil craft can exceed 70 knots (81 mph)!

## Hovercraft

Matching the hydrofoil in speed over water is the *hovercraft*. The hovercraft glides over the surface of the water on a cushion of air, and is, in fact, often called an *air-cushion vehicle*. It operates equally well over land as over water, but it is for water transport that it has been primarily developed.

The air cushion is maintained by a powerful fan, which blows a constant stream of air through inward-pointing nozzles around the underside perimeter of the body. Sea-going hovercrafts are fitted with flexible 'skirts' all around to reduce the rate at which the air leaks away from the cushion.

Hovercraft are propelled by means of huge, backward-pointing propellers. Steering is effected by swivelling the propellers themselves or by deflecting the air stream from them by means of tail fins. In large hovercraft, such as the *SR-N4*, the power to drive the fans and the propellers comes from gas-turbines. The *SR-N4* is a car ferry operating the English Channel crossing. It can carry 30 cars and 250 passengers between Dover and Boulogne in about a third of the time (35 minutes) it takes conventional car ferries. Its top speed is 78 knots (90 mph).

A hydrofoil boat at speed, showing the way the foil lifts the hull clear of the water. This boat has what is called a surface-piercing hydrofoil. One great advantage of using hydrofoils on inland waterways is that they make relatively little wash.

'Princess Margaret', one of British Rail's Seaspeed hovercraft leaving its Dover base for the Channel crossing. It is powered by four gas-turbine engines, which drive 19-foot propellers. Note the hinged rudder sections on the tail fins, which act in the same way as an aeroplane rudder.

# Navigation

Mariner's Compass

These days *navigating,* or guiding a ship across the oceans, is a very precise matter. Although ships' navigators still use traditional methods of navigation, they are relying more and more on sophisticated electronic equipment which gives almost 'pinpoint' accuracy. *Radar* is perhaps the best-known of all the navigational aids.

The most essential piece of navigation equipment on any ship, however, is still the *compass,* which always points in the same direction. The traditional mariner's compass is a special kind of magnetic compass, so mounted that it is little affected by the pitching and rolling of the ship. But most ships these days have a much more accurate instrument called a *gyrocompass,* which contains a rapidly spinning gyroscope wheel. Once the wheel is set spinning, nothing will deflect it.

The compass cannot really help a ship's navigator to his destination unless he knows where he is to start with. He can gain a rough idea of where he is by *dead reckoning.* He plots on a chart the distance and direction the ship has travelled. The compass reading tells him the direction, and the ship's *log* (a kind of marine 'speedometer') tells him the speed and the distance travelled. He can make a more accurate estimation of his position by observing the Sun by day and the Moon by night. He takes compass bearings of several stars, observes the angles

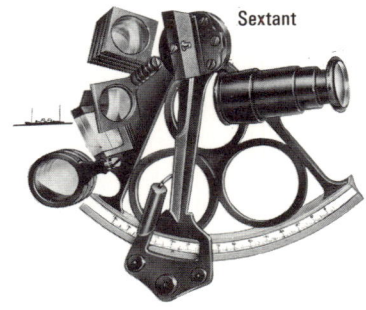

Sextant

Two basic instruments that sailors have used for navigation for centuries, the compass (above) and the sextant (right). The compass needle always points to magnetic north, which is close to but not coincident with the north geographic pole. The sextant is used for measuring the angle of the Sun above the horizon. From this the sailor can estimate local time. From his chronometer he notes Greenwich mean time. From the difference in time he can calculate his distance from Greenwich, that is, his longtitude.

Lighthouses and lightships (below) aid navigation by warning the sailor of dangerous shoals, quicksands, rocks or wrecks. Lightships are used where it is not practical to build lighthouses.

they make with the horizon and notes the precise time on the very accurate *chronometer* (see page 202). By consulting a book of astronomical tables called the Nautical Almanac he can estimate his position quite accurately.

On cloudy days and nights, this kind of 'celestial' navigation is of no use. That is when electronic methods of navigation show their superiority. The most widely used method involves the use of *radio beacons* on shore which beam out radio signals. Each beacon has its own identifying call-sign. A *radio-direction finder* on board ship can obtain bearings from two or more beacons. Where the bearings cross gives the location of the ship.

Other radio navigational aids depend on the measurement of the difference in time it takes a radio wave to travel from two fixed stations on shore to the ship. Two systems are well known. One, called *Loran,* is a long-range aid that covers the Pacific and the West and far North Atlantic. The other, the *Decca,* is a relatively short-range aid that covers the British Isles, the English Channel and the North Sea. Aircraft navigate by these systems too.

Navigation underwater is very difficult as you can appreciate. The latest nuclear submarines have a very complex *inertial* guidance system. Sensitive electronic devices sense every change in direction or speed and feed the information into a computer. The computer can work out the submarine's position almost instantly and with extreme accuracy.

# Docks

Ports provide sheltered moorings and all the facilities ships require to load and unload their cargo and passengers. The larger ports also have facilities for ship repair and maintenance.

The design and layout of the port depends to a large degree on the tidal range there, that is, the difference between high and low tides. Ports on inland and land-locked seas, such as the Mediterranean and the Baltic, have scarcely any tides at all. This considerably simplifies port practice. Ships enter port and tie up at *quays,* built-up areas of the waterfront, or *jetties,* structures built out into the water.

At many other ports, however, there is an appreciable difference between high and low tides. At Bristol, in the west of England, for example, the tidal range is at times as great as 40 feet. You can imagine the difficulties involved in trying to unload a ship moving upward and downwards 40 feet every 12 hours with the tide! Simple quays and jetties are therefore no good, and docks are used instead.

*Docks* are enclosed basins fitted with massive, watertight lock gates. At high tide the gates are opened and ships sail in. Then the gates are sealed shut. When the tide falls outside the dock, the ships inside are not affected. Loading and unloading the ships can therefore proceed unhindered.

A wide variety of mechanical handling equipment is used on the quayside to handle cargo. There are many kinds of cranes: some travel on rails along the quay: some, so-called *portal* cranes, span roads and railway tracks; some are fixed; others float on pontoons. Elevators and conveyors handle bulk cargo such as grain, ore, and coal. Fork-lift trucks are also widely used now.

Ship repair in port is carried out in *dry docks,* also called *graving docks.* They are concrete chambers which can be filled with water and then emptied. The ship to be serviced floats into the filled dock, the lock gates are closed behind it, and the water is pumped out. After work has been completed, the dock is flooded and the ship can sail out. A *floating dock* may be used at a port and also for emergency repairs at sea. It is a large steel cradle with a U-shaped cross-section. By letting water into hollow tanks in its base, it can be sunk low in the water. In this position the ship for repair is towed in. Water is blown from the tanks to make the dock rise gradually from the water, taking with it the ship.

Above: An aerial view of Liverpool Docks, showing several of the docks in use with their gates closed. Docks such as these are necessary when the rise and fall of the tide at a port exceeds more than about 12 feet.

Left: A ship in floating dock at Amsterdam. Floating docks are attached to many of the world's major ports. They have the advantage over ordinary dry docks that they can be trimmed to accommodate damaged vessels with a pronounced list.

# Important Minerals

**Asbestos** Some minerals, called *asbestos*, are found naturally in the form of tiny grey, silky fibers. The longest fibers can be spun into yarn and then woven into cloth, just like cotton and wool. Because it is literally made of rock, this cloth is fireproof. Firemen wear suits of asbestos cloth when fighting fierce fires. Short-fibered asbestos is used to make other products that must withstand heat, such as automobile clutch and brake linings. Mixed with a suitable binder it is used to make roofing material and insulation for boilers and furnaces. The heat shield which protects astronauts returning to Earth in their space capsule contains asbestos, too. Asbestos is also a good electrical insulator, and resists acids and alkalis. The varieties of asbestos belong to two mineral groups: *chrysotile* asbestos and *amphibole* asbestos. Chrysotile, or white asbestos, accounts for 90 per cent of the world's production.

**Azurite** is one of the carbonate ores of copper. It has a spectacular deep rich blue color.

**Barites** (barium sulphate) also called *baryte*, is the chief ore of barium. One of the main uses of barites is as an extender in paints. It is mixed with zinc sulphide to form the pigment lithopone. Barites may be colorless, white, or tinted yellow or red. It is quite heavy for an earthy mineral. Its name means 'heavy', and the natural ore is sometimes called *heavy spar*.

**Bauxite** is the only profitable ore of the lightweight metal aluminum. Common clay also contains aluminum, but it cannot be extracted easily. Bauxite should really be considered a rock rather than a mineral because it is a mixture of minerals, rich in *alumina* (aluminum oxide).

**Calcite** The great chalk and limestone cliffs seen at the seaside and elsewhere are made up almost entirely of this one mineral. These forms of calcite are slightly impure. Chalk, the whitest, is a pure form. But limestones are stained yellow, brown, and even red by impurities. The crystal-clear form of calcite is called *Iceland spar*. It literally makes you 'see double'. Place a piece of Iceland spar over a line of type in a book. You will see two lines when you look through the crystal.

**Cassiterite** (tin oxide) Practically all the tin we use comes from one ore – *cassiterite*, also called *tinstone*. The word cassiterite comes from the name ancient traders gave to the British Isles. They called them Cassiterides – the 'Tin Islands' – because of the valuable tin mines there. Cassiterite is usually dark brown in colour and is hard and very heavy – almost as heavy as lead. Most cassiterite is found in placer deposits in gravel and stream beds and mined by dredging.

**Chalcopyrite** or copper pyrites, is the most common source of copper. It is a sulphide of iron and copper. It has a beautiful deep yellow color, which, due to alteration, may flash all colors of the rainbow like a peacock's feathers. For this reason, chalcopyrite is sometimes called 'peacock ore'. It is sometimes confused with iron pyrites, which is very similar to it. Both minerals have often been mistaken for gold and have earned the name fool's gold.

**China Clay** Most clays are grey, brown, or red, but pure *china-clay*, or *kaolin*, is white. It is the clay used in making all our fine pottery. It is also used in paper-making as a filler to produce smooth paper and in the manufacture of paints, soaps, and medical poultices.

Most important minerals are mined underground. Here miners are drilling the shot holes for the explosives used to break up a vein.

**Corundum** (aluminum oxide). You would hardly think that there was much in common between a fiery red ruby, a brilliant blue sapphire, and a grinding wheel. But there is. They are all different forms of the very hard mineral corundum which is second only to diamond in hardness. Ruby and sapphire are rare and valuable crystal forms of corundum which are prized as precious stones. The grinding wheel is made up of the common dull-grey form of the mineral. Corundum makes an excellent abrasive because it is so hard. Sometimes it is found mixed with magnetite, and the mixture is called *emery*. Powdered emery is glued onto paper and cloth to make the familiar emery paper or emery cloth we use to smooth down metal surfaces.

**Cryolite** (sodium aluminum fluoride). The main use of cryolite is in making aluminum. It is used in molten form to 'dissolve' the aluminum ore before smelting. Cryolite is a fascinating mineral. Its name means 'frost stone'. If you drop a piece of it into a glass of water, it practically disappears. It does this because it bends light rays in almost exactly the same way as water does. This means that, as far as your eye is concerned, there appears to be nothing but water in the glass. Another interesting feature about cryolite is that it melts easily – even in a match flame. Cryolite is found in large quantities only in southern Greenland, near Ivigtut.

**Cuprite** (copper oxide) is one of the many ores of copper. It is reddish-brown and is sometimes called 'ruby copper' because of its appearance.

**Diamond**, the pure crystal form of carbon, is the hardest of all the minerals and apart from the finest emeralds and rubies, the most valued of all the precious stones. When cut and polished, diamond is the most brilliant of them all. As it occurs naturally, deep underground or in placer deposits, diamond is covered in a dull grey film. In this form it is called a *rough* diamond. The rough diamond is cut with a great many flat sides, or facets, to produce the greatest possible brilliance. The 'cutting' and polishing of the facets is done by grinding with discs coated with diamond dust. Diamond is so hard that the only material which will cut it is itself! Not all rough diamonds, however, are suitable for cutting and polishing into fine gems. They are used in industry as abrasives and for the bits of rock drills in mining operations. They are called *industrial* diamonds.

**Feldspar**, or *felspar*, is the most common and widespread of all minerals. Several kinds of feldspars are found in the form of crystals in granite and other igneous rocks. The crystals may be white, pink, green or blue. Large amounts of feldspar are used for making pottery and glass. Feldspars are the commonest of the complex silicate minerals.

**Fluorite** (calcium fluoride), also called *fluorspar*, is a beautiful and interesting mineral that can be made to glow in the dark. Some specimens glow a light blue in the dark when ultra-violet light is shone on them. Fluorspar lends its name to this effect – *fluorescence*. Fluorite crystallizes in perfect cubes, which may be colorless, yellow, brown, green, purple, and blue. A banded blue variety is called *blue john*. It was once used for carving vases. Today, most fluorite is used in steel-making furnaces to make the materials melt more easily. Some is used to put the 'fluoride' in some kinds of toothpaste and in drinking water.

**Galena** (lead sulphide) is by far the richest source of lead. Often it contains valuable traces of silver as an 'impurity'. Galena is very soft and very heavy. It can easily be broken into tiny cubes. It has almost the same colour as lead itself.

**Graphite** The black 'lead' of the pencils you write with are made not from lead at all but from *graphite*, mixed with clay. Graphite is crystalline and one of the forms of carbon. It is one of the softest minerals and feels 'greasy' to the touch. Some machinery is in fact 'greased' with it. The really interesting thing about graphite is that it is chemically the same as diamond, which is the hardest substance known!

**Gypsum** (hydrated calcium sulphate) Blackboard 'chalk' is made of *gypsum*. Like chalk, gypsum is white and quite soft. If you have broken an arm or a leg, you will also have met gypsum in another guise – *plaster of Paris*. This is used for making plaster casts of all kinds. It is also used by builders to make many kinds of plasters and wall coverings. Plaster of Paris is made by heating gypsum. When water is added to it, it can be moulded into any shape. But it soon dries and hardens.

**Haematite** (iron oxide) is an iron ore with one of the highest iron contents. It is so called because of its blood-red color. Sometimes it is found in rounded lumps and is then called 'kidney-ore'. Often it occurs in the form of grey, mirror-like crystals, and is called *specular* (looking-glass) ore.

**Halite** (sodium chloride) Most of our common table salt comes from massive deposits of rock salt, which consist almost entirely of the mineral *halite*. We use salt in the home for cooking and for preserving meats, for example. But most salt is used in industry to make a wide variety of chemicals, such as soda (for making glass) and chlorine (for purifying water).

**Ilmenite** (iron and titanium oxide) is a heavy, black mineral that is the chief source of the valuable metal titanium. It is sometimes called *titaniferous iron ore*. Vast amounts of ilmenite are used to make the paint pigment titanium white (titanium dioxide).

**Limonite** (hydrated iron oxide) is a yellowish-brown ore of iron. It is often found in marshes and shallow lakes and is therefore sometimes called *bog iron ore*. The yellowish-brown colour of sand, clay, and many other rocks is caused by 'staining' by limonite.

**Magnetite** (iron oxide) is the iron ore richest in iron. It is black and metallic looking. It is a powerful natural magnet, which can pick up pieces of iron and steel. Also, a piece of magnetite can be used as a compass needle. This property gives the mineral its other name – *lodestone* ('guiding stone').

**Malachite** is one of the carbonate ores of copper. It is a vivid deep green. The green deposit, or *patina*, which you often see on copper and bronze statues and ornaments is a thin coating of malachite.

**Mica** occurs naturally in transparent, flaky sheets which can easily be separated. Mica sheet is widely used in electrical equipment as an insulator. It is by far the best insulator there is. The thin sheets of mica are very flexible and elastic. Thick crystals of mica are made up of many of these sheets and are called 'books'.

Superb crystals of galena, or lead sulphide, also called lead glance. It is often found with zinc blende.

'Kidney ore', an interesting variety of the iron-oxide ore haematite. It is easy to see how this ore got its name.

A purple banded variety of fluorspar, or fluorite. The mineral occurs in a variety of forms and may fluoresce.

**Pitchblende** is the most important source of the radioactive metals radium and uranium. It is a brownish-black mineral that has been likened to pitch, hence its name. Pitchblende is actually a mixture of several minerals including uranium oxide.

**Pyrites** (iron sulphide) is one of the minerals which has often been mistaken for gold on account of its brassy yellow appearance. For this reason, it is called 'fool's gold'. It contains too much sulphur to be a useful iron ore, but it is a valuable source of sulphur. Also, many deposits of pyrites are rich in copper and gold.

**Quartz** (silicon dioxide, or silica) is one of the commonest of all minerals. It is one of the main minerals in the massive granite rocks which make up a great part of the Earth's surface. Sand and gravel are nearly all quartz. Crystals of quartz are particularly beautiful, and many of them are valued as semi-precious stones. The purest form of quartz is *rock crystal*, which is perfectly clear. It has certain electrical properties which make it suitable for use in radio and telephone instruments, and in precision clocks.

**Siderite** (iron carbonate) *Clay-ironstones* are rock-like masses which contain the iron-ore mineral *siderite*. They do not contain nearly as much iron as the other iron ores. But they are important ores in countries such as Britain and other countries in Europe which have few rich ore deposits.

**Sphalerite** (zinc sulphide) is the main ore of zinc. It is also called *zinc blende*. Both of the words *blende* and *sphalerite* mean 'deceiving' or 'treacherous'. The mineral is so called because it is difficult to recognize and may resemble more valuable minerals, such as galena, a rich source of lead and silver. It is usually brown or black.

**Sulphur** is a soft yellow mineral which is one of the chemical elements. It is unusual because it melts readily at a temperature just above that of boiling water. Also, it burns, and gives off choking 'sulphurous' fumes. These are the fumes you can smell when you go near a volcano. And sulphur is, indeed deposited in and around the rocks near volcanoes, often in the form of beautiful crystals. Sulphur is one of the most valuable raw materials for the chemical industry. It is made first into sulphuric acid and then into many products, including large amounts of fertilizers. Sulphur itself is used for *vulcanizing* rubber.

**Talc** We come into contact with this mineral at a very early age in the form of *talcum powder*. It is ideal for a baby's skin because it is one of the softest of all minerals. Talc is so soft that it is given the lowest grade of mineral hardness. Baby powder accounts for only a small part of talc production. Very large amounts are used in making pottery, paint, and paper. The great softness of talc makes it feel 'soapy'. For this reason the solid form of talc is often called *soapstone*.

# Mining and Quarrying

Vast quantities of minerals and rocks are taken from the ground each year to satisfy the demands of modern industry. Removing minerals, including coal, from the ground is known as *mining;* removing rock from the surface is known as *quarrying.*

Deposits of minerals may occur in massive beds between strata (layers) of rocks or in veins that cut through these strata. Such deposits may lie on or only just beneath the surface of the ground, or thousands of feet down. The closer the deposits are to the surface, the easier and cheaper they are to extract.

## Surface Mining

The technique of surface mining is known variously as *open-cast, open-cut,* or *strip* mining. What happens is this. Huge power shovels strip off the layers of earth, or *overburden* covering the mineral deposit. Then the deposit is shattered into smaller pieces by blasting with explosives. Power shovels scoop up the shattered rock and drop it into trucks or freight cars for removal. The excavators cut down into the ground or into a hillside in a series of steps, or terraces, which are a very noticeable feature of open-cast mines. Iron ore, bauxite, and coal are among the common mineral deposits mined in this way.

## Quarrying

Quarrying is similar in many respects to open-cast mining, but explosives are used much more carefully. This is particularly true when valuable building stone such as marble is being quarried. Where possible, quarrymen split up the massive outcrops of stone into slabs simply by drilling or cutting and then splitting with wedges. They

Quite a number of minerals are mined at the surface, this being termed opencast working. The technique is well illustrated here. The massive dragline excavator on the higher level strips off the earth overburden to expose the mineral deposit. Then a power shovel below scoops up and loads the ore into trucks.

use natural faults or 'grain' in the rock where they can. In very hard rock, such as granite, however, explosives do have to be used to split the rock. But they are used so sparingly, in conjunction with drilling, that the rock splits neatly without shattering.

Most rock, however, is quarried for purposes other than for building. Vast amounts of chalk (for cement-making) and limestone (for steel-making and so on), for example, are needed in industry. And it does not matter whether the rock is shattered or not. Quarrying techniques for these kinds of rocks therefore make much freer use of explosives. The quarrymen drill holes deep into the rock all along the quarry face and pack explosives into them. When the explosives are fired, thousands of tons of rock are brought down and shattered at one go. The same kind of technique is used for quarrying the rock needed in the construction of roads, dams, and so on.

No explosives are, however, needed for quarrying such deposits as clay, sand, gravel, and shale. They are naturally soft or loose and can usually simply be dug out by excavators. Sand and gravel pits soon fill with water, and dredgers must then be used to remove the deposits.

## Dredging

Dredging is a technique that is also used in another kind of surface mining called *placer* mining. Placers are deposits of heavy minerals which form in the beds of lakes or streams and also on the sea-shore. Gold and precious stones are to be found in stream placers. So is cassiterite, the ore of tin. Huge mechanical dredgers which can dig up 15,000 tons or more of ore-bearing gravel a day are employed in tin-mining. The gravel is washed with powerful jets of water which separate the lighter gravel from the heavy tin ore. This is an up-to-date version of the method of *panning* which gold miners traditionally used.

## Underground Mining

Many mineral deposits, however, occur deep underground, and extraction is much more expensive – and often dangerous, of course. The principle of ordinary *underground mining* is to sink vertical shafts down into the ground and then drive horizontal tunnels from this shaft towards the mineral deposits. These tunnels must be supported, or 'shored-up' along their length to reduce the risk of them collapsing. This may be done with wooden or metal *props* or by lining with concrete. The deposits are

A marble quarry. The stone is extracted carefully by splitting and drilling so as to cause the least damage to it. In other quarries, such as those quarrying limestone, explosives may be used to break up the rock.

In underground mines the ore body is usually broken up by explosives. Pneumatic rock drills are used to drill the shot holes for the explosives.

generally blasted loose with explosives and hauled along the tunnels in railway wagons to the shafts, from where they are removed to the surface. Underground mining has become much more mechanized in recent years. In coal mining, for instance, automatic coal-cutters and conveyor systems are a feature of most modern mines.

Copper ores, gold, and diamonds are among the other valuable minerals mined underground. The deepest mine in the world is the East Rand gold mine near Johannesburg in South Africa, which has a shaft 11,246 feet (over 2 miles) deep. Gold is already being worked at 9,800 feet there. Mining at any depth underground is dangerous. There is constant danger of rock falls. There may well be explosive gases such as firedamp (mainly methane) present. And deep down the temperature is extremely high. At about 9,000 feet, it may be higher than 40°C. The air for ventilation must therefore be refrigerated. You can appreciate what a problem it is to provide adequate ventilation.

Some minerals are mined underground in a totally different way. Take sulphur, for example. Superheated steam is used to melt the sulphur beneath the ground. Then compressed air forces the molten suphur to the surface. Rock salt – that is, ordinary common salt in massive form – may be mined in quite a similar manner. Water is pumped down into the deposit, where it dissolves the salt, and back again to the surface. At the surface the water is boiled off, leaving solid salt. Other soluble minerals, including potash, may be extracted like this. Drilling for oil and natural gas is really a special method of mining, too (see page 77).

# Extracting and Refining Metals

## Extraction

Metals form the basis of our present industrial civilization. Over 400 million tons of steel alone are used throughout the world every year to make an incredible variety of products, from automobiles, trains, ships, and all kinds of machinery to 'tin' cans, knives, forks, and spoons. You can see from the section 'Metals and their Uses', beginning on page 143, just how important metals are in our everyday lives.

Metals make up quite a large part of the Earth's crust, but only a few of them are found *native,* that is, in their pure metal form. These metals include gold, platinum, silver, and copper. They can be found native because they do not readily combine with other chemical elements. But most metals, such as iron, lead, zinc, and aluminium, are far too reactive to be found native. They occur in the ground as chemical compounds called minerals. When these minerals can be profitably processed to provide the metal, they are known as *ores*.

It is seldom that the metal ores are pure enough to be processed straightaway. They are almost always found mixed with a lot of unwanted rock, clay, dirt, stones, and so on. Before the ore can be processed this unwanted material, called *gangue,* must be removed. This may well account for 99 per cent of the deposit mined from the ground! The removal of gangue as far as possible from the ore is known as *mineral dressing*.

Most iron ores are exceptional in that they contain comparatively little gangue, and they can be processed more or less as they are taken from the ground. The main iron ores (haematite, magnetite and limonite) are oxides, that is, combinations of iron with oxygen. They can be reduced to metallic iron by heating with coke in a *blast furnace,* so called because hot air is blasted through it to produce a high temperature. The oxygen in the ore combines with the carbon of the coke to form carbon dioxide gas, which

Concentrating cassiterite, or tin ore by hydraulic treatment at a Malayan mine. The crushed ore is carried by a stream of water over tables. The heavy cassiterite settles out while the lighter earthy minerals are washed away. The crude ore is then roasted, smelted and refined to give tin 99 per cent pure.

### Blast Furnace

The blast furnace used for iron smelting is a huge steel tower 100 feet or more high. It is lined with refractory brick to withstand the tremendous heat developed during the firing of the furnace (see page 85). The temperature may be as high as 1700°C.

Coke, iron ore, and limestone are charged into the top of the furnace through a specially shaped valve, called a *double-bell* valve. This prevents the loss of heat and furnace gases. The air blast is heated in brick stoves before it enters the bottom of the furnace through nozzles called *tuyeres*. The stoves are themselves heated by burning waste gases from the top of the furnace. The coke burns in the air to yield carbon monoxide gas, giving out great heat. Carbon monoxide acts as a reducing agent, removing oxygen from the ore to leave iron metal. The function of the limestone is to remove earthy impurities, forming a slag. Molten iron and slag run down to the bottom, or *hearth* of the furnace, where they separate into two layers, the slag floating on top of the iron.

Every few hours the molten iron is drawn off, or *tapped*. It is either poured into moulds or taken while still molten for further refining. The iron produced in the furnace is called *pig iron*. It contains quite a high proportion of carbon and other impurities, such as silicon and phosphorus. Removal of most of the impurities, or *refining* is necessary before the metal is useful.

Blast furnaces operate continuously day and night until the furnace lining wears out. Many of the larger furnaces produce well over a quarter of a million tons of pig iron a year.

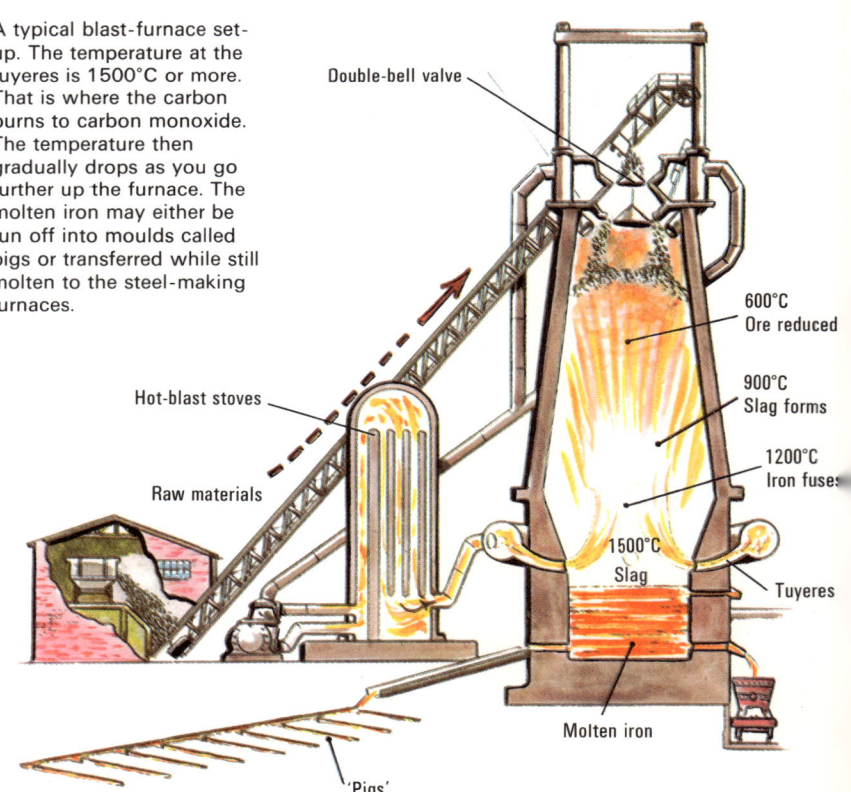

A typical blast-furnace set-up. The temperature at the tuyeres is 1500°C or more. That is where the carbon burns to carbon monoxide. The temperature then gradually drops as you go further up the furnace. The molten iron may either be run off into moulds called pigs or transferred while still molten to the steel-making furnaces.

escapes. Molten iron remains. Limestone is added to the furnace to help remove the impurities. It combines with them to form a molten slag, which can be separated readily from the molten metal. This process of extracting an impure metal from its ore by heating is known as *smelting*.

Heating with coke is a standard method of smelting oxide ores. But not all metal ores, of course, are oxides. Zinc and lead, for example, occur as sulphides, which cannot be smelted directly. What happens in this case is that the sulphide ores are *roasted* in air, a process that converts the sulphides to oxides. They can then be smelted as before. Carbonate ores are roasted, too, to convert them to oxides.

As we have already noted, many ores, oxides included, have to be concentrated before they can be processed further. And this reduces them to a finely divided state. In this state they are not really suitable for smelting because they could well 'gum up the works'. What is done, therefore, is to

Pouring molten gold into small ingot moulds. Each bar of gold so formed weighs about 30 lb and is currently worth something between $15,000 and $80,000 depending on fixed price or floating price.

heat them strongly enough so that the separate particles fuse together into larger more manageable lumps. This process is called *sintering*.

Smelting is not the only method of extracting metal from its ore. There are many others, which are more complicated. One way that does not involve heating is *leaching*. The metal is dissolved from the ore with chemicals. The solution formed is then treated in some sort of way so that the metal is *precipitated*; that is, it comes out of solution. A leaching method is used to obtain gold from its ore. A solution of sodium cyanide is used to dissolve the gold.

A metal may be extracted from solution by means of electrolysis, too, that is, by passing electricity through it. Aluminium is extracted by electrolysis from a solution of its ore, alumina, in molten cryolite. This produces pure metal first time. Magnesium is prepared by electrolysis, too, from a solution of magnesium chloride in molten potassium chloride.

A group of blast furnaces at a steel-making plant. Note the hot-blast stoves on the left of the picture.

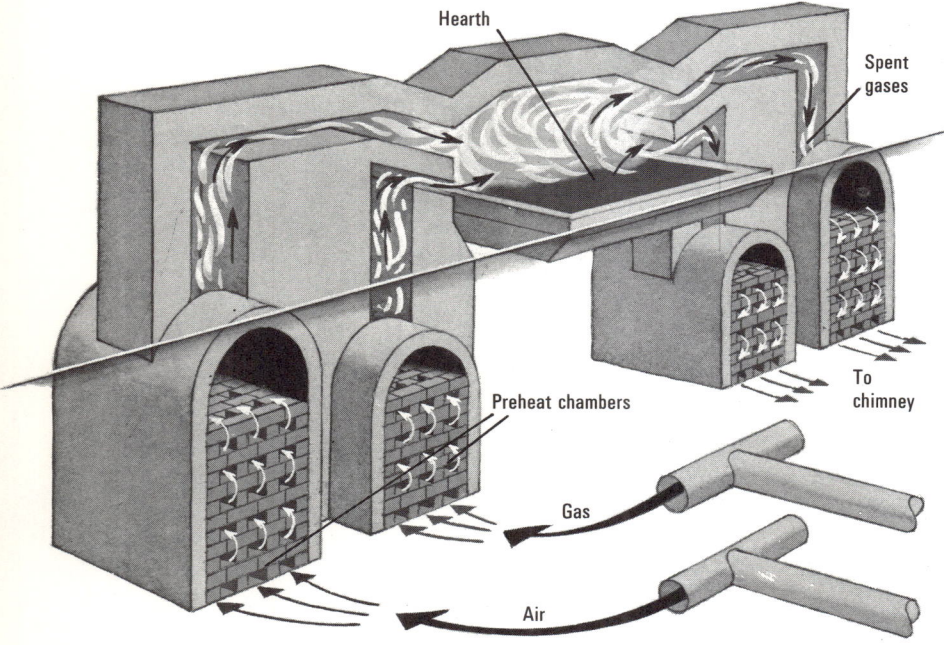

The open-hearth process. Pig iron is refined in a shallow hearth open to the furnace flames. The fuel gas and air enter the furnace through one set of brickwork chambers, and the spent gases leave through another, which they heat. Periodically the gas flows are reversed so that the ingoing fuel and air are preheated.

Ordinary smelting of a metal ore, however, produces impure metal. Before it can be used it must be purified, or *refined*.

## Refining Metals

Metals that have been extracted from their ores by such methods as smelting invariably contain some kinds of impurities. The use for such crude metal is limited because the impurities generally cause weakness, brittleness or other undesirable properties in the metal. Mostly, then, metals are refined, or purified after smelting. In some cases there may be another purpose in refining. Sometimes the impurities in the crude metal are more valuable than the metal itself! For example, the crude lead and zinc obtained by smelting galena and zinc blende often contain valuable traces of silver and gold. Refining in such cases is therefore done very carefully to preserve the impurities removed from the crude metal.

In metal-refining generally, however, the main object is to obtain pure metal. The impurities are discarded. One of the commonest methods of removing impurities is literally to burn them out. This is the method that is used in making steel from pig iron. Here we shall take steel-making to illustrate this kind of refining. Another basic method of purifying crude metal is by electrolytic refining in which electricity is used to separate the pure metal from the impurities. Copper, for example, is refined in this way.

## Steel Making

The main impurity in pig iron is carbon (up to 5 per cent), which was introduced during the smelting process in the blast furnace. In addition there are small amounts of manganese, phosphorus, silicon, and other elements. The metal does not become generally useful until the carbon content has been reduced to about 1·5 per cent or below. That is when it becomes known as *steel*.

The most important steel-making process is the *open-hearth* process. It is so called because the shallow hearth of the furnace is open directly to the flames which melt the steel and burn out the impurities. Steel scrap, iron ore, and the pig iron are charged into the furnace, together with a substance like limestone which helps to form a slag. Oxygen from the iron ore and from the furnace gases oxidizes, or 'burns out' the impurities. The carbon is converted to carbon monoxide gas, which bubbles through the molten metal and slag. This part of the process is known as the 'carbon boil'. Other impurities pass into the slag.

Another important process is the *Bessemer* process. Molten pig iron, together with, say, limestone is charged into a converter, which looks like a giant concrete-mixer. Air is then blown under high pressure through the molten metal and oxidizes the impurities. During this 'blow' sparks and flames shoot high into the air in a truly spectacular fashion. The impurities pass into the slag as before. Similar kinds of processes which use oxygen rather than air are rapidly replacing the Bessemer process. In the *basic-oxygen*, or *L-D* process, for example, a jet of pure oxygen is blown at supersonic speed into the molten pig iron. In the *Kaldo* process a jet of oxygen is blown onto the surface of the molten iron in a rotating converter.

An electric-arc furnace refining scrap steel. Electrodes are lowered to the surface of the charge and then gradually withdrawn. The electric arc that develops creates a very high temperature, which melts the charge.

One of the most recent and interesting methods of refining pig iron is by what is called *spray steel-making*. In this method a ring of jets of oxygen at high pressure break up a falling stream of molten pig iron into a spray of fine droplets. Lime to form a slag is introduced at the same time. The particles of metal are so small that refining takes place almost instantaneously, while the particles are falling into a receiving vessel.

*Electric furnaces* are used for refining the highest-quality steels. In these furnaces the charge is heated by electricity and therefore there is no danger of contamination of the metal by burning furnace gases, as there is in the other processes. The charge for the electric furnace is not pig iron, but steel scrap of known composition, together with the usual lime and sometimes iron ore. Only the best grades of steel are made in electric furnaces because of their heavy consumption of electricity, which is expensive.

The most important electric furnace is the arc furnace. Carbon electrodes in the roof are lowered to the charge, and an arc is struck between the electrodes and the charge. The heat produced melts the charge. In the induction furnace alternating current is passed through a coil surrounding the charge. Eddy currents are set up in the charge, causing heat to be produced.

## Electrolytic Refining

The electrolytic method of refining, which is used, for instance, to purify copper, works by the principle of *electrolysis*. Two metal rods are placed in a solution of a compound of the metal and a direct electric current is passed from one rod (the *anode*) to the other (the *cathode*). The passage of electricity causes pure metal from the anode to be deposited on the cathode. In copper refining, therefore, crude smelted copper is made the anode of an electrolytic tank, containing copper sulphate solution, while the cathode is made of pure copper. When current is passed, pure copper from the impure anode is deposited, via the solution, on to the cathode. The impurities form a slime beneath the anode.

# Steel and Other Alloys

Pure iron is quite a weak metal, and it is not very hard, either. But add a little carbon as well as traces of certain other metals, and it becomes both strong and hard. According to what is added, the iron can be given other desirable properties as well. Adding other substances to a metal to alter its properties is called *alloying*, and the product an *alloy*.

Most of the alloys of iron are better known to us as *steels*. Many other metals form useful alloys, too. *Brass* and *bronze* are common alloys of copper. *Solder* and *pewter* are well-known lead alloys. Aluminum and zinc alloys are also in widespread use. However, it is the iron alloys that are by far the most important.

## Cast and Wrought Iron

The pig iron that is made in the blast furnace can be considered one kind of iron alloy, but it is a very crude one. It contains far too many impurities and other materials as well as being uneven in composition. By only slight refining it can be made into *cast iron*. The pig iron is remelted with coke in a *cupola*, which is really a small blast furnace.

The molten metal, which is very fluid, is then poured into shaped moulds, or casts (see page 136). Cast iron is an ideal material for making engine blocks and machine frames. It is strong, hard, rigid, and absorbs shock well. Its main disadvantage is that it is brittle.

*Wrought iron* is a more refined form of pig iron which is quite pure iron (0·1–0·2 per cent carbon) with threads of slag running through it. It is made by heating the pig iron with iron oxide in a 'puddling' furnace. The oxygen in the oxide combines with the impurities, which either boil away as gas or form a slag. But the temperature of the furnace is not high enough to make the metal fluid. That is why it contains threads of slag. Wrought iron is little used today, except for such things as gates and chains.

Refining pig iron in the various steel-making furnaces reduces the carbon content and removes other unwanted substances. The steel-maker stops the refining process when the metal has reached the carbon content he wants. Then he adds controlled amounts of other elements to bring the steel to the desired composition.

## Carbon Steels

There are two principal kinds of steels, carbon steels and alloy steels. The properties of *carbon steels* depend mainly on the amount of carbon present. *Mild* steel (up to 0·25 per cent carbon) is the ordinary kind of steel that is used for girders, automobile bodies, bicycle frames, and so on. *Medium-carbon* steel (0·25–0·45 per cent) is stronger than mild steel and is used for bridge members, nuts and bolts, and tools of many kinds. High-carbon steel (0·45–1·5 per cent) is hard and tough, and is used for cutting tools, drill bits, saws, and so on.

## Alloy Steels

The properties of alloy steels depend not on the carbon they contain, but on other alloying elements. One of the most familiar alloys is *stainless steel*. As we all know, ordinary steel corrodes, or rusts if it is left out in the rain, is stained by fruit juices, and so on. But adding chromium and nickel to steel makes it resist corrosion and stains. Both chromium and nickel resist corrosion well, and they tend to impart that property to their alloys. One of the most common kinds of stainless steels contains about 18 per cent chromium and 8 per cent nickel.

Nickel by itself in steels makes them strong, hard and shock-resistant. One parti-

Bronze is one of the most widely used alloys. It is ideal for casting purposes since it flows freely when molten. The Peter Pan statue in London's Kensington Gardens (above) is cast in bronze.

---

**Familiar Alloys**

**Brass,** an alloy of copper and zinc, is one of the best-known alloys. There are many different kinds of brass, each with different amounts of copper and zinc. The brasses get harder and stronger as the amount of zinc increases. Aluminum and iron are sometimes added to brass to make it stronger.

**Bronze** is an alloy of copper and tin. It is hard, wears well, and does not corrode easily in the air. Statues are often cast in bronze. The so-called 'copper' coins are made of bronze, but they contain only a tiny amount of tin, with some zinc. Some so-called bronzes, such as aluminum bronze, contain no tin at all. Aluminum bronze is an alloy of aluminum with copper. It is very strong and resists corrosion well. Silicon bronze is a useful alloy which withstands attack by chemicals well.

**Cupronickels** are alloys containing copper and nickel. The most familiar is that used for 'silver' coins. It contains 3 parts of copper to 1 of nickel. Cupronickels have good resistance to corrosion and can be worked easily cold.

**Duralumin** is a lightweight alloy of aluminum, copper, and a little magnesium used widely in aircraft construction. It displays the interesting phenomenon of 'age-hardening'. After heat treatment, it does not reach its maximum hardness for some days.

**Pewter** is an alloy that contains mainly tin and lead. It is quite soft and can be shaped easily. It was once widely used for making tableware. Today, pewter is used in decorative metal-work and for making such things as tankards.

**Solder** is an alloy, also mainly of lead and tin, that melts at a low temperature. It is widely used to join electrical wires and for sealing tin cans.

**Type metal** is an alloy used for making printers' type. It contains lead, antimony, and some tin. In a printing machine every letter is cast in a tiny mould. The lead in the alloy makes it easy to cast. The antimony hardens it and makes the castings sharp and accurate.

cularly interesting nickel alloy steel is called *invar*. It is remarkable because it scarcely expands or contracts with heating or cooling like other metals do. It is used for accurate measuring tapes, tuning forks, pendulums, and in precision instruments.

One of the most common alloying elements in steels is manganese. It is added to practically all of them, including the carbon steels, to a greater or lesser extent in order to prevent brittleness. The so-called manganese steel itself is very strong and hard-wearing. It is used, for example, for excavators, rock-breaking machinery, and armor plating.

High-speed tool steels keep their hardness and edge at very high cutting speeds when the metal may become red-hot. They generally contain tungsten and chromium as alloying elements, and sometimes vanadium, molybdenum, or cobalt.

The properties of both alloy and carbon steels can be modified and often greatly improved by what is known as *heat treatment* – reheating them and cooling them at a controlled rate. Other alloys besides steel respond to heat treatment in the same way.

Newly refined steel from the furnace is poured from huge ladles into a row of ingots. This is termed 'teeming'. The ingots are then allowed to solidify, and taken to another part of the plant where they will be shaped.

Before shaping, the steel ingots are 'stripped', or removed from their moulds and placed in the soaking pit (below). Here they are heated to a uniform temperature to relieve any stresses set up when the molten metal cooled in the ingot mould. After soaking, they are ready for the shaping treatment.

# Shaping Metals

There are quite a variety of ways of shaping metals. The method chosen in a particular case depends on what size and shape are required, what metal is being used, and so on. Some metals are shaped hot, others cold. Molten metal is shaped by casting. Hot, but solid metal can be shaped in several ways. It can be pressed into shape by forging. It can be formed into sheets by rolling. Rods and tubes can be made by forcing the metal through holes. Wires are made by drawing cold metal through holes. One of the latest techniques of shaping involves the use of powdered metals. Cutting blocks of metal to shape on machine-tools is very widely practised.

## Casting

Casting is one of the simplest ways of shaping a metal. A casting is made by pouring molten metal into a mould of the required shape. The metal cools in the shape of the mould. This is exactly like making a jelly in a mould. Probably the best-known metal used in castings is the so-called *cast iron*. Aluminium and zinc alloys are frequently shaped by casting, too.

The mould for metal castings may be made from sand or from metal having a higher melting point than the metal being cast. The first stage in making a sand mould is to make a model of the finished object. Then damp sand is packed firmly around it. The sand used must be a special sort that binds together well. When the model is removed the impression remains. A channel, or *runner* is made in the sand down to the cavity to let the metal in. Another, called the *riser,* is cut to let the air escape from the mould. Sufficient metal is passed in so that the channel and riser are filled. This ensures that the mould itself is completely filled. Often extra metal must be added to the mould after the main bulk has been poured. This is necessary because of the shrinkage of the metal inside as it cools.

The big disadvantage of the sand mould is that it can be used only once. It must be broken up to release the casting. This disadvantage has been overcome by using metal moulds that can be used over and over again. The metal mould is called a *die*. The method of casting in it is called *die-casting*. The die is made in two halves, which are separated to release the casting. The molten metal can simply be poured into the mould, as in the sand mould. Or it can be forced into the mould under pressure. This method of pressure diecasting can be used to produce castings automatically on a mass-production basis. Then the dies have hollow channels inside them through which cooling water circulates. This quickly chills the molten metal into a solid state, and the castings can be ejected from the mould with the minimum of delay.

Castings in sand moulds or dies are made one at a time. But there is a method of continuous casting that produces simple shapes, such as slabs and rods, continuously. Molten metal from the furnace is poured directly into water-cooled moulds. As it solidifies, the piece of metal is gradually and continuously withdrawn from the bottom of the mould.

Pipes and similar shapes can be cast, too, by a somewhat different technique. Liquid metal is poured into a rapidly rotating cylindrical mould. The metal is flung by the centrifugal ('away-from-the-center') force onto the walls of the mould and takes up the shape of a pipe.

Casting is an essential preliminary pro-

Continuous casting is a method often used for shaping metals such as steel and aluminium. The picture shows a continuous-cast aluminium slab ready for despatch to the rolling mill.

Pressing is a very common shaping technique used throughout industry. Massive hydraulic presses are used for the purpose, many of which are capable of exerting thousands of tons pressure. The picture shows a hydraulic press used in the aircraft industry for hot-forming titanium sheet. The sheet is pressed between heated ceramic dies.

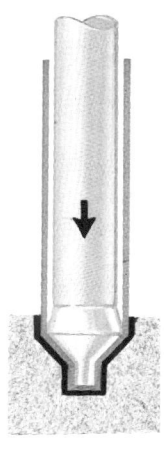

Soft tubes for holding toothpaste for example may be made by impact extrusion. The metal is put in a die and a punch is forced down quickly onto it. The impact causes the metal to flow, and it shoots upwards to form the tube.

Casting in molten metal has been practised for centuries and it is still a very important method of shaping. Sand moulds are still widely used. The molten metal is poured into the mould through a channel called the runner. A channel called the riser is also necessary to allow air and excess metal to escape.

cess for the other methods of shaping, too. The molten metal from the metal-making furnace is poured into rectangular moulds to form blocks called *ingots*. Because of the size of the ingot, cooling is uneven throughout it, giving rise to stresses. Also, flaws and hollows may develop because of shrinkage. To bring it into a more uniform condition, the ingot is therefore reheated to white heat in a brick-lined furnace called a *soaking pit* for several hours. After this treatment the ingot is squeezed into a suitable shape for subsequent processing.

### Forging

Forging is one of the oldest ways of shaping metals. The blacksmith used to practise hand forging at the village forge. He would first heat the metal until it was red-hot and soft. Then he would hold it in a pair of tongs and hammer it into shape on a stand called an *anvil*. Sometimes he would shape the metal by hammering it into a metal die.

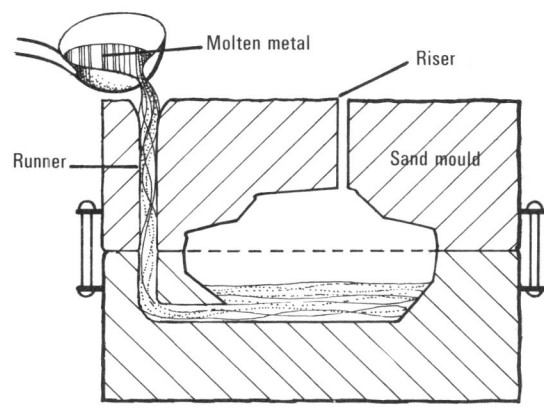

Modern forging works on the same principles. But now most forging is done by heavy power machinery. The main methods of forging are drop forging and pressing.

In *drop* forging, the hammer blows needed to shape the hot metal come from a *drop hammer*. This is a machine with a heavy metal ram that drops under its own weight. Sometimes the ram is forced down even harder by a steam hammer, which works by steam pressure.

The ram carries the upper, or male half of the metal-shaping die. The metal is placed on the floor of the machine, which also forms the bottom, or female half of the die. The metal is forced into shape when the hammer falls and the two halves of the die come together.

*Pressing* is used for shaping many things, such as automobile bodies from sheet steel. They are shaped not by a sharp hammer blow but by gradual squeezing. The massive hydraulic presses are incredibly powerful and press down with a force of tens of thousands of tons.

### Rolling

Metal that is required in sheet form to make such things as car bodies and tin cans is made into sheets by *rolling*. The metal is first rolled hot because it can be rolled more easily in this state. The rolling mill is like an enormous mangle. The white hot ingot of metal passes through pair after pair of hard, heavy rollers, which squeeze it and reduce its thickness. Each set of rollers is slightly closer together than the previous one. Eventually the thick, fairly short block

of metal is reduced to a long, thin sheet. After hot rolling, the thin sheet may be rolled cold. Cold rolling gives the metal a more accurate and harder finish. Sometimes the metal is rolled in specially shaped rollers. It can be shaped by this means into such things as railway lines and girders.

### Extrusion

One simple way of making metal rod is by forcing a piece of the hot, softened metal through a die hole of the right size. This is somewhat like squeezing toothpaste out of a tube. This process, called *extrusion,* is widely used for making rods and tubes. The die is made of very hard, polished steel to withstand wear. Sometimes the metal is forced through the die; sometimes the die is forced through the metal. To make a tube, the metal is squeezed between the die and a pointed steel rod called a *mandrel*.

Thin-walled tubes, such as toothpaste tubes, are made by *impact* extrusion. A punch hits a piece of metal in a die very hard.

**Working Properties of Metals**

Whenever we think of metals, we think of substances that are strong, hard, and tough. And these are three of their most important properties. Other important things to know about metals are how much they can stretch, how easy they are to shape, and how brittle they are. These physical properties vary from metal to metal. Steel is strong but is difficult to shape and bend cold. Copper is much weaker but can be bent and shaped easily when cold. The properties of a metal can be altered as we have seen (page 134) by adding other metals or elements to it to form an alloy and by heat treatment.

The methods of shaping and using metals depend on their physical properties. Metals with great *strength* and *hardness* are naturally difficult to form or cut. They must be shaped by grinding. Metals with good *toughness* resist continual bending and pulling.

A metal with high *ductility* can be stretched a lot without breaking. This makes it easy to shape. For example, wire is made by drawing such a ductile metal through progressively smaller holes. Ductile metals also have good *plasticity*. They can be easily formed into more or less any shape. Other metals only become plastic, or easy to shape when red hot. This is why steel, for example, is generally worked when hot.

Another related property is *malleability*. It is a property that enables metals to be hammered or rolled into thin sheets without cracking. Aluminum is an example of a malleable metal. It can be rolled into foil less than a thousandth of an inch thick. Thicker aluminum foil constitutes the 'silver' paper in sweet and cigarette wrappings. But the most malleable metal of all is gold. It can be beaten into sheets so thin that they are practically transparent.

*Brittleness* in a metal such as cast iron makes it stiff and suitable for rigid structures. But a brittle metal snaps easily if it is bent or deformed even slightly. It therefore cannot be shaped by such methods as forging, rolling, and so on, which involve considerable deformation. A tendency to brittleness is introduced into a metal by casting.

---

Top left: A red-hot steel ingot about to enter the rollers of a hot-rolling mill. It will go through a series of such rollers until it is reduced to the required thickness. Left: Pouring molten iron into sand moulds at the foundry.

The metal becomes plastic and flows upwards between the punch and the die to form the tube.

### Drawing

Another way of making thin rod is by *drawing* cold metal through the hole in a die. The process simply involves drawing a thicker rod through progressively smaller holes. It has to be done in stages because the metal is too strong to be pulled straight away through a tiny hole. By an extension of the process, fine wire can be made. Cold drawing gives the metal a shiny surface and may make it stronger. Often rods and tubes made by extrusion and other methods are cold-drawn through a slightly smaller die to produce a more accurate finish and to make them stronger.

### Powder Metallurgy

The method of making solid metal pieces from small metal grains, or powder is known as *powder metallurgy*. Some metals, such as platinum and tungsten, are obtained commercially in small grains. But they cannot be melted and cast into solid blocks in normal industrial furnaces because their melting points are too high. But when grains of the metal are pressed together and heated in a furnace, they form a solid piece.

Powder metallurgy is also used for shaping small articles from many other kinds of metal. Powdered metal is put into a die to give it the required shape and subjected to very great pressure. Then it is *sintered,* or heated strongly in a furnace to bind the grains together. No metal is wasted in this process.

One of the most outstanding examples of this technique is the production of tungsten carbide. This material is one of the hardest substances known to man. It is used for the tips of cutting tools and for making dies for shaping metals. The manufacture of tungsten carbide involves heating tungsten powder with finely divided carbon at a temperature of about 1500°C.

Forging the head flange of a pressure vessel for a nuclear-fuelled power station, using a mandrel. The massive flange began as a 180-ton vacuum-cast alloy steel ingot. After forging and machining its weight was reduced to only 40 tons.

# Joining and Cutting Metals

Many products can be fashioned out of a single piece of metal by one of the methods we have already discussed. But this is not usually possible where large or complicated shapes are concerned. In these cases it is usually cheaper and easier to build up the shape by joining simple pieces together. You obviously cannot forge a 1,000 foot-long hull of a ship out of a single piece of metal in one operation!

One simple means of joining metal parts is by *screwing* or *bolting* them together. Another is by *riveting*. Rivets are metal plugs with a rounded head at one end. They are pushed through holes in the parts to be joined (which must overlap, of course) and are hammered until a head forms at the other end. This joins the parts together. Often the rivets are hammered home while they are red hot. Then, as they cool, they shrink

Riveted joint

Below: Arc welding is used on a vast scale in the heavy engineering industries, such as shipbuilding. An electric arc is struck between two electrodes, which results in a temperature high enough to fuse steel. The picture shows typical arc-welding equipment. Note the worker is wearing a head mask. This has a dark glass visor to protect the worker's eyes from the glare.

slightly and therefore squeeze the parts together more tightly.

The other common method of joining metals are welding, soldering and brazing. They involve the localized melting of the metals themselves or of a binding metal. They are preferable to the above mechanical methods because the parts to be joined do not have to be drilled first. Drilling naturally weakens the metal.

## Welding

Welding produces much stronger joints than either soldering or riveting. There is an important weight advantage over riveting, too, because the parts to be joined can meet end to end and need not overlap. Welding is widely used in industry because it is both quick and cheap. Automobiles, for example, have a very large number of welded joints. A ship is constructed of literally thousands of welded plates. Iron and steel are both very easy to weld. Aluminium and other important metals and their alloys can now also be welded satisfactorily.

In welding the edges of two pieces of similar metals are heated until they melt, or *fuse* together. They cool down in one piece to form a strong joint. The metal may be heated by electricity or by a gas flame. The metals are cleaned during welding by a compound called a *flux*. What is equally important, the flux also prevents the air getting to the joint and oxidizing the metal. The presence of dirt or a layer of oxide on the metal would naturally result in a weak joint.

Sometimes the metal parts are pressed together as they melt to help form the joint. This is called *pressure* welding. When no pressure is applied, the process is called *fusion* welding.

One simple way of pressure welding is to hammer the pieces of red hot metal together. This was the method blacksmiths used. Modern methods of pressure welding are butt welding and spot welding. They are forms of what is called *resistance* welding: the heating effect comes from resistance to the passage of electricity.

In *butt* welding, the ends of the metals to be joined are placed together and heated by a strong electric current. When they soften, they are pushed firmly together to form a weld. In *spot* welding, overlapping metal sheets to be joined are gripped between two metal rods, or *electrodes* carrying electric current. The current passing through the electrodes softens the metal in between. Then the electrodes press the sheets together to form a welded spot. *Seam* welding welds by a similar method, using circular

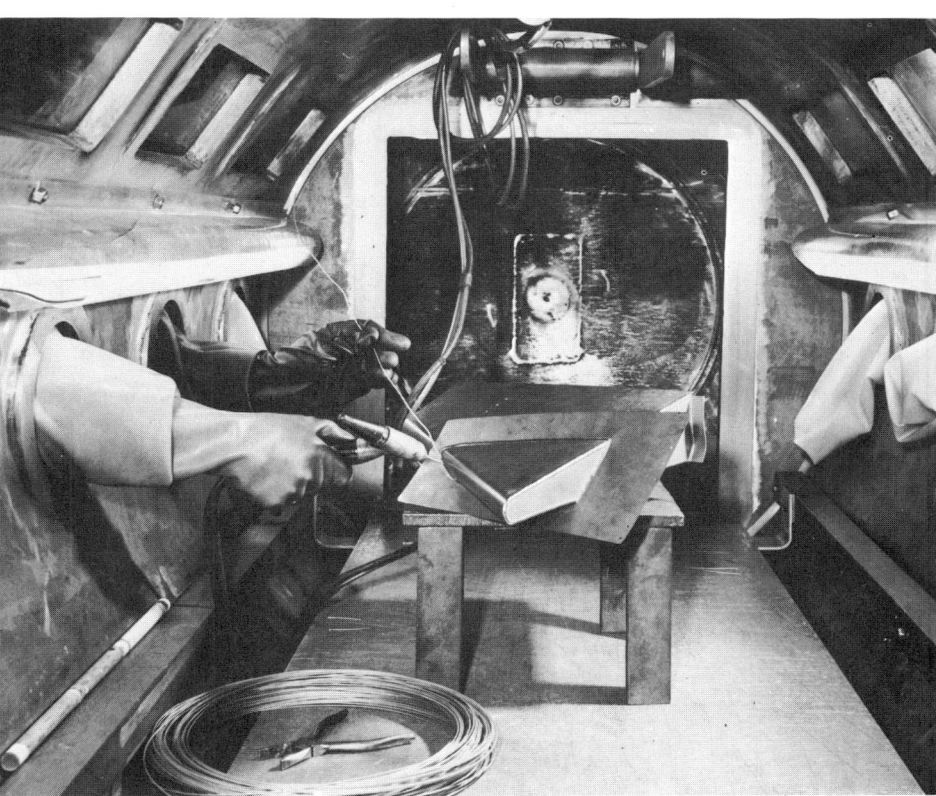

Above: Seam welding a steel drum. Note the circular electrodes. They apply a heavy electric current to soften the metal between them, and press together to make a neat, continuous welded joint.

Above right: Inert-gas welding of titanium in a glove chamber. The sealed chamber is filled with inert gas, such as argon. Such a technique prevents the possibility of oxidation and contamination of the weld.

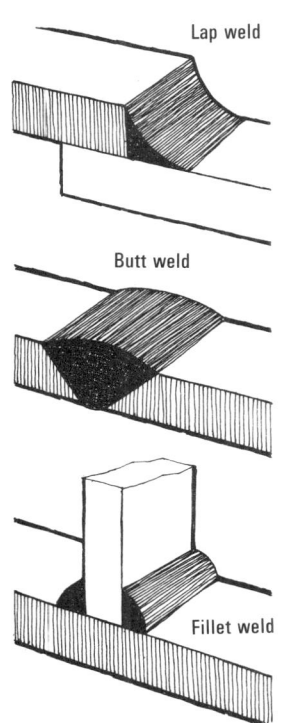

Lap weld

Butt weld

Fillet weld

discs to carry the electric current and to apply the pressure.

In fusion welding, extra metal is often added to the joint from a *filler rod*. The rod is melted at the same time as the metals and runs into the joint. It helps to bind the metals together. There are two main methods of fusion welding – arc welding and gas welding.

In *arc welding,* the metal is fused by the intense heat of an electric arc. The arc is struck between an electrode and the base metal. In *carbon-arc* welding a carbon rod is used as the electrode, and a separate filler rod is required. In *metal-arc* welding the filler rod is used as the electrode. Practically all metal-arc welding is done with an electrode that is coated with a fluxing material. In *inert-arc* welding an unreactive gas such as argon or helium is fed through nozzles surrounding the electrode. It prevents the oxygen in the air attacking the fused metal surface.

In *gas welding* the metal is heated by a very hot flame from a *blowpipe,* or *gas torch*. The most common gas-welding torch is the *oxyacetylene* torch. A high temperature (more than 3,000°C) is produced by burning a mixture of acetylene and oxygen. The gases are kept under high pressure in heavy iron cylinders. Pipes carry the gases to the welding torch, where they mix. The oxyacetylene mixture is ignited at the tip of the torch to produce an intense flame.

Acetylene is not the only gas used in gas torches. Sometimes hydrogen is used instead. The so-called *oxyhydrogen* torch is more suitable than the oxyacetylene torch for welding some metals. Propane, one of the heavy gases obtained from natural gas, is also sometimes used.

### Cutting

Oxyacetylene torches are used not only to join metals by welding but also to *cut* metals. Workers in many industries shape metal parts by cutting them out of solid metal. Scrap merchants and demolition experts need to cut girders and other large metal

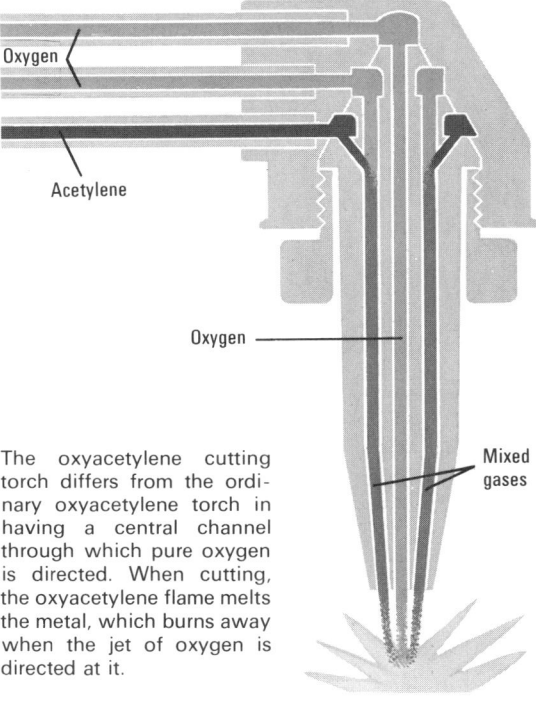

The oxyacetylene cutting torch differs from the ordinary oxyacetylene torch in having a central channel through which pure oxygen is directed. When cutting, the oxyacetylene flame melts the metal, which burns away when the jet of oxygen is directed at it.

parts into more easily manageable pieces.

The oxyacetylene cutting torch is somewhat different from the welding torch. What happens is that the oxygen-acetylene flame is used to pre-heat the metal to be cut and then a jet of pure oxygen cuts clearly through the heated metal. The oxygen causes the metal to burn away rapidly. The head of the torch has a central hole, from which the oxygen issues forth. Around it are a number of openings for the pre-heating oxyacetylene mixture.

Above: One of the more recent welding techniques is electron-beam welding, in which a high-speed stream of electrons does the actual heating. The beam can be precisely focused and carefully manipulated. The technique is ideally suited to the welding of small components and those metals that would become contaminated by gaseous impurities during ordinary welding. The picture shows two electron-beam welding machines used to fabricate intricate components in nickel and titanium alloys for the Concorde supersonic airliner.

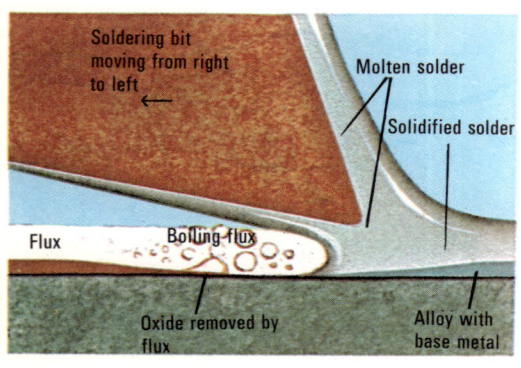

Left: Illustration of the process of soldering. Flux is applied to remove the dirt and oxide from the metal surface, thus ensuring better adhesion for the solder.

### Soldering and Brazing

Soldering and brazing are processes of joining two metals by melting another metal or alloy between them. The temperature at which the alloy melts is far below that at which the metals themselves would melt. In soldering the alloy used, *solder,* contains lead and tin. It is called *soft* solder because it has quite a low melting point. Soft solders are used in plumbing to mend and joint pipes. They are also widely used for joining electrical wires in televisions, radios, and so on, and for sealing tin cans. Rosin is a common flux used in soldering.

In brazing the alloy used, which often is brass, has a much higher melting point than soft solder. It is known as *hard* solder. The joint formed is naturally much stronger than that formed by soft solder. The joints in bicycle frames are formed by brazing. Another hard solder, called *silver solder,* is an alloy of silver, copper and zinc. Borax is a common flux used for brazing operations, often as a mixture with boric acid.

# Metals and their Uses

**Aluminum (Al)** is a particularly valuable metal because it is so light. It is much lighter than steel, and some of its alloys are as strong as steel. Aluminum and its alloys are therefore widely used in railway carriages, ships, and aeroplanes, where lightness and strength are needed. Also, aluminum does not rust as ferrous metals do when exposed to air. It does oxidise, but the oxide film prevents further corrosive action.

There is a wide range of aluminum alloys. 'Duralumin', an alloy with copper, is often used for aircraft parts. Other important aluminum alloys contain copper, magnesium, and silicon.

In the home, aluminum and its alloys are used for window frames, venetian blinds, appliances, and saucepans. Milk bottle tops are made from thin aluminum foil. Aluminum is suitable for saucepans because it conducts heat well. It conducts electricity well, too. Because of this property and its lightness, it is widely used for overhead power cables.

There is more aluminum in the Earth's crust than any other metal. It is found, for example, in common clay. But removing it from the clay is too difficult. The only suitable ore is bauxite, which contains aluminum oxide, or alumina. Producing aluminum from bauxite is very different from the usual methods of obtaining metals from their ores. First the bauxite is crushed and heated with caustic soda. Alumina dissolves and leaves behind the impurities. The pure alumina obtained is then dissolved in molten cryolite. Aluminum is produced when electricity is passed through the solution.

**Antimony (Sb)** A hard, brittle metal that is widely used for hardening lead and making it more resistant to corrosion. Type metal is a lead-antimony alloy with small amounts of tin and sometimes copper. The chief source of antimony is the mineral stibnite, or grey antimony ore (antimony trisulphide).

**Arsenic (As)** Not a true metal, but one of its forms has metallic properties. It is steel-grey and brittle. It is used in some alloys and in making lead shot. Its compounds are among the most deadly poisons known.

**Barium (Ba)** A soft, silvery-white metal that bursts spontaneously into flame in air. It is occasionally alloyed to nickel to improve the strength and hardness. It occurs as the sulphate, barites, or heavy spar, and the carbonate witherite. It is one of the alkaline earth metals. Its compounds colour a flame a beautiful apple-green, and are used in fireworks.

**Beryllium (Be)** The lightest of the structural metals, strong, hard, and with a high melting point. It is alloyed with copper, aluminum, iron, and nickel to improve strength, hardness, and resistance to corrosion. The chief source of beryllium is the silicate mineral beryl, the transparent green form of which is the precious stone emerald.

**Bismuth (Bi)** A greyish-white crystalline metal with a slight reddish tinge to it. Its main use is in alloys of low melting point, such as solders and the so-called fusible alloys. Wood's metal, for example, which is an alloy of 50 per cent bismuth with lead, tin, and cadmium, melts at about 70°C.

**Cadmium (Cd)** A soft, silvery-white metal that is used for electroplating. It is alloyed with copper for making power cables, and it confers added strength. It is a common constituent of fusible alloys of low melting point. With nickel it is used in storage batteries. With mercury it forms a cell, the Weston cell, that is used as a standard in voltage measurements. The only natural compound of cadmium is the sulphide, greenockite.

**Caesium (Cs)** A silvery-white alkali metal resembling sodium. It is highly reactive and decomposes water. It is used in photo-electric cells. Its compounds are rare.

The vaporized metal is very important for it is used in that most accurate of timepieces the atomic clock (see page 203).

**Calcium (Ca)** A soft, white, alkaline earth metal that tarnishes rapidly in air. It is occasionally used in alloys with copper and with lead. Its compounds are widespread, especially calcium carbonate (chalk, limestone, and marble).

**Chromium (Cr)** We handle chromium every day in one of its forms. This may be as chromium plating or as stainless steel cutlery. Chromium is plated on other metals to protect them. It is hard and does not lose its shine through corrosion. Steel is made stainless by the addition of chromium and nickel. Nickel and chromium together form important alloys. Chromium is found as chromite, or chrome iron ore.

**Cobalt (Co)** The main use of this tough metal is in the manufacture of alloys. With aluminum and nickel it forms an alloy (alnico) which is excellent for making powerful permanent magnets. With chromium and tungsten, it forms stellite, an alloy that is remarkably resistant to high temperatures. Most of the world's cobalt is obtained as a by-product from ores that are worked for other metals. Most comes from the copper mines.

**Copper (Cu)** One of our most useful metals. It is also one of the few metals that can be found native. But most of our copper comes from ores mined deep underground.

Copper has an attractive reddish-brown color. It can be easily beaten into thin sheets and complicated shapes and drawn into wire. Also it does not corrode quickly when exposed to the air. These properties have made it valuable for thousands of years for making jewellery, ornaments, jugs, vases, and cooking vessels.

Copper is particularly suitable for cooking vessels because it conducts heat extremely well. It also conducts electricity well. This is why copper is used in such things as electric cables, telephone wires, electric motors, and television sets. Two of the best-known alloys — brass and bronze — contain copper. Brass contains zinc; bronze contains tin.

The main ore of copper is chalcopyrite, or copper pyrites, a sulphide.

Left: A fascinating variety of copper utensils displayed in a small French town famous for its copperware. Copper has been used in similar ways down the ages. It is easy to shape, conducts heat well, and has excellent resistance to corrosion.

Right: Aluminum, like many other metals, is very widely used in sheet form. Cast first as an ingot, the aluminum is progressively reduced in thickness by a succession of rollers. The picture shows the final stage, the aluminum sheet being now only a fraction of an inch thick.

**Gallium (Ga)** A rare grey metal with a low melting point (30°C) that is similar in some respects to aluminum. It was once known as ekaaluminum.

**Germanium (Ge)** A metal used to make the tiny transistors in our pocket radios. It will form much larger crystals than most other metals. These crystals have the special electrical properties needed for radios. The transistors are very much smaller and tougher than the radio valves they replaced. They also need less power to operate. Germanium is usually obtained from the flue dust of power stations.

**Gold (Au)** One of our most precious metals. Its beautiful 'golden' color never becomes dull. It never corrodes. It can be shaped more easily than any other metal. These properties have made gold the most sought-after metal for thousands of years.

One of the widest uses of gold is for ornaments and jewellery. It can be made into delicate webs, drawn into fine wire, and beaten into thin sheets. Pure gold is too soft for day-to-day jewellery like wedding rings. A little copper or silver is therefore added to harden it. Gold was once widely used for making coins, such as the English sovereign. But few gold coins are now made.

Gold is one of the few metals mined in native form. It is also found in minute traces in silver and copper ores. It can be obtained during the refining of other metals.

**Hafnium (Hf)** A rare white metal that resembles titanium and zirconium. It has a few uses in the electrical industry, but is too expensive to be widely used.

**Indium (In)** A rare silvery metal that is even softer than lead. It is used to coat engine bearings to make them withstand wear and corrosion better. It is sometimes used in solders.

**Iridium (Ir)** A rare, precious metal that resembles platinum in some respects. It is extremely hard and melts only at a very high temperature (2,450°C). It is generally used as an alloy with platinum for making chemical and electric equipment, jewellery, and surgical instruments. It is extremely corrosion-resistant. For this reason, it is used for the tips of fountain-pen nibs.

**Iron (Fe)** One of the first metals used by man and still one of the most popular. It is seldom used in a pure state. Its main use is as a basis for steel. Iron and alloys containing iron are called *ferrous* metals. Other metals are called *non-ferrous* metals. One way in which iron is used by itself is in magnets because it can be easily magnetized.

The iron obtained by smelting iron ore in a blast furnace is called *pig iron*. It contains many impurities. Steel is made by refining pig iron. Pig iron is remelted and poured into moulds to produce *cast iron*. Cast iron is cheap and widely used in industry for such things as machine frames and engine blocks. Another form of iron, called *wrought iron*, is a kind of refined pig iron with lumps of slag in it. It is not much used today except for ornamental gates and chains.

Iron ores are to be found in most parts of the world. The chief ones are haematite, magnetite, and limonite, all oxides.

**Lead (Pb)** One of the heaviest of the common metals and also one of the weakest. It is soft and can be shaped easily. It does not corrode quickly like iron, for example. Because of these properties, lead has been used for hundreds of years piping water and for lining roofs. Large amounts of lead formerly were made into chemical compounds for use in paints. It is widely used in alloys. Solders are alloys with tin. Type metal contains antimony. Another lead-antimony alloy is used in car batteries. Lead is found combined with sulfur in the ore galena.

**Lithium (Li)** The lightest of all metals. It is an alkali metal like sodium, and burns vigorously when placed in water, giving off hydrogen gas. It is occasionally used in alloys.

**Magnesium (Mg)** Even lighter than aluminum, magnesium is normally used in alloys with other metals. A common alloy contains aluminum, zinc, and manganese. These make the magnesium stronger and more resistant to corrosion. Being so light, magnesium alloys are widely used for aircraft parts. Magnesium burns in air with a brilliant white flame. But it can be melted and cast safely and easily. It is, in fact, the easiest metal to shape and machine.

Magnesium is one of the most plentiful metals on Earth. Magnesite and dolomite are two common ores. Magnesium can also be obtained by processing sea water.

**Manganese (Mn)** is used almost entirely in alloys. In general, it makes them stronger and tougher. Manganese is added to practically all steels. It improves their quality and increases their strength. It is added as ferromanganese, an alloy with iron. Manganese is found combined with oxygen in such ores as pyrolusite and psilomelane.

**Mercury (Hg)** is the only metal that is liquid at ordinary temperatures. It is often called *quicksilver* because of the way it flows. The most familiar uses of mercury are in thermometers and barometers. It is also used as a vapor in some street lamps. Most metals dissolve in mercury to form *amalgams*. Dentists use gold and silver amalgams to fill holes in teeth. Mercury is obtained from the sulphide ore cinnabar.

**Molybdenum (Mo)** is a strong, hard metal with a high melting point (2,620°C). It is widely used in the manufacture of steels for cutting tools, drills, aircraft parts, and so on. It is obtained from two minerals, molybdenite (molybdenum sulphide), which is by far the most important, and wulfenite (lead molybdate).

**Nickel (Ni)** is best known for its uses in electroplating and, with chromium, in stainless steels. It forms a great variety of alloys. The copper alloys include the cupronickels used for many 'silver' coins. One of its alloys with chromium is used for the wires of electric fires. Similar alloys are used in the high-temperature parts of jet engines. Nickel's main ore is pentlandite, a sulphide of iron and nickel.

**Niobium (Nb)** A rare grey metal, once called *columbium*. It is used for making the high-temperature steel alloys in jet engines and rockets. It is found with tantalum, which has similar uses, in the ores columbite and tantalite.

**Osmium (Os)** The heaviest of all known elements, it is twice as heavy as lead. It is similar to platinum. It is also used in alloys with platinum and iridium. It is completely resistant to acid attack. Osmium is found, associated with other elements of the platinum group, in alluvial deposits and sometimes in gold ores.

**Palladium (Pd)** A hard, white metal similar to platinum although it is not as heavy nor as costly. One of its main uses is in industry as a catalyst for promoting chemical reactions. It occurs native with platinum.

**Platinum (Pt)** is a precious metal more valuable than gold. It is similar to gold in many ways. It does not lose its colour (silvery white) or shine, and it can be shaped easily. But it is harder and heavier than gold. One of its main uses is in jewellery. In industry platinum is very valuable as a catalyst that helps certain chemical processes to take place. It is used in oil refining and in making nitric acid. Platinum is found scattered in small grains in some gravels. It is expensive to extract.

**Potassium (K)** One of the very reactive alkali metals that decomposes water. It is extremely soft and can be cut with a knife. The pure metal has few uses. Its compounds, however, are widespread.

**Radium (Ra)** is a very rare metal. It is found in small traces in ores that contain uranium, such as pitchblende. Like uranium, radium is radioactive and it gives off powerful radiation. This radiation is widely used in medicine for treating cancer.

**Rare Earths** This name is given to a group of metals which are very similar chemically, and which are usually found together in Nature. The mineral monazite is the chief source.

**Rhenium (Re)** A rare metal with a high melting point (3,170°C). It is occasionally used in alloys and in thermocouples, but it is too expensive for widespread use.

**Rhodium (Rh)** A hard precious metal occurring with and resembling platinum. It is used for plating other metals, especially silver, and, alloyed with platinum, for making thermocouples.

**Rubidium (Rb)** One of the highly reactive alkali metals. It is soft and white and resembles sodium. It is slightly radioactive.

Above: A Tanzanian silversmith at work decorating a silver tray. Decoration is comparatively easy since silver is malleable and ductile.
Left: Stainless steel in an unfamiliar setting, as jewellery. Nickel and chromium are the most important alloying elements in stainless steel.
Right: A nuclear power station, which uses uranium as "fuel". Enormous quantities of energy are released when uranium atoms split.

**Ruthenium (Ru)** A hard, brittle metal of the platinum group. It occurs with platinum in Nature. It is alloyed with platinum and confers hardness.

**Silver (Ag)** One of the precious metals, silver has long been used for making ornaments and fine jewellery. Some of the finest tableware has been made in silver. Like gold, silver can be shaped easily and corrodes hardly at all. But in towns silver 'blackens', or tarnishes, because of sulphur in the atmosphere. Silver conducts heat and electricity better than any other metal. But it is too expensive to use on a large scale.

Silver was once used to make coins. But now most of the world's 'silver' coins are made from copper and nickel. Much of our 'silver' tableware is only plated with silver, for cheapness. For jewellery, a little copper is added to silver to harden it. By far the biggest use of silver today is in chemical compounds. Silver forms compounds that are sensitive to light. They are used in the photographic industry.

**Sodium (Na)** A soft, silvery white alkali metal that reacts violently with water. It is used in the molten state as a cooling agent in some nuclear reactors. As a vapor it is used in sodium vapor street lamps which give out a vivid yellow light. Its compounds are widespread in Nature. Common salt (sodium chloride) from the sea or from massive deposits of rock salt is the main source of the metal.

**Strontium (Sr)** A pale yellow alkaline earth metal that is similar to calcium. Its compounds color a flame a deep crimson, and are used for this purpose in fireworks.

**Tantalum (Ta)** A rare metal that is used in alloys for jet and rocket engines that have to resist high temperatures. It is also used for lamp filaments and in some cutting tools. It occurs with a similar metal, niobium, in columbite and tantalite.

**Tellurium (Te)** A semi-metallic element that is sometimes alloyed with lead to give improved hardness and strength.

**Thallium (Tl)** A white, relatively soft metal resembling lead. It has few uses at present. Its compounds are deadly poisonous.

**Thorium (Th)** A dark grey radioactive metal that resembles platinum in some respects. It has important uses in the atomic energy industry. Its compounds are used in the manufacture of gas mantles. The two most important sources of thorium are monazite and thorite (thorium silicate).

**Tin (Sn)** The most familiar use of this metal is in making tin cans. The can is made from thin steel plate coated with tin. It is made in this way because steel itself would rust, and tin by itself would not be strong enough. The tin plate is made by dipping the steel in molten tin or by electroplating. Tin forms important alloys, including bronze (with copper), pewter, and solder (both with lead). Another important use is in metals for machine bearings. These alloys contain mainly tin with some antimony and copper. Tin ore, cassiterite, is an oxide.

**Titanium (Ti)** is a strong, lightweight metal that is used mainly in alloys. Some parts of chemical plants are made of titanium because it is very resistant to corrosion. It is used in bone surgery for the same reason. Titanium and its alloys, especially steels, are particularly useful in aircraft and their engines, where lightness and strength are important. Most titanium is used not as a metal but in the form of its oxide in white paints. The metal is obtained from ilmenite, a mixed oxide of titanium and iron. Its extraction is costly.

**Tungsten (W)** melts at a higher temperature than any other metal (3,380°C). This is why it is used for the filaments in light bulbs and in radio valves. Tungsten is also very hard. High-speed cutting tools are made from steel containing tungsten. They keep their hardness even though they get red-hot. One of the best cutting materials of all is tungsten carbide, in which the metal is combined with carbon. Tungsten is found in a few rare minerals, including scheelite.

**Uranium (U)** is the metal used as a 'fuel' in atomic power stations. Under certain conditions, uranium atoms split into smaller pieces. When this happens, an enormous amount of energy is released as heat (see page 33). This process, called *fission*, makes it possible for uranium to be a fuel. Uranium ores are found in many parts of the world, but only in minute traces. Concentrating them is a costly business. Pitchblende is one of the main ores.

Uranium is one of the radioactive elements which continually give off tiny penetrating particles, or radiation from inside their atoms. By losing these particles, they gradually change into certain other elements which have smaller atoms. For example, uranium changes into radium. Then radium changes to lead.

The reverse process also occurs. The uranium atom can take in atomic particles to form a bigger atom. The new element formed is called an artificial element because it is not normally found on Earth. This new element is a metal known as plutonium. It, too, is radioactive. Further artificial elements can be obtained by bombarding plutonium with atomic particles.

**Vanadium (V)** A very hard white metal that is used extensively in steel alloys, to which it gives greater strength and elasticity. Vanadium steel is used for locomotives and automobile parts such as the chassis, axles, and springs that must absorb shock well.

**Zinc (Zn)** is a metal that is widely used in alloy form. Its most important alloy is brass, in which it is combined with copper. Alloys of zinc and aluminium are used extensively for diecasting. Many parts of cars and appliances such as washing machines are made of these diecast alloys. Zinc resists corrosion quite well. It is therefore used as a coating for protecting steel. This is called *galvanizing*. Dry batteries contain thin zinc plates. The main ore of zinc is zinc blende, or sphalerite, a sulphide.

**Zirconium (Zr)** A rare, soft white metal. It ignites at a low temperature and is used in photographic flash-bulbs. It is occasionally used in alloys. Its oxide zirconia is an excellent refractory material and is used in rocket motors.

# Machining Metals

Sometimes metal articles can be shaped into their final form in only one operation, such as forging. But usually they must be finished to the exact shape and size by machining operations, such as drilling and grinding, which remove excess metal. For example, the engine block of an automobile is first made by casting. Then holes have to be drilled into it. Some parts have to be flattened; others have to be polished; and so on.

The machines that carry out machining operations are called *machine-tools*. All of them work by some form of cutting action. The actual cutting edges must naturally be made of specially hardened steel because metal itself is so hard. Also, they cannot slice off a huge chunk of metal at one go. They must remove it little by little.

To do this quickly, the tools have to turn at very high speeds. They are usually driven by electric motors. To prevent excessive wear on the cutting edges, the tools are generally lubricated by a watery emulsion of oil called a cutting oil. The constant stream of liquid past the cutting edge also serves to remove the heat generated by cutting.

The machine-tools that carry out machining operations have several things in common. They have a means of clamping the *workpiece* (the part to be shaped) and moving or rotating the cutting tool against it. (Sometimes it is the other way round.) They have adjustments to move the workpiece and the tool up and down, in and out, and sideways. Sometimes this is done by hand, sometimes by power.

The basic machining operations are turning, drilling, planing, milling, and grinding.

In *turning,* the workpiece revolves against a cutting edge. The edge can be moved along and in and out to give various shapes. The workpiece revolves in a *lathe*. This is one of the most important machines in any work-

A numerically controlled machine tool in action. Positioning and multiple drilling of the workpiece is entirely automatic once the machine is set in operation. Instructions are fed to the machine by a tape played on the console seen at right of the picture. It is termed "numerical control" since initially directions for the machine are translated into numbers.

## The Lathe

The ordinary lathe found in most workshops is what is known as a *centre* lathe because the workpiece is held and revolves between centres. The main frame of the lathe, called the *bed*, carries the *headstock* and the *tailstock*. In the headstock is the *chuck*, a kind of vice that holds one end of the workpiece. The chuck can be driven at various speeds by an electric motor. The tailstock supports the other end of the workpiece and can be slid towards or away from the chuck as necessary to accommodate different-sized workpieces.

The cutting tool or tools are mounted in clamps, or *tool-posts* on the *cross-slide*, which can be moved towards or away from the rotating workpiece across the bed. Supporting the cross-slide, is the *saddle*, which can move along the bed. The lathe operator moves the cross-slide in and out and the saddle along by means of hand-wheels. He controls the chuck speed by means of levers on the headstock.

The *turret*, or *capstan* lathe has a number of cutting tools mounted on a special tool-holder called a *turret*. When brought into use, it operates automatically — one tool follows another in rapid succession.

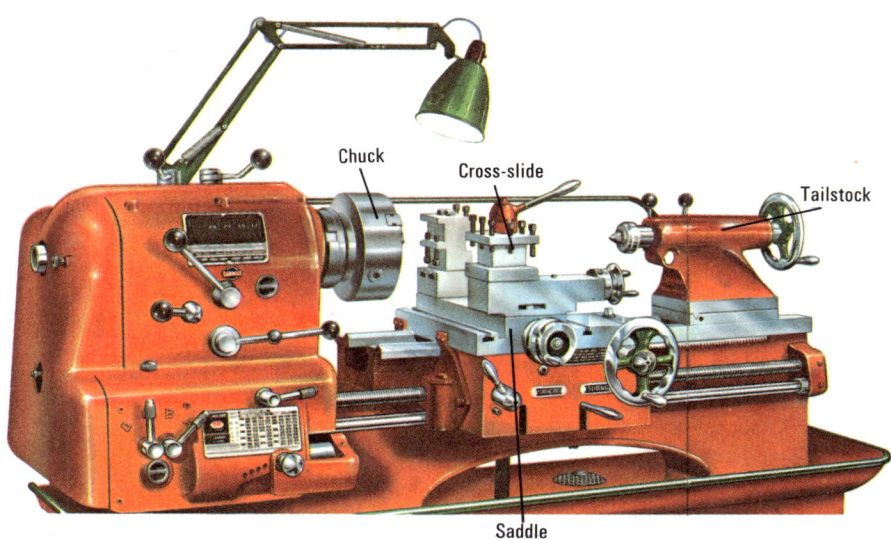

shop. It can also be used for drilling and cutting screw threads.

In *drilling*, a rotating drill cuts a round hole in the metal. Drills are generally mounted in *drill presses*, by means of which they can be lowered and pressed onto the workpiece and then removed. The usual type of drill is called a *twist drill* because of the shape of the groove in it. It cuts at its tip only. *Boring* is a similar operation to drilling, but the tool has only one cutting edge, whereas a drill has several. *Reaming* is a way of finishing a drilled hole very accurately. The tool it uses looks somewhat like a drill, but cuts all along its edge.

*Planing* produces large, flat surfaces. On castings, planing is often the first operation to produce accurate vertical and horizontal surfaces. The cutting tool remains stationary while the workpiece travels beneath it.

Related to the planer are the shaping and slotting machines. The *shaping machine* produces small, flat surfaces. The cutting tool travels horizontally on a ram that moves back and forth, shaving a thin slice of metal off on each forward stroke. The workpiece remains stationary. The *slotting machine* does the same kind of thing, but acts vertically. The tool shaves the metal at each downward stroke of the ram.

*Milling* is a way of cutting slots and grooves. The tool that does this is a sharp, toothed wheel. The wheel spins round rapidly and bites into the metal as the workpiece traverses beneath it. By using a wide wheel a surface can be 'scraped' flat.

In *grinding*, a rough-surfaced, spinning wheel is brought into contact with the metal and gradually rubs it away. The wheel is covered with very hard grains of an abrasive material such as carborundum, or silicon carbide. For grinding exceptionally hard metals, diamond wheels may be used. On larger machines the grinding surface may take the form of a continuous abrasive belt.

Grinding is often used as a finishing operation to bring a workpiece to accurate dimensions. It can be made an accurate operation because only a little metal is removed at a time. *Honing* is a technique in which a revolving abrasive head is inserted into a drilled hole. It is used to finish, say, the bores of engine cylinders with complete accuracy. For a really fine finish a very fine grinding technique called *lapping* is employed.

Many of the machining operations these days are carried out automatically on complicated machines that work electronically. Instructions are fed into the machine, a piece of metal goes in one end, and the finished product comes out the other.

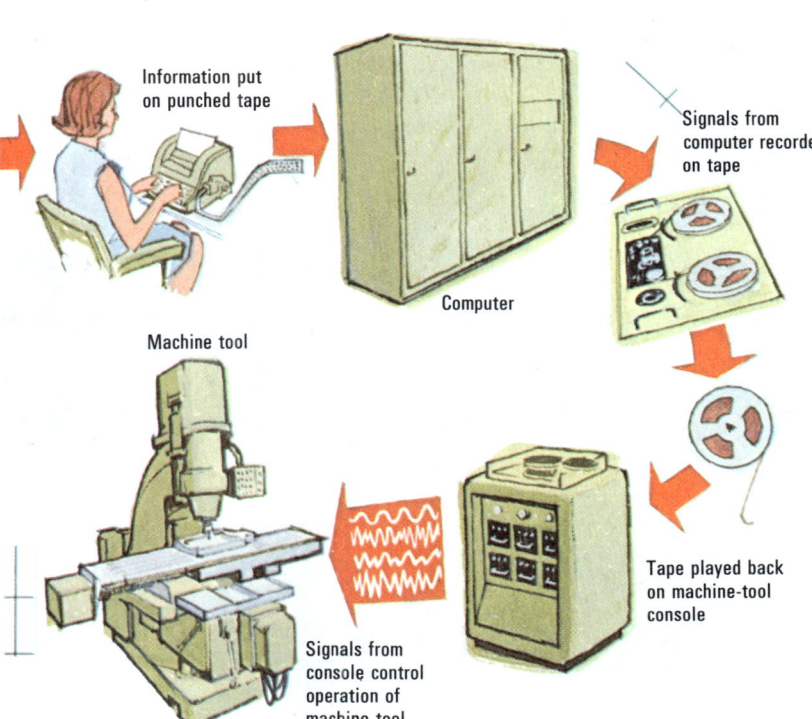

A typical set-up for automatic numerically controlled machine-tool operation. Instructions to guide the machine tool are fed into a suitably programmed computer, which converts them into electronic impulses that the machine tool can act upon.

147

# Civil Engineering Construction

Dams and bridges are among the most spectacular structures built by man. Massive dams up to a thousand feet high block deep river gorges and store water for producing electricity or for irrigation. Slender bridges suspended from cables carry traffic hundreds of feet over wide river estuaries. Many of them form part of a network of motorways that carry high speed traffic the length and breadth of the country.

Designing and constructing dams, bridges, and roads forms an important part of *civil engineering*. But the civil engineer is concerned with much more than just roads, bridges, and dams. He also designs and builds tunnels, harbours, canals, railways, sky-scrapers, and any large buildings, such as factories and power stations.

The life of a civil engineer is varied and exciting. One year he may be building a dam in the freezing cold of the Canadian Rockies. The next year he may be designing a harbour in the sweltering heat of Africa.

But no matter what kind of construction is involved and where it is, the project must be properly designed and carefully planned. Otherwise the structure being built may collapse, or the work may take much longer than expected and be very costly. The various stages of designing and planning are basically similar for most kinds of projects. Let us take building a dam as an example.

A team of engineers visit the general region where the dam is to be built and select a suitable *site*. Then they decide upon a *design* for the dam and calculate what *materials* will be needed in its construction. They prepare detailed *engineering drawings,* which show the design of each part of the structure and how each part fits into the whole.

Local materials are used as far as possible for construction. But for other materials *transport* will be required. Many miles of new roads or railways may have to be built so that materials can be readily transported.

One of the biggest planning tasks on a large project is to ensure that labour, machines, and materials are in the right place at the right time. Engineers now sometimes use computers to help them with this planning, together with techniques such as *critical path analysis*. This technique analyses all the operations that have to be carried out and shows how they are related to one another.

When all the plans for a project have been drawn up, the actual work can begin. The exact method of construction the civil engineer uses will depend on what is to be built. But the preparations for construction follow much the same pattern in practically all cases. The site for building must be surveyed, the soil must be tested, and large amounts of earth must be moved.

Accurate *surveying,* or the measuring of distances and heights, is essential if construction is to be accurate. It must be carried out both before and during construction.

*Soil-testing* is very important. The engineer cannot build satisfactorily unless he knows what soil is present beneath the surface of the ground. This is because the methods he uses depend to a great extent on the strength and other mechanical properties of the soil.

*Earth-moving* is often one of the biggest jobs during construction. Using powerful machines, the civil engineer levels large areas of ground, cuts through hillsides, and digs great pits. He prepares the ground in this way so that he can lay firm *foundations*. Then and only then can building commence.

Dam-building is one of the biggest tasks of the civil engineer. Millions of cubic yards of concrete are often required to give the dam the stability it needs to withstand the pressure of the water stored behind it. The picture shows the Grande Dixence dam in Switzerland in an advanced stage of construction. This dam, at 932 feet high the highest in the world, was completed in 1961.

## Surveying

In all kinds of construction work it is very important to be able to measure distances and heights accurately. In bridge building, for example, construction generally begins at several points at the same time. Obviously, the various parts cannot be joined together if they are built at different levels or out of line! Enormous sums of money would be wasted if that occurred. By measuring, or *surveying* the land accurately, the engineer can make sure that everything goes according to plan.

The simplest method is *chain surveying*, but it is only suited to the measurement of small areas. The basis of this method is to build up a 'framework' of straight lines covering the area to be surveyed and measure their length. The positions of objects and natural features are then measured in relation to this standard framework. The lengths are measured by means of a standard-sized *chain*, which consists of a number of steel *links*, joined by rings. Every so many links along the chain there is a *tab*, or *tally* to indicate how far it is from the end. One common chain, the *engineer's chain,* has 100 links and measures exactly 100 feet. The *metre chain* is also widely used and may be either 20 or 25 metres long.

For accurate surveying over a large area the surveyor must use instruments to help him. The most important are the theodolite and the level. The *theodolite* is a precision instrument that measures horizontal and vertical angles between points on the ground. It consists of a telescope mounted on a tripod so that it can be swung round in the horizontal plane and also tilted up and down in the vertical plane. The difference in angles between one position and another can be read from horizontal and vertical scales. Before any readings can be taken, the horizontal axis must be made precisely horizontal. This is done by adjusting three or more levelling screws beneath the telescope.

The surveyor uses the theodolite to determine the positions of various points by means of a technique known as *triangulation*. He forms a framework of triangles over the region he is surveying. With an accurate steel tape he measures the length of one side of a triangle as a *baseline,* on which his subsequent calculations will be based. From each end of the baseline he sights with his theodolite the third point of the triangle. From these sightings he obtains all the angles of the triangle. By simple trigonometry, knowing the length of one side (the baseline) he can calculate the lengths of the other sides.

The *level* is an instrument for determining the relative heights of points on the ground. It is a telescope with a bubble tube attached, again mounted on a tripod. It has similar adjustments to the theodolite to ensure that the instrument is level. When it is level, it can be used to find the height of a distant point relative to its own height. The surveyor looks through the telescope at a staff held on the point. This levelling staff is about 14 feet long and has a scale marked on it. By observing the scale, the surveyor can calculate the height of the distant point very accurately.

Far left: Bridge-building tests the ingenuity of the civil engineer to the full. Suspension bridges are the rule where large gaps have to be spanned, such as river estuaries and straits.

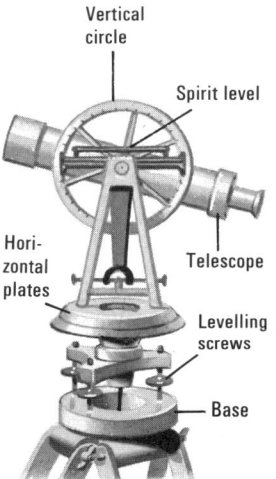

The essential elements of a surveyor's theodolite. The levelling screws are used in conjunction with the spirit level to ensure accurate levelling of the instrument before use. The horizontal plates are engraved with scales that permit reading of angles around the horizon. Angles of elevation are read from a scale on the vertical circle.

# Preparing for Construction

In any kind of construction work the ground must be prepared in one way or another. In building roads, for example, preparation will include clearing the ground, cutting through hills, filling in hollows, building embankments, and so on to give as level a surface as possible. The machines used for clearing and excavating the ground and shifting the excavated material are called *earth-moving*, or *muck-shifting* machines. They include tractors, scrapers, and excavators.

## Tractors

Tractors are used both to shift earth and to pull or push other machinery, such as the scrapers. They are driven by petrol or diesel engines. Some have wheels with huge rubber tyres. Others have *crawler*, or *caterpillar* *tracks* instead of wheels. They are called *track-laying* vehicles. The tracks consist of an endless chain of metal plates that form a surface over which the tractor can travel. Caterpillar tractors can travel easily over rough and soft ground because their weight is more evenly distributed than it is, say, in wheeled vehicles. They are steered, not by a steering wheel, but by means of levers which control the movement of each track. When the tractor driver wishes to turn, he cuts out the drive to the inner track, allowing it to freewheel. The outer track continues to move, and therefore the tractor swings round.

A caterpillar tractor with a curved steel blade set squarely in front is called a *bulldozer*. As the bulldozer moves forward, the blade bites into the ground and strips off a layer of earth. The curve of the blade causes the earth to 'roll up' in front of it. The bulldozer is widely used in all forms of construction work to shift earth and to move obstacles. Small bulldozers are called *calf-*

A general construction scene, with foundations being laid and reinforcement in position for the walls. As on any construction site, bulldozers, cranes, and rollers are much in evidence.

*dozers*. In most machines, the earth is pushed straight ahead. In some, called *angle-dozers*, the blade is set at an angle so that the earth is pushed to one side. Bulldozers are all very well for moving earth up to about 50 yards. But beyond this distance pushing becomes uneconomical, and scrapers are generally used instead.

## Scrapers

Scrapers are used particularly in road construction, where large areas of ground have to be levelled. They may be self-propelled or towed by tractors. Scrapers consist basically of a huge bowl mounted on large wheels. On the underside of the bowl there is a steel cutting edge. The steel blade slices into the ground as the scraper moves forward, and the layer of earth stripped off is forced into the bowl.

## Excavators

Excavators can remove earth from above or below their own level. They can slice into a hillside, excavate deep cuttings, or dig trenches. There are several kinds of excavators, each designed for a particular task. But they have the same basic parts. The base consists of a *chassis,* or frame, fitted with tyres or caterpillar tracks. The engine and control cab rests on a *turntable,* which allows the top part of the machine to swing round in a circle. The digging tool is attached to or suspended from a long metal arm, called a *boom,* or *jib*. The boom can be raised or lowered and the digging bucket opened or closed by means of wires or a hydraulic mechanism controlled from the cab.

Among the most common excavators are the face shovel, the drag shovel, the skimmer, the dragline, and the grab. The *face shovel* is used to excavate earth above the level on which the machine stands. A toothed bucket mounted on the boom bites into the earth and moves upwards. The *drag shovel* is similar to the previous machine, but acts downwards and backwards. It excavates below the level on which it stands. It can be used for digging trenches, too. The *skimmer* is designed to remove only a thin layer of soil on its own level. It excavates with the boom and bucket horizontal.

The *dragline* can make fairly shallow excavations over a wide area. The bucket is suspended by a long cable from the boom and is 'cast' forwards much like a fishing line is cast. It is dragged back again over the ground, scooping up earth as it does so. A *grab* is used to make deep excavations and for excavating loose ground or gravel under water. In the grab a bucket is dropped from a height and 'grabs' a pile of earth in its jaws.

To provide firm foundations in weak ground piles are sunk. Sometimes they are cast in situ in reinforced concrete. Often they are driven in with hefty pile-drivers. The picture shows a 200-foot pile being driven. Such lengths are usual if the ground is very weak.

Right: Sheet piles being driven. These piles are so shaped that they interlock. They are often used for shoring up the sides of excavations, as here.

Below: Bulldozers are invaluable in construction work, being able to move soil and rubble at an appreciable rate. They are powered by diesel engine, have hydraulic-lift systems to control dozer or bucket, and have crawler tracks so that they can travel easily over rough and waterlogged ground.

All of these excavators have single buckets. Another type has a series of buckets on an endless travelling belt which can be angled down into the ground. It is used for cutting trenches.

## Cranes

Cranes can be seen on any construction site lifting heavy loads or transferring materials from one part of the site to another. There are many different kinds of cranes, but generally they each have certain things in common. The load is lifted by a hook attached to a *pulley block*, which is a device for increasing lifting power. The pulley block can be moved up and down by winding a wire cable, or *rope* on a *winch*. In most cranes the block hangs from a steel boom. *Mobile* cranes are constructed in much the same way as the excavators mentioned earlier. They simply have different attachments.

A *tower* crane has a horizontal swing jib on top of a high tower. The jib has a counterweight at one end and can swing round in a circle but not up and down. The load hangs from a device called a *crab*, which travels along the jib. Tower cranes are a familiar sight in most cities these days since they are used extensively during the construction of tall office blocks, hotels, and so on.

## Foundations

Any large structure, such as a bridge, a dam or a skyscraper, must have firm foundations. If they are not strong or firm enough, they will eventually give way or sink into the ground. The structure will be damaged and may even collapse. Most foundations these days are made of reinforced concrete.

An engineer does not know what kind of foundations to build and where to build them until he has examined the ground. He must know what kind of soils lie beneath the surface, their strength, whether they will sink under heavy loads, whether they let through water, and so on. The study of these properties is known as *soil mechanics*.

Samples of the soil are obtained by boring holes in the ground at various points. They

In highway construction the ground needs to be rolled and compacted firm. Heavy sheepsfoot rollers are generally used for this purpose. Here two crawler-tractor drawn rollers are working together.

A self-propelled scraper with a full load being given a helping push by a bulldozer. Scrapers are able to remove large volumes of soil quickly and efficiently. They unload by tipping.

are taken to a soil laboratory to be classified and tested. From the results of these tests, the engineer can estimate how the soils will behave when built upon, and so design the foundations accordingly.

The kind of soil is very important. Some soils, for example, can be made stronger by *compaction,* that is, forcing the soil particles together. This can be done by heavy rolling, flooding, or even by blasting. Other soils can be made more solid and water-tight by the injection of certain substances, such as cement *grout,* a watery mixture of cement; hot asphalt (for large hollows), and fine clay.

It is best, of course, to build on rock. For heavy structures such as bridges, it is almost essential. Large amounts of earth may have to be excavated before the foundations can be laid. Or steel or concrete piles may be sunk down to rest on the rock.

## Piles

Piles must be sunk into firm soil if there is no rock near the surface. Then it is the friction of the pile against the soil that provides most of the support. Some piles have to be sunk a hundred feet or more to reach firm soil.

There are several different piling methods. One is to force precast concrete or steel piles directly into the ground by hammering. Another is to make a hole in the ground and then fill it with concrete. The piles thus formed are known as *cast in situ* piles. The hole may be made by boring with a large screw device called an *auger.* Or it may be made by driving a steel casing into the ground. Concrete is poured into the casing, which is then withdrawn. The method used depends on the nature of the soil present. The 'hammers' used in piling are called *pile-drivers.* Some work by steam, some by compressed air.

## Underwater Foundations

Special techniques must be used to build foundations for bridge piers, and so on, beneath the water. There are two main methods. One is to construct a temporary dam called a cofferdam around the site of the pier and build up from the river bed. Another is to sink a prefabricated structure called a *caisson* down into the bed. Cofferdams can be used only when the water is fairly shallow. Caissons must be used in deeper water.

The usual type of cofferdam consists of steel *sheet piles,* which are driven into the river bed and lock together to form a watertight barrier. Then the water enclosed by the piles is pumped out, and work on the pier can begin. The cofferdam is removed when the work is completed.

Construction of bridge foundations often takes place within a cofferdam. The cofferdam holds back the water so that workmen can excavate inside. Where cofferdams are impractical because the water is too deep, caissons have to be used instead.

### Caissons

There are three main types of caissons, the pneumatic caisson, the open caisson, and the box caisson. They are usually rectangular or circular in shape and are made of concrete, steel, or both.

The pneumatic caisson is designed so that men can work inside it on the river bed. Its lower end has a cutting edge which helps it to sink into the bed. The men work inside a working chamber, which is supplied with compressed air to keep the water out. The caisson sinks downwards as the workmen remove material from around the cutting edge. Excavation stops when the caisson reaches solid rock, which provides a firm support for the structure. The caisson is then filled with concrete and provides excellent foundations.

The workmen enter and leave the chamber through air locks to maintain the air pressure. Materials and equipment pass through similar locks. The men cannot pass directly from the high-pressure section to the outside. They must spend a certain time under decompression when the air pressure is gradually lowered to normal. Otherwise they would get an attack of what is called caisson disease, or 'the bends'. It is a kind of very severe cramp caused by the formation of gas bubbles in the blood.

The *open caisson* is open at both ends. A mechanical grab removes material from the cutting edge, and the caisson sinks down until it rests on firm bed-rock. The *box caisson* is a huge box which is floated out to the desired position and then sunk by filling with sand or water. Sand or concrete must sometimes be put down first to make the bed level.

# Bridges

Bridge building is one of the most exciting tasks of the civil engineer. Using modern materials and modern methods of construction, he can build very long, slender bridges across wide river estuaries and deep valleys. He no longer has to build roads and railways to follow the rise and fall of natural features.

There are four basic types of bridges, beam, the arch, the cantilever, and the suspension bridge. The choice of a particular type to carry a road or railway over, say, a stretch of water depends on many factors. They include the *span*, or width to be crossed, the depth of the water, the nature of the ground on each side and under the water, and whether ships have to pass underneath.

Suspension bridges are among the most graceful structures built by the civil engineer. Elegant, but extremely functional nevertheless, suspension bridges can span a gap of three-quarters of a mile with ease. The Severn bridge shown below is interesting because its deck is aerodynamically designed to counter the effects of wind loads.

Each type of bridge has certain features that make it suitable for a particular crossing. A suspension bridge, for example, is used when the water to be crossed is wide and deep. The supporting towers for this type of bridge are usually built quite near the shore, where the water is shallower. If the river is wide but shallow, a simple beam or arch bridge with a number of short spans may be more suitable.

But no matter what type of bridge is chosen, it must have firm foundations both on land and in the water. The bridge designer must calculate the load which the foundations must support. This includes the *dead* load of the bridge itself, the *live* load of the traffic it carries, and additional loads caused by wind, for example. He can estimate wind load by testing scale models of the bridge in a wind tunnel.

When the foundations have been laid, and the bridge piers have been built, work can begin on the upper part of the bridge. This *superstructure* includes the bridge deck, which carries the road or railway tracks. The exact kind of superstructure will depend on the type of bridge being built. It may consist of a network of steel girders, welded steel plates, precast and reinforced concrete, or a combination of these materials. Work proceeds from the bridge piers outwards until the structures meet in the middle. Often some temporary form of support, such as steel scaffolding, is needed to hold up the structures until they meet. This is particularly necessary in arch bridges which have no strength until they are completed.

## Beam Bridge

The beam bridge is the simplest kind of bridge of all. It consists of a long beam of timber or steel, supported at the ends by piers. But the span cannot be too great otherwise the sheer weight of the beam will make the bridge collapse in the middle. One way of increasing the strength, and therefore the span, of a beam bridge is by building a truss on it. A truss is a simple framework of steel girders.

A long beam bridge can be built by supporting the beam with several piers so that each span is not too wide. This is usually practical only in shallow water where the piers can be easily built.

## Arch Bridge

For greater spans, an arch bridge can be used. In an arch bridge, the weight of the bridge is transmitted at an angle down both sides of an arch to supporting abutments. These abutments must be carefully designed to support loads pressing on them at an angle. Arch bridges can be built of stone, brick, concrete and steel, or a combination of these materials. Steel-arch bridges can span the greatest width.

## Cantilever Bridge

Most kinds of bridges are supported evenly along their length by arches, piers, or cables. But the cantilever bridge is somewhat different. It is made up really of two identical, but separate parts. Each part consists of a beam that is supported by a pier about half way along its length. One end of the beam is anchored to the bank. The other end merely projects outwards towards an identical beam projecting from the other bank.

The parts of the beams which overhang into the center are called *cantilever arms*. In most cantilever bridges, the two cantilever arms do not meet. They have between them a *suspended span,* which is usually about the same length as the cantilever arms.

## Suspension Bridge

The suspension bridge is a spectacular structure: the road literally hangs in the air from a pair of cables carried by two towers, one on or near each bank. The suspension cables supporting the bridge are made up of tight bundles of strong steel wires. They are anchored firmly behind the high suspension towers. The road deck hangs from the cables by steel rods called *suspenders.* Suspension bridges can span much greater widths than any other kinds of bridges.

The Verrazano-Narrows bridge across the entrance to New York Harbor is the world's longest span suspension bridge, with a span of 4,260 feet. The suspension towers for this bridge are no less than 690 feet high. The length of wire used for the suspension cables was enough to circle the Earth's equator more than five times!

---

**Other Kinds of Bridges**

*Swing bridges* move about a central pier and swing through a right-angle to allow vessels to pass through to either side.

*Bascule bridges* move up and down rather like a drawbridge. One end of the bridge is hinged to the bank, while the other can be raised or lowered. A heavy counterweight attached to the bridge makes movement easier. Tower Bridge in London is a bascule bridge with a lifting arm on each side of the bank.

*Pontoon bridges* are temporary bridges that rest on flat-bottomed boats called pontoons anchored to the river bed. Permanent bridges may have to be built in a similar way if a river bed is too soft for normal bridge foundations. The pontoons are then made of reinforced concrete.

*Bailey bridges* are temporary bridges, designed originally for army use. They consist of a number of steel trusses and panels that are connected to one another by steel pins at the corners.

Above: A slender stone-arch bridge in the Swiss Alps. It looks remarkably fragile, but its arch shape gives it the requisite strength.
Left: The Europa bridge in Austria is a fine example of a modern beam bridge. Tall concrete pillars raise the roadway hundreds of feet above the valley floor.
Below: The world's most famous bascule bridge, Tower Bridge in London.

# Dams

Large quantities of water are required in the home for drinking and washing, in agriculture for irrigation, or watering crops, and in industry for cooling and manufacturing. To meet these needs, water must be stored. This is especially important in countries when rain falls only at certain times of the year. The main way of storing water is to build a barrier, or dam, across a river to form an artificial lake, or reservoir.

The stored water behind the dams can also be used to produce electricity. Water taken from the bottom of the dam passes through pipes and spins huge water turbines, which generate electricity. Power produced in this way is called *hydroelectric power*.

The first thing a dam builder must do is to decide where to build the dam. He will probably investigate several possible sites along the course of a river before deciding on one and choosing a suitable type of dam. The choice of site will depend on the purpose of the dam: whether it is for water supply or for electricity generation. For water supply wide, flat river valleys are favored to give a large capacity, if this is possible without building over-long stretches of embankment. The depth, or *head* of water stored is not really important. Local materials such as clay available on these sites are suitable for building earthen dams and embankments. For hydroelectric schemes, however, the needs are different. What is required is a large head of water to drive the turbines. Narrow, rocky gorges are therefore preferred. These sites are also suited to the construction of high concrete or masonry dams.

Before dam building can begin, the river must be diverted away from the site of the dam. This is often done by building diversion tunnels to take the water around the site. The river is blocked by dumping earth and rock into it or by building a cofferdam. The foundations of the dam can then be built in the dry. Then, when the dam has reached a certain height, it can be used to store water. The river is unblocked, and the diversion tunnels are closed. The dam is then completed as the water level continues to rise.

A classic arch dam with the curve toward the water. Arch dams derive their strength from their shape not their mass, though for safety they are often thickened at the base. Then they are called a gravity-arch dam.

## Earthen Dams

All the world's biggest dams are *earthen*, or *embankment* dams. But they may contain rock, sand, and gravel as well as earth. Those containing shattered rock fragments are called *rock-fill* dams. They often have an earthen core. The Fort Peck Dam across the Missouri River in Montana, USA, is the biggest earthen dam in the world. It contains some 125 million cubic yards of earth and rock.

A typical earthen dam has a very wide base and gradually slopes inwards and upwards on both sides to become fairly narrow at the top. It is made watertight in one of two ways. A thick core of clay or concrete may be built into it. Or the side of the dam facing the water may be lined with sheet steel, a layer of concrete, or stone. The clay used in the core is called *puddle clay*. The core is built up from what is called a 'cut-off' trench across the valley under the dam. The trench, which is subsequently filled with clay or concrete, is necessary to prevent water seeping beneath the dam and flowing downstream.

The material for the dam, earth, rock, and so on, is tipped into position by trucks, or moved in by scrapers and bulldozers. These heavy vehicles travel slowly along the top of the rising dam and help to press down the material already placed.

Some dams are built up from silt or a mixture of sand, clay and gravel in water, which is pumped into position through pipelines. The pipelines are arranged so that the clay settles near the centre of the dam to form a watertight core. These dams are called *hydraulic-fill* dams.

## Concrete Dams

Concrete or masonry dams are suitable when rock lies reasonably close to the surface and excavation is comparatively easy. There are three main types: gravity, arch, and buttress. The *gravity dam* depends for its strength on weight alone. Every part of the dam is sufficient to withstand the water pressure upon it. The cross-section of the dam is approximately that of a right-angled triangle with the upright face towards the water.

*Arch dams*, however, are much thinner in section than gravity dams. They are built against the sides of a valley and curved in the shape of an arch towards the water. It is the shape that gives the arch its strength. The pressure of the water travels down the arch to the sides of the valley. Sometimes, the arch is thickened at its base so that it becomes partly a gravity dam, too. This is then known as a *gravity-arch dam*.

*Buttress dams* are a form of gravity dam which consist of a comparatively thin facing supported from behind by a series of buttresses. There is a very great saving in weight over the gravity dam.

Millions of tons of cement, rock, and gravel are needed in large concrete dams. The rock and gravel are usually quarried locally to avoid heavy transport costs. The cement may also be made in a special plant set up near the construction site if the necessary raw materials are available.

## Hydroelectric Schemes

In dams designed to produce hydroelectric power, huge pipes called *penstocks* carry water from the bottom of the dam to the *powerhouse,* which contains the generator turbines. Sometimes the penstocks are tunnels driven through rock. Even the powerhouse itself may be blasted out of rock. The powerhouse of the Kitimat hydroelectric scheme in British Columbia (Canada), for example, is a vast cavern some 1,100 feet long, 120 feet deep, and 80 feet wide, hewn out of solid rock. Dams are built with a by-pass channel called a *spillway*. When the lake behind the dam gets too full, the water overflows down the spillway instead of over the dam.

Above: An ideal site for an arch dam, at the end of a narrow, rocky valley. The sides of the valley provide firm support for the arch. Note the pylon at the top left carrying power lines from the generating station linked with the dam.

Left: An unusual view of the Grand Coulee dam on the Columbia River, in the American state of Washington. The biggest concrete structure in the world, the Grand Coulee contains over 10 million cubic yards of concrete and has a weight of about 20 million tons.

# Roads

We have come to rely heavily on motor transport in our everyday lives. There are already well over 100 million automobiles in the world, and the number is increasing rapidly. As a result, many new roads must be built and existing roads reconstructed to carry the increasing volumes of traffic. Roads designed for safe, high-speed motoring are being built in most countries. They are known variously in different parts of the world as motorway (Britain), autoroute (France), autobahn (Germany), autostrada (Italy), and so on. It was in Italy, in fact, that the first motorway was built in 1925.

Motorways are designed so that traffic going in opposite directions is clearly separated. They are made with gentle curves and easy gradients so that traffic does not have to slow down. There are no junctions or traffic-lights on a motorway. Other roads pass over or underneath it. Traffic from a crossing road enters by means of a *flyover* junction.

The first stage in road building is to survey the countryside to find the best possible route to take, which will avoid as many obstacles as possible. In this way, expensive operations such as bridge-building and tunnelling can be kept to a minimum. When the route has been chosen, bulldozers, excavators, and scrapers move in to prepare the road bed.

The majority of roads these days are surfaced with asphalt, or macadam. It is spread hot by machine, and then rolled while still hot by a heavy roller.

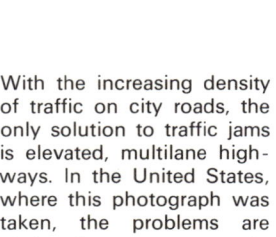

With the increasing density of traffic on city roads, the only solution to traffic jams is elevated, multilane highways. In the United States, where this photograph was taken, the problems are acute, since one person in three owns an automobile.

The properties of the soil beneath the road, the *subgrade,* are very important. If the soil is weak, materials such as sand and gravel must be mixed with it to strengthen it. The subgrade can also be improved by heavy rolling or ramming. On top of the subgrade goes a layer of crushed stone, which forms the base for the top-surface or *pavement.*

There are two basic kinds of pavement, rigid and flexible. A *rigid* pavement is made of concrete and will not give even if the soil beneath it sinks a little. The concrete is laid on the base between parallel metal forms. Spreading and finishing machines, which travel on rails alongside the forms, smooth and compact the concrete.

A *flexible* pavement consists of small broken stones, usually held together by tar. The mixture of stones and tar is called *tarmacadam,* or *tarmac*. It is spread hot by finishing machines that squeeze it evenly over the road and compact it. Heavy rollers then compact it further into a hard-wearing surface. *Asphalt* is somewhat similar.

# Skyscrapers

Tall skyscrapers and large blocks of flats and offices are being built in towns throughout the world. It is on these building sites that most of us see civil engineering in action. There we see the great machines at work – bulldozers, excavators, cranes, pile drivers and so on, and observe the different stages of construction.

The first stages, as for any other form of construction, are to survey the site and to test the soil. If there is firm soil near the surface, bulldozers and excavators remove the upper layers of earth. Concrete foundations can then be laid on a firm base. But sometimes firm soil is a long way down, and excavating would not be practical. Then piles are driven down from the surface to firm soil or rock to support the foundations. This is especially important with really tall buildings, which we appropriately call *skyscrapers*. The firmness of the soil will determine the maximum height to which a building can be constructed in safety. Where there is solid rock close to the surface, as on Manhattan Island, New York, there is virtually no limit to the height of building, at least as far as the strength of foundations is concerned. That is why Manhattan has some of the tallest buildings in the world.

When the foundations have been laid and have set, work can begin on the upper part, or *superstructure* of the building. The first job is to build the *frame* that will support it. The frame may consist of precast reinforced concrete or steel beams, which are lifted into position by cranes and bolted together. Or it may consist of reinforced concrete cast in position in wooden moulds, or *shuttering*. Because of this frame, the walls take no load at all, unlike a conventional, two- or three-storey house in which the walls support the whole building. The walls may therefore be made thin, light materials such as glass or aluminium sheet. They are known as *curtain* walling.

The walls and floors of the building may also be cast in position. But a faster method is sometimes used. The floors and walls are precast in concrete in a factory and merely assembled on the building site.

More modern and quicker methods of construction include the lift-slab and jackblock methods. In the *lift-slab* method, all the floors are cast at ground level and lifted into position on supporting columns. In the *jackblock method* the roof and each floor are cast at ground level and gradually jacked up into place.

Above: High-rise, skyscraper apartments and offices dominate many of the world's big cities. When city land is limited and expensive the obvious solution is to build upwards. From an engineering point of view the height of the skyscraper is limited by the strength of the soil. Piles are invariably used for foundations for high buildings if rock is not present close to the surface of the ground.

The United States leads the world in skyscraper construction. Chicago (above) occupies a special place in skyscraper history, for it was there that the first metal-frame skyscraper was built, 10 stories high, in 1885. In 1974 Chicago boasted the world's tallest building, the Sears Building, 1,550 feet high.

Right: More modest skyscrapers are built in San Francisco, but not just for aesthetic reasons. There is an ever-present danger of earth tremors.

# Tunnels

It is often necessary when building roads or railways to drive tunnels through hillsides and mountains in order to avoid a steep winding route. The world's longest road and railway tunnels run through the Alps in Europe. Beneath the streets of London, Paris, Moscow, New York and many other cities underground railways travel in a network of tunnels many miles long. In dam-building, too, tunnels are often required to divert rivers while the dam is being built or to convey water to the powerhouse. But the longest and most numerous tunnels of all are the aqueducts, or water-supply channels which carry water to our cities.

The simplest method of building tunnels close to the surface is known as *cut-and-cover*. This has been used extensively in the construction of sub-surface underground railways. A wide trench is dug down to the level desired. Then walls are built on each side, and the excavation is covered over with a roof. The top soil is replaced to restore the surface to normal.

Before any deep tunnelling can begin, engineers must sink holes or make other tests to determine the nature of the ground and to find the best site for the tunnel. The method of tunnelling chosen will depend on whether the ground is hard, soft, or likely to let water through. Most tunnels must be lined with steel or concrete to prevent them from collapsing.

The Mont Blanc road tunnel extends for more than 7 miles through the Alps between France and Italy. The picture above shows the drilling platform, called the jumbo, which was used to bore a series of shot holes into the rock face simultaneously.
Right: The mouth of the finished tunnel, which has been given a concrete lining. Note the excellent roof lighting.

This was the circular digger used to bore through London clay during the construction of London Transport's new Victoria Line Underground system. Note in the picture the steel ribs that are put in place as excavation proceeds to prevent collapse of the tunnel.

In hard rock, tunnels are driven by blasting with explosives. Powerful rock drills driven by compressed air bore holes into the rock face. Sometimes they are mounted together on a frame called a *jumbo* that runs on rails. Explosives placed in the holes blast away the rock. The shattered rock is loaded into trucks and removed. Then the process is repeated until the tunnel is completed. In soft rock, such as sandstone and chalk, it may not be safe to use explosives. Then the rock must be broken up by pneumatic picks.

In soft ground such as clay, a *tunnelling shield* is used to prevent the tunnel from collapsing during construction. The shield is like an open-ended tin, with a cutting edge at the open end. Powerful jacks force the shield forwards into the soil. Workmen then dig out the earth from inside the shield, and the process is repeated. Tunnel linings of concrete, steel, or cast iron are put into place behind the shield as it moves forward. In some types of shield the earth is removed by rotary diggers.

If there is danger of water seeping through during tunnelling, compressed air is fed to the shield to keep the water out. The workmen and materials enter and leave the shield through an air lock to maintain the pressure.

Another method of building tunnels is to sink prefabricated sections into a deep trench and join them together. This method may be used, for example, for crossing a river estuary where the bed is too soft for normal tunnelling methods.

# Canals and Harbors

Rivers and seas have long provided us with a vital means of communication. But rivers do not always go where we want to go. Therefore we build artificial waterways, or *canals* to take us. And where transport by sea is concerned, we find that there are not enough natural harbors around the coast to provide protection when we are loading or unloading our ships. Therefore we build artificial harbors.

## Canals

As for other construction work, building canals raises its own special problems, although in many respects it is similar to building roads. There is a lot of earthmoving to be done; cuttings must be excavated; and embankments must be raised so that the route will be level. If existing lakes or rivers are included in the canal route, they may have to be widened or deepened by dredging. Or a dam may have to be built to deepen them.

The problem with canals is that, unlike roads, they must always be built on the level. They must go up and down slopes in a series of horizontal steps. A vessel passes from one level to another through a *lock,* which is a chamber with gates at each end. Going up, water flows into the lock from the higher level and raises the vessel to that level. Going down, water flows out of the lock and drops the vessel to the lower level.

## Harbors

Artificial harbours are formed by building a protective wall called a *breakwater* out into the sea so that it encloses an area of water. A gap in the breakwater serves as the harbor entrance. There are two main kinds of breakwater. A *blockwork* breakwater is made of massive stone or concrete blocks that weigh up to 50 tons. The blocks are lifted into position by massive, powerful cranes. A *rubble-mound* breakwater consists of lumps of rock, sometimes topped with concrete slabs. The rocks are dumped into the sea from barges or railway trucks, which run along the breakwater as it is built.

The breakwater and harbor works need a great deal of maintenance because of the destructive force of the sea. Dredging must also be done regularly to prevent the harbor channels from becoming blocked by sand and mud shifted by the sea or by silt deposited by nearby rivers. Harbor engineers try to allow for the effects of silting, wave action, and so on, when they design the harbor. They estimate these effects by making tests on scale models of the harbor, simulating as closely as they can prevailing local conditions.

Some of today's giant oil tankers have to be unloaded offshore, because they are too big to use existing harbors. A structure called a *jetty* is built in deep water to provide berthing space for them. It consists of a reinforced concrete or steel deck built on piles, which extend down to rock beneath the sea bed.

An oil tanker discharging its cargo at an offshore jetty. These huge ships draw too much water to be able to use ordinary harbor installations.

Canals of course cannot go up and down hills. Where a change of level occurs, locks are installed for raising or lowering canal craft. At international ship canals such as Panama the locks may be 1000 feet or more long.

# Facts and Feats of Construction

## Suez Canal

The Suez Canal links the Mediterranean and the Red seas across the Isthmus of Suez. It runs for about 100 miles between Suez in the south to Port Said in the North. It reduces the distance between Europe and the Far East by several thousand miles.

The Canal does not have locks, because there is almost no difference in levels between the Red Sea and the Mediterranean. It is about 35 feet deep, but can take only one-way traffic for most of its length. It has to be dredged continually to prevent the desert sand from blocking it. Ships, which travel in convoy through the Canal, take about 15 hours to make the passage.

Ferdinand de Lesseps, a French diplomat and engineer, took 10 years to build the Canal, which was opened in 1869. Until July 1956 the Canal was under the control of the Suez Canal Company, in which Britain and France had significant holdings. Then President Nasser of Egypt seized control, provoking armed intervention by France, Britain and Israel in the following October. During the fighting, the Canal was put out of action for several months by sunken vessels. It was closed again as a result of the six-day Arab-Israel war in June 1967, leaving several ships trapped in the Bitter Lakes.

## Panama Canal

De Lesseps also planned the route for the Panama Canal, which links the Pacific and Atlantic oceans across the Isthmus of Panama in Central America. The Canal runs for about 50 miles from Colon on the Atlantic to Panama on the Pacific. It saves ships sailing between the two oceans a journey of about 8,000 miles round Cape Horn at the tip of South America.

Work started on the Canal in 1881, but in 1889 De Lessep's company went bankrupt after completing a quarter of the project. An American company eventually took over, but it was 1914 before the Canal was opened to traffic. In all, 55,000 workmen died before the completion of the Canal, mainly from tropical diseases.

Construction of the Canal involved damming a river to make the 32-mile-long Gatun Lake; excavating the 9-mile-long cutting, Gaillard, or Culebra Cut; and building three sets of locks to take the ships from the sea up to lake level and down again. One set of locks — the Miraflores Locks at the Pacific end of the Canal — is the largest in the world. Each lock has a length of more than 1,000 feet.

## St. Lawrence Seaway

This seaway between Canada and the United States is a modern feat of canal engineering. It stretches about 180 miles from Montreal to the mouth of Lake Ontario. It opened up the Great Lakes — the largest inland waterway in the world — to ocean-going vessels. Ships can now sail 2,500 miles inland from the Atlantic Ocean.

Construction of the canal involved cutting many new channels and widening others and building a number of locks and dams. Some of the dams were built both to control the water level and to produce electricity. The United States and Canada began work on the project in 1954, and it was officially opened in 1959.

## Eiffel Tower

The Eiffel Tower dominates the skyline of Paris. It was designed by the French engineer Gustave Eiffel for the Universal Exhibition of Paris in 1889. Its huge, wrought-iron skeleton rises 986 feet into the air from a base 330 feet square. Lifts and spiral staircases carry visitors to an observation platform at the top of the Tower. There is also a restaurant and post office at this level. The Tower is now used to transmit television programmes. With its television mast, added in 1957, it has a height of 1,052 feet.

---

One of the most famous, and one of the most beautiful bridges in the world, the Golden Gate, in San Francisco Bay.

## Empire State Building

The world's most famous skyscraper is the Empire State Building in New York City. It reaches a height of 1,250 feet, with 102 storeys. The building was completed in 1931. Twenty years later, a television mast was added to give it a total height of 1,472 feet. It was the world's tallest building until 1970, when the World Trade Center topped it by 100 feet.

## Snowy Mountains Scheme

One of the largest power and irrigation projects in the world is the Snowy Mountains Scheme in south-eastern Australia. Three rivers rise in the Snowy Mountains — the Murray, the Murrumbidgee, and the Snowy. The main object of the scheme is to divert the waters of the Snowy to the Murray and the Murrumbidgee. The Snowy flows down the southern slopes of the mountains, where there is adequate rainfall. The Murray and Murrumbidgee flow north and west where the rainfall is low. Diverting the Snowy into the drier regions will greatly increase crop production in the area.

The scheme includes 200 miles of tunnels and aqueducts to carry the water through the mountains; a large number of dams; and nine hydro-electric power stations.

## Mont Blanc Tunnel

The Mont Blanc Tunnel under the Alps is the largest road tunnel in the world. It carries the road for just over seven miles thousands of feet under Mont Blanc, highest mountain in the Alps. It is open all the year round, unlike the other alpine roads, which are blocked by snow for half the year.

Blasting the tunnel through solid rock took from 1959 until 1962. Teams of French and Italian tunnellers started work from opposite ends of the tunnel and met near the middle with perfect accuracy. This was a considerable feat after tunnelling for 7 miles through a mountain.

The tunnellers were in constant danger from rock bursts and from water trapped in the rock. Often they had to work up to their knees in thick mud. They were not safe even outside the tunnel. Several of them were killed by avalanches cascading down the mountain sides.

## The Greatest Bridges

| LONGEST OVERALL | Feet |
|---|---|
| Pontchartrain Causeway (really a viaduct) in Louisiana, USA | 126,000 |
| Maracaibo Lake, in Venezuela. | 25,400 |
| Chesapeake Bay, on the eastern coast of the United States. | 22,970 |
| Mackinac Straits, between Mackinaw City and St. Ignace, Michigan, USA. | 19,200 |
| Lower Zambesi, at Sena in Mozambique, Africa. | 12,060 |
| Tay, over the Firth of Tay, near Dundee on the eastern coast of Scotland. | 11,650 |

| LONGEST SPANS | |
| *Suspension Bridges* | Feet |
| Verrazano Narrows, from Staten Island to Brooklyn across New York City harbour, USA. Its length between supports is 6,690 feet. | 4,260 |
| Golden Gate, across the entrance to San Francisco Bay, USA. Its length between supports is 6,450 feet. | 4,200 |
| Mackinac Straits, Michigan, USA, is the longest suspension bridge with a distance between supports of 8,340 feet. | 3,800 |
| George Washington Bridge, New York, the strongest suspension bridge. | 3,500 |
| Salazar, across the River Tagus, or Tejo at Lisbon, Portugal. It has an overall length of about 10,000 feet. | 3,323 |
| Forth Road, across the Firth of Forth, in eastern Scotland. Its overall length is 5,980 feet. | 3,300 |
| Severn Road, across the Severn estuary, in south-western England. | 3,250 |

| *Cantilever Bridges* | Feet |
| Quebec, Le Pont de Québec, across the St. Lawrence River in Quebec, Canada. Its overall length is 3,240 feet. | 1,800 |
| Forth Rail, across the Firth of Forth, in eastern Scotland. It has two main spans of equal length. | 1,710 |
| New Orleans cantilever bridge | 1,575 |
| Howrah, across the River Hooghly, Calcutta | 1,500 |

| *Steel-Arch Bridges* | Feet |
| Bayonne, across the Kill van Kull, connects Bayonne, New Jersey, to Staten Island, New York. | 1,652 |
| Sydney Harbour, in New South Wales, Australia. Its span is 25 inches shorter than the Bayonne Bridge, but its total length (3,770 ft.) is greater. | 1,650 |
| Thatcher Ferry across the Panama Canal in central America. | 1,120 |
| Widnes-Runcorn, from Widnes, Lancashire, to Runcorn, Cheshire. Overall length is 3,490 feet. | 1,082 |

| *The Longest Tunnels* | Miles |
| Seikan Tunnel, beneath Tsugaru Strait, Japan (completion 1973) | 38½ |
| East Finchley to Morden, via Bank (London Transport Board Underground railway system). | 17¼ |
| Golders Green to South Wimbledon (Underground). | 16 |
| Simplon II, between Switzerland and Italy, | 12½ |
| Simplon I, earlier than II and 22 yards shorter. | 12½ |
| Apennine, in the Apennines, in northern Italy. | 11 |
| Rokko Tunnel (rail), between Osaka and Kobe, Japan | 10 |
| St. Gotthard, in the Alps, in southern Switzerland. | 9¼ |
| Lotschberg, in the Bernese Alps, in central Switzerland. | 9 |
| Mont Cenis, also known as the Fréjus, between southern France and Switzerland. It was the first tunnel to be cut through the Alps (1871). | 8½ |
| Cascade, in Cascade Mountains, in the State of Washington, United State's longest. | 7¾ |
| Mont Blanc, between France and Italy, longest road tunnel. | 7 |

The longest tunnel of any kind is the 85-mile-long water-supply tunnel which supplies Hillview Reservoir in New York City, USA.

## The Tallest Structures

| | Feet |
|---|---|
| Ostankino television tower, in Greater Moscow, USSR, is the tallest self-supporting structure in the world. | 1,762 |
| World Trade Center, New York, topped out in 1970. | 1,353 |
| Empire State Building, New York. | 1,250 |
| John Hancock Center, Chicago, topped out in 1968. | 1,105 |
| Tokyo Television Tower, Japan (self-supporting). | 1,092 |
| Chrysler Building, New York, completed in 1931. | 1,046 |
| Eiffel Tower, Paris. | 986 |
| Wall Tower, New York. | 950 |
| Bank of Manhattan, New York. | 927 |
| Rockefeller Centre, New York. | 850 |

| *United Kingdom* | Feet |
| Post Office Radio Tower, in London. It has a concrete structure 579 feet tall, extended to full height by the radio antenna. | 619 |
| The National Westminster tower block, Bishopsgate, London. | 600 |
| Blackpool Tower, in Lancashire. | 518 |
| Euston Tower, London. | 407 |
| Salisbury Cathedral, in Wiltshire. | 404 |
| Co-operative Insurance Society Building, in Manchester. | 400 |

The Post Office Tower, now a prominent London landmark. Measuring 619 feet to the tip of the radio antenna it carries, it is Britain's tallest building.

Dominating the skyline of Paris is the 986 foot high Eiffel Tower, completed in 1889. It is located on the Left Bank of the Seine on the Champs de Mars.

# Cement and Concrete

Cement is one of the most important raw materials used in construction work because it is the essential ingredient in concrete. Millions upon millions of tons of concrete are used every year in building roads, bridges, dams, and so on. It is cheap, easy to make, and is waterproof and fireproof. It can also be used under water. When concrete is first mixed, it forms a pasty mass that can be cast into any shape. But it quickly sets rock-hard.

Concrete is made by adding water to a mixture of cement, sand, and gravel or stone. The sand is called *fine aggregate* and the stones are called *coarse aggregate*. The sand fills in the gaps between the larger stones so that there will be no air trapped between them. The presence of air would weaken the concrete.

The most common type of cement is *Portland* cement, which is a fine, grey powder made by roasting a crushed mixture of limestone, clay, and other materials. Large lumps of the chalk or limestone rock from a quarry are broken down into smaller lumps by powerful mechanical crushers. Then they are mixed with the other materials and water to form a *slurry*. The slurry enters a ball mill where tumbling steel balls grind it finely. From there it passes into a long, cylindrical kiln that slopes slightly downwards and rotates so that the materials gradually move along it. In this kiln the materials are first dried and then heated white-hot to form a hard, coke-like mass, called *clinker*. After cooling, this clinker is finely crushed into the familiar fine powdery form of cement. A certain amount of gypsum is added to control the rate at which the cement will eventually set.

Reinforced concrete is used on a massive scale in the construction industry. The reinforcing bars can be clearly seen here. The photograph is interesting however because the concrete members, which form part of a bridge, are being glued together! An epoxy resin adhesive is being used that will be at least as strong as the concrete it is bonding.

Reinforced concrete is also a common material for road surfaces. The surface is laid by what is termed a concrete "train". A series of spreading machines and vibrators follow one another along the path of the road, travelling on side rails. Reinforcement is generally put down between successive layers of wet concrete.

Concrete is usually mixed on the site where it is to be used in machines called concrete mixers. They have a rotating drum which mixes the ingredients thoroughly together. The wet concrete may be taken to where it is needed in huge buckets, or *skips* carried by cranes, or it may be pumped into place through a pipeline. Sometimes concrete is supplied ready-mixed in huge mixer trucks, which pour it directly into place.

Concrete may be cast in its final position in moulds made of wood. Or it may be *precast* into shape and then taken to the building site to be assembled.

### Reinforced Concrete

Ordinary concrete is strong when it is being squeezed, or under compression. But it is weak when it is being stretched, or under tension. If, say, a concrete beam is supported only at its ends, this weakness may cause it to crack and collapse. To prevent this weakness steel wires or rods are cast into the concrete. It is then known as *reinforced* concrete. The strength can be further increased by stretching the wires or rods either before or after they are placed in the concrete.

In *pretensioning*, the concrete is cast around rods that are being stretched. When the concrete sets, the rods are released and compress the concrete, making it stronger. In *post-tensioning,* the concrete beam is cast with holes along its length. Steel wires are inserted in the holes, pulled out, and then anchored so that the beam is under compression. This again strengthens the beam.

# Water Supply and Treatment

With the rapidly expanding population and the increasing demands of industry, the provision of a plentiful supply of clean, fresh water has become quite a problem. And, associated with it, is the problem of disposing of equivalent quantities of *sewage* – the dirty water containing animal waste, decaying vegetable matter, soap and detergents, and all kinds of rubbish.

Not many towns have enough water locally to supply the day-to-day needs of the population and industry. The water must be carried, often over great distances, from a river or lake. *Reservoirs* are constructed to ensure a constant supply of water all the year round. They are made by building a dam across a river at a suitable point. The water is carried from its source in open channels or pipes called *aqueducts*.

Water straight from the river is dirty and generally contains harmful bacteria and other organisms which can cause disease and give rise to dangerous epidemics. Typhoid, para-typhoid, cholera, and dysentry are among the commonest water-borne diseases. Before the water arrives at the taps in our homes, it therefore has to undergo purification in water-treatment plants.

The first stage of purification takes place in storage reservoirs into which the dirty river water is pumped. Much of the dirt settles out, and many harmful organisms die. But the water is still generally cloudy when it is pumped from the reservoir to the water-treatment plant. At the plant it is mixed with certain substances which bring down the tiny dirt particles causing the cloudiness as a sediment. This is allowed to settle in *sedimentation tanks*. The process of *filtration* which follows clarifies the water further and also removes many polluting organisms. Filtering is done through beds of sand or other similar media. Finally the water is made completely sterile by *chlorination*, the adding of chlorine or compounds containing chlorine (bleaching powder and sodium hypochlorite). After being tested for purity, the water is pumped into the public water supply. Some local authorities add fluoride to the water to help reduce tooth decay in children. But the practice has not been whole-heartedly applauded, despite convincing evidence of its effectiveness.

## Sewage Disposal

Clean water goes into a town in the water mains. Dirty water leaves it through the sewers. To prevent serious pollution of our rivers and lakes and reduce the danger of disease, sewage is generally treated at *sewage farms* to make it harmless.

First of all it is passed through screens which remove much of the solid matter in it. The screened sewage then passes through a *detritus tank* or *channel*, in which heavy bits of grit and gravel settle out. Afterwards it flows into the *sedimentation tank*, where the fine solids suspended in it settle out as a liquid *sludge*. The clarified water then goes through processes to oxidise any organic matter remaining in it. The commonest way of doing this is by spraying the water onto porous filter beds through revolving sprinklers. Organisms within the bed oxidise the decaying organic matter into harmless products. And the water is then pure enough to be pumped back into the river.

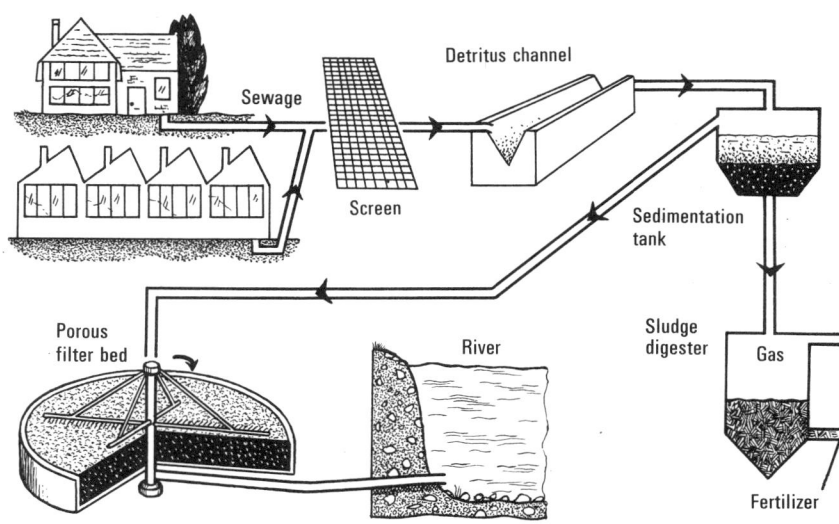

The main elements of the sewage disposal process. Waste from house and factory travels via the sewers to the sewage treatment plant. First it is screened of large matter. Then it passes along a detritus channel, where grit settles out, and into a sedimentation tank. There fine particles settle as a sludge that can be processed to give fuel gas and fertilizer. The clear water is sprayed over filter beds and any remaining organic matter in it is removed.

Beneath our cities are huge tunnels like this which carry away rainwater from the roads and sewage from the houses to sewage plants or into rivers or the sea.

# Chemical Combination

Chemistry is the science that is concerned with the composition and properties of substances and the changes they undergo. Substances with fixed and definite properties can be termed chemical substances. Specimens of gold and salt, for example, always have the same properties, no matter where they are obtained from. Substances like wood and brass (an alloy), however, do not. Their properties—hardness, colour, and so on—vary from specimen to specimen.

Some chemical substances, such as gold and iron, can be subjected to all kinds of treatment—heat, cold, electricity, acids, and so on—and either remain unchanged or else form new chemical substances of which they form only part. Substances which cannot be broken down into simpler substances are called *elements*. Other substances break down under such treatment to yield two or more elements. Salt, for example, when it is melted and an electric current is passed through it, will split up into the elements sodium and chlorine.

Certain basic laws describe chemical combinations. A chemical compound always contains the same elements in the same fixed proportions by weight (*law of constant composition*). When two elements combine to form two or more compounds, the weights of one that combine with fixed weights of the other are in simple ratio (*law of multiple proportions*). These are two of the basic laws. John Dalton (1766–1844) put forward his celebrated atomic theory of matter in 1803 in an attempt to explain laws such as these. Chemical elements, he said, are made up of basic particles or *atoms*, which cannot be further subdivided in the course of a chemical change. Chemical compounds are formed by the combination of atoms of different elements in simple proportions.

Joseph Gay-Lussac (1778–1850) found that the volumes of gases taking part in a chemical change bear a simple relation to one another, if measured at the same temperature and pressure (*law of gaseous volumes*). For example, one volume of hydrogen will combine with one volume of chlorine to give two volumes of hydrogen chloride. Dalton and others realised that there must be some simple relation between the numbers of atoms present in equal volumes of a gaseous substance. He thought that perhaps equal volumes contained the same number of atoms. But this did not work: 1 atom of hydrogen + 1 atom of chlorine cannot give 2 compound atoms of hydrogen chloride because the atoms of hydrogen and chlorine are not divisible.

It was left to Amadeo Avogadro (1776–1856) to provide the answer. He supposed that the basic physical unit of a gas was not a single atom, but a cluster of atoms, which he called a *molecule*. Avogadro put forward his

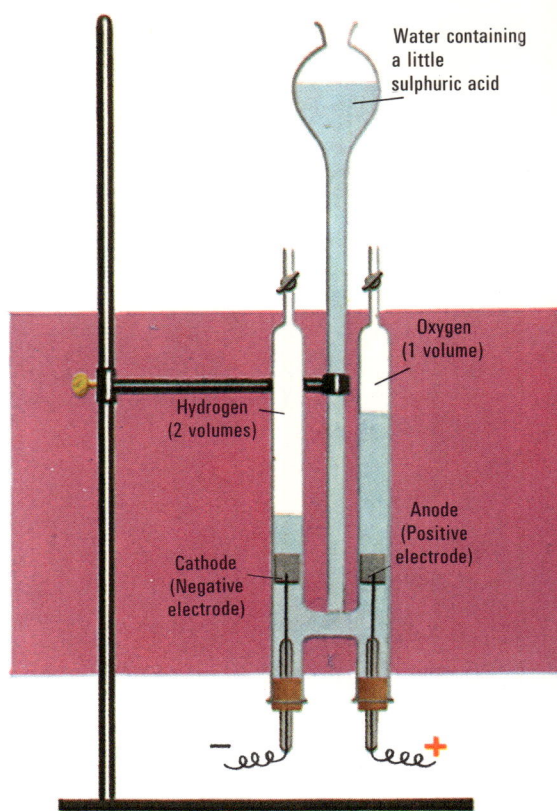

Apparatus for the electrolysis of water. Water is split up by an electric current into its component hydrogen and oxygen. In the reaction two molecules of water are split to give two molecules of hydrogen and one molecule of oxygen. This accounts for the fact that the volume of hydrogen produced during electrolysis is twice that of oxygen. (Equal volumes of all gases at the same temperature and pressure contain the same number of molecules—Avogadro's hypothesis).

A great many materials will under suitable conditions form beautiful crystals of regular shape. The picture shows a fine specimen of calcite (calcium carbonate) crystals. The regular external appearance of a crystal hints at the regularity of the crystal lattice at the atomic level.

famous hypothesis that equal volumes of all gases under the same temperature and pressure contain the same number of molecules. Just as we talk of an atomic weight (see page 26), so we talk of a *molecular weight* which is the weight of a molecule of a substance compared with that of an atom of carbon, which is taken as 12.

## Chemical Change

When you heat chalk (calcium carbonate) fiercely, you drive off carbon dioxide and leave lime (calcium oxide). When you add dilute sulphuric acid to zinc, hydrogen is given off and zinc sulphate is formed. When you add silver nitrate to sodium chloride, silver chloride is precipitated as a white solid, leaving sodium nitrate in solution. These are common examples of chemical changes. They are changes in which the atoms in the reacting molecules have re-grouped to form different molecules.

The above chemical reactions take place quite readily. Heat is often required before many reactions will proceed at a reasonable pace. Sometimes pressure is necessary, too. And some reactions will not proceed unless there is a catalyst present. This is a substance that promotes the chemical reaction, but does not take part in it. Heat, pressure, and a catalyst are required to make ammonia from hydrogen and nitrogen, for instance (see page 169).

### Valency

The number of atoms of each element present in a molecule of a compound is determined by the valency of the elements. Valency is, broadly speaking, the property of an atom to enter into chemical combination, that is, to form chemical bonds, with other atoms. The valency of an element can be defined as the number of hydrogen atoms which an atom of the element will combine with or displace. Carbon will combine with four hydrogen atoms (to make methane $CH_4$) and therefore has a valency of 4. Oxygen combines with two hydrogen atoms (to make water $H_2O$) and has a valency of 2. An atom combines with other atoms to form molecules by means of some of the electrons surrounding the nucleus of the atom (see page 26). It is the electrons in the outermost 'shell' which are available for bonding atom to atom. There are three kinds of valencies, that is, basic ways by which atoms form chemical bonds: electrovalency, covalency, and co-ordinate valency. Electrovalency occurs in salts, for example, such as common salt, sodium chloride. The sodium atom loses an electron to become a positive ion (*cation*), indicated as $Na^+$. The chlorine atom gains an electron to become a negative ion (*anion*), indicated as $Cl^-$. In their compounds the atoms are held together by electrical forces. This kind of bond is called an *electrovalent* bond. Most metals and hydrogen form positive ions. Most non-metals form negative ions.

In covalency atoms share a pair of electrons, one coming from each atom. They do not form ions. Carbon, for example, forms covalent bonds with four hydrogen atoms in methane. Co-ordinate, or *dative* valency is a special kind of covalency in which only one of the atoms provides the pair of electrons for sharing.

Not only atoms, but groups of atoms within the molecule have their own valencies. Such groups are called *radicals*. The sulphate radical, which forms a negative ion, depicted as $SO_4^{--}$, is a common one.

The most convenient way of representing chemical changes is by means of symbols and formulae in what are called chemical *equations*. Every element has a certain symbol that identifies it—oxygen is O, hydrogen H, zinc Zn, and so on. Some symbols are straightforward like those quoted, but others are derived from the Latin names for the elements. For example, Cu is copper (for *cuprum*); Pb is lead (after *plumbum*), and so on. (For a complete list of the elements and their symbols see page 27.)

The formula of a compound represents the number and kind of atoms in its molecules. Thus a molecule of oxygen is made up of two atoms of oxygen and is represented as $O_2$. Water contains two atoms of hydrogen to one of oxygen and is represented as $H_2O$. Calcium carbonate is $CaCO_3$—one calcium, one carbon, and three oxygen atoms.

We construct a chemical equation by writing the reacting substances, or the reagents on the left and the products on the right, as in simple arithmetic. The reaction between one molecule of hydrogen and one molecule of chlorine to yield two molecules of hydrogen chloride can be represented thus:

$$H_2 + Cl_2 = 2HCl$$

Since no atoms can be formed or destroyed in a chemical reaction, the number of atoms each side of the equation must balance.

Below: Representations of sodium and chlorine atoms and ions. Sodium loses an electron to become a positive ion. Chlorine captures an electron to become a negative ion.

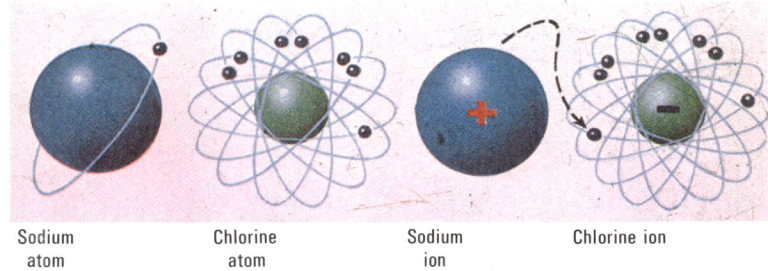

Sodium atom  Chlorine atom  Sodium ion  Chlorine ion

Right: An impression of the crystal lattice of common salt. This arrangement is an example of a simple cubic lattice.
Below: A molecule of carbon dioxide. It is a covalent compound, with the three atoms sharing some of each others' electrons.

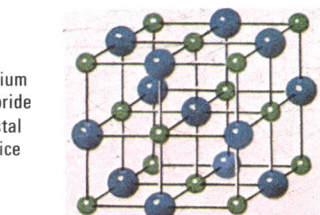

Sodium chloride crystal lattice

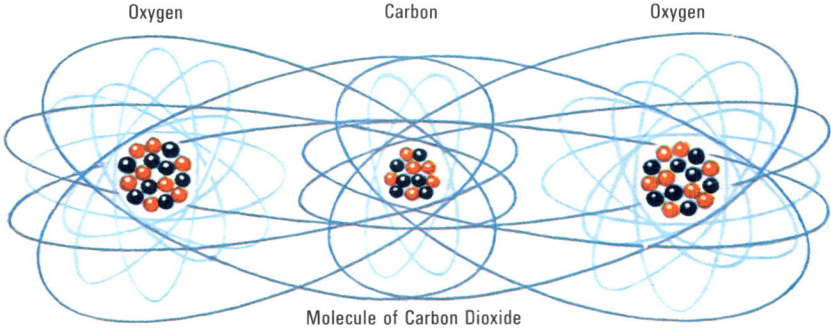

Oxygen  Carbon  Oxygen

Molecule of Carbon Dioxide

# The Chemical Industry

The modern chemical industry can be regarded for simplicity as comprising the *chemical-producing* sector, and the *chemical-processing* sector. Products of the first sector include petrochemicals, acids, alkalis, drugs, synthetic resins, and plastics powders. Products of the second include paint, rubber, medicines, foodstuffs and cosmetics.

The basic chemical manufacturing processes involved are generally known as 'unit processes' and are really laboratory processes which have been scaled up for factory operation. The major unit processes employed in chemical production include oxidation, nitration, hydrogenation, esterification, cracking, and polymerization. Any one of these processes may require the presence of a *catalyst,* a substance which causes a chemical reaction to occur without itself undergoing any change. Generally catalysts last a very long time before needing any replacement. In chemical processing, physical operations like mixing, blending, distilling, dissolving, evaporating, drying, boiling, filtering, extracting, and refining are carried out.

It must not be thought, however, that these operations are special to either sector. Generally both the chemical-producing and chemical-processing industries use a combination of chemical and physical operations during manufacture.

## Chemical Operations

*Oxidation* in its simplest terms means adding on oxygen to another substance. But it has come to mean much more than this. It is used to describe reactions in which not only oxygen, but elements chemically similar to oxygen (for example, sulphur and chlorine) are added to a substance, or in which hydrogen is removed from a substance. And modern electro-chemical theory considers that any process by which the positive valency of an element is increased should be called oxidation. An oxidation reaction is always accompanied by an opposite reaction – *reduction.* The terms thus cover a wide field of activity and embrace reactions of widely different types.

*Nitration* is, again, quite simply, the addition of nitrogen or a nitrogen-containing group (for example, the nitrate group – $NO_3$) to a substance.

*Esterification* is the formation of organic compounds by the reaction of organic acids and alcohols. Esters are in organic chemistry what salts (formed from acids and alkalis) are in inorganic chemistry, but they have different properties.

*Hydrogenation* means adding a hydrogen atom to what is known as an unsaturated organic compound. It is used to produce more useful fuels and edible fats (for making margarine, for instance) and requires very high temperatures, pressures, and usually the presence of a catalyst.

*Cracking* is the decomposition by heat and pressure, sometimes in the presence of a catalyst, of less valuable hydrocarbons (thick oils) into more valuable lower-boiling

Research plays a vital role in the chemical industry. Most major chemical companies have their own research staff. A great deal of research is also carried out at governmental research establishments and universities.

products such as gasoline and gases like ethylene. (See page 79.)

*Polymerization* means causing a simple molecule to crosslink with other similar or different simple molecules (called *monomers*) to form a *polymer*; this is done to give products with better tensile strength, flexibility, electrical or cold resistance, and so on. It often requires high temperatures, pressures and a catalyst. A well-known example is the conversion of styrene monomer into the very useful moulding plastic polystyrene.

**Physical Operations**

*Mixing* and *blending* are closely allied operations and may deal with either liquids or solids or a mixture of the two. The apparatus used obviously depends upon the job – the type of product being handled, the degree of mixing required, the viscosity (thickness) of the mix (for liquids), and so on. Mixing machinery thus varies from a simple rotating barrel to a heavy-duty beater capable of preparing several tons of anything as thick as dough.

*Distilling*, as applied to liquids, means heating the liquid until it vaporizes and then cooling and condensing it. It is a process used to purify liquids, separate different liquids and to yield liquids of accurately known boiling point. Another form of distillation is known as *destructive* distillation which, in the case of coal, for example, simply means heating until all the volatile substances are driven out, leaving behind coke (see page 74).

*Dissolving* uses similar equipment to mixing, but is concerned with dissolving a solid in a liquid, for instance, sugar in water. Heat may be needed and so the machinery may have additional parts, such as immersion heaters, steam jackets or pressure lids.

*Evaporating* and *drying* are similar operations and usually require shallow pans with a large surface area to allow liquid to evaporate away quickly in currents of warm air. Faster operation is often obtained by placing the trays in heated vacuum ovens.

*Filtering* and *extracting* are generally done in a *filter press*, which is simply a machine with lots of very big filter papers fitted inside it. Sometimes it is the clear liquid that is wanted and sometimes the solid collected on the papers. Another form of extracting uses hot solvent to dissolve something out of a solid (for instance, oils out of nuts). The solution obtained is evaporated, and the solvent distilled for reuse.

*Refining* really means getting rid of unwanted ingredients present in a material and will utilize any of the processes described so far, depending on the type of im-

Mixing is one of the commonest physical operations carried out in the chemical industry. This picture shows a battery of mixers at a paint factory in which the various ingredients are intimately blended to form a stable finished product.

purity to be removed. Often several processes are used one after the other. This is especially so in manufacturing and processing of chemicals for the food and drug industries. Chemicals used in laboratories for analytical work are also treated by refining, and you can usually tell if this is the case since the letters 'C.P.', meaning 'chemically pure', appear somewhere on the label.

**Heavy Chemicals**

The chemical processing industry is subdivided into two major sections – *heavy* and *fine*. It is a division more of convenience than

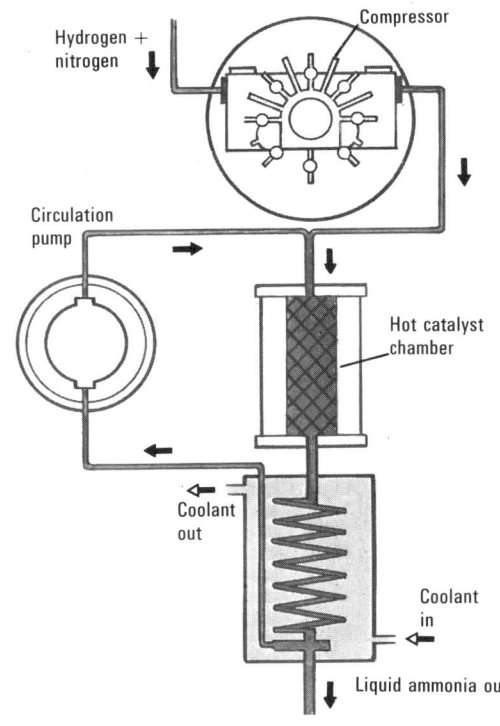

Outline of the production cycle for the manufacture of ammonia. Hydrogen and nitrogen gases are compressed and fed into a heated chamber containing a catalyst. Some of the gas combines to form ammonia, which is then condensed in a cooler. The remaining gas is recycled back to the catalyst chamber.

169

Tall distillation, or fractionating towers are a common sight at chemical plants. Distillation is an extremely common way of separating liquids of different boiling points. Where this is not possible, methods such as liquid-liquid extraction—treating the liquids with another which preferentially dissolves one—have to be utilized.

of scientific accuracy. Caustic soda, vitamin B, and radium bromide are each products of the chemical industry. But whereas the world output of caustic soda is measured in millions of tons, that of vitamin B probably only reaches tens of tons, and that of radium bromide only a few grams!

Caustic soda, then, is the heaviest weight production of these three examples and has come to be known as a 'heavy' chemical. Other products in this category are sulphuric, nitric, hydrochloric and phosphoric acids, soda, ammonia, saltpetre (sodium nitrate), and fertilizers.

There are in the world relatively few plants engaged in making the vast tonnages involved, but they are highly efficient. One aspect of this efficiency has been the need to reduce waste and fume to a minimum to avoid atmospheric pollution. This has resulted in the waste products being converted into valuable by-products, and heavy chemical factories often produce quite a lot of other materials as well.

### Sulphuric Acid

Sulphuric acid has been called the 'lifeblood' of the chemical industry because so many chemical processes depend on it. In Britain alone about four and a half million tons are consumed annually. The modern method of manufacture is known as the 'contact process'. Sulphur dioxide gas and air are drawn over certain catalysts in a hot (400°C) 'contact chamber' to form sulphur trioxide. Sulphur trioxide dissolves in water to give sulphuric acid.

The sulphur dioxide is obtained either by burning sulphur in air or by roasting pyrite, or iron sulphide. Either vanadium pentoxide, iron oxide, or platinized asbestos may be used as catalysts.

### Soda

The manufacture of soda, by the 'ammoniasoda' process, provides an interesting example of what is known as a *cyclic* process. In the process salt (sodium chloride) solution is saturated with ammonia gas and then with carbon dioxide gas. Sodium bicarbonate is precipitated, and ammonium chloride (sal ammoniac) remains in solution. Roasting this very pure sodium bicarbonate yields soda, or sodium carbonate, and carbon dioxide gas, which is piped back and used to precipitate more bicarbonate. The ammonium chloride solution is heated with slaked lime to liberate ammonia gas, and this also is piped back. The principle of a cyclic process, therefore, is to use reagents over and over again.

## Caustic Soda

Caustic soda, which used to be made chemically from sodium carbonate, is today prepared by an electrolytic process. Strong brine solution is passed through a series of electrolytic cells, which have carbon rods round the sides as anodes and liquid mercury on the bottom as the cathode. The sodium chloride dissociates, or splits up to give chlorine gas at the carbon anodes and metallic sodium at the mercury cathode, where it forms an alloy called an *amalgam*.

The chlorine is led away and compressed into cylinders to be used in other chemical manufacture. The sodium-mercury amalgam is run into water, where the metallic sodium reacts violently to form caustic soda solution, leaving pure mercury to be used again (another example of recycling). The solution is then concentrated by evaporation. This process can be halted at any desired point to give solutions, called *lyes,* of known concentration. Lyes are used extensively in soap-making (see page 172).

## Ammonia

Ammonia is invariably prepared by synthesis (building-up), the basic ingredients being nitrogen and hydrogen. Three volumes of hydrogen and one volume of nitrogen are compressed to more than 200 atmospheres and passed through towers packed with red-hot catalyst (iron oxide). During passage through the tower, something like 4% of the nitrogen and hydrogen is converted to ammonia. The mixture of ammonia, hydrogen, and nitrogen then passes into very efficient cooling chambers, where the ammonia liquefies. The nitrogen-hydrogen mixture passes on for recycling.

The nitrogen for the process can be obtained by distilling liquid air or from producer gas. (This is made by passing air over red-hot coke, yielding a mixture of about two-thirds nitrogen and a third carbon monoxide.) Hydrogen may be obtained by electrolysing water (or usually a caustic soda solution) or from water gas, which is a mixture of hydrogen and carbon monoxide.

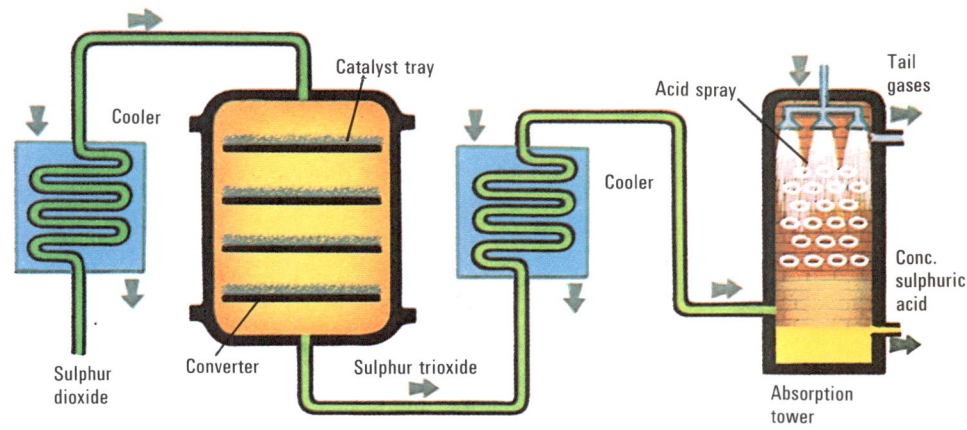

Outline of the contact process for the manufacture of sulphuric acid. It superseded the old lead chamber process. It involves converting sulphur dioxide to sulphur trioxide by heating with a catalyst and then absorbing the sulphur trioxide in dilute acid sprayed in an absorption tower. Concentrated acid results.

Milling and grinding of pigments and powders has to be done in many industries, especially paint manufacture. Only in a finely divided state will the pigment remain suspended within its vehicle.

### The Chemical Reactions

The reactions taking place during the production of the heavy chemicals described here can be summed up by the following chemical equations.

*Sulphuric acid*
sulphur dioxide + oxygen = sulphur trioxide
$$2SO_2 + O_2 = 2SO_3$$
sulphur trioxide + water = sulphuric acid
$$SO_3 + H_2O = H_2SO_4$$

*Soda*
carbon dioxide + salt (solution) + ammonia (solution)
$$CO_2 + NaCl + NH_4OH$$
= sodium bicarbonate + ammonium chloride
$$= 2NaHCO_3 + NH_4Cl$$
sodium bicarbonate = sodium carbonate + carbon dioxide + water
$$2NaHCO_3 = Na_2CO_3 + CO_2 + H_2O$$

*Caustic soda*
sodium chloride (salt) = sodium + chlorine
$$2NaCl = 2Na + Cl_2$$
sodium + water = sodium hydroxide + hydrogen
$$2Na + 2H_2O = 2NaOH + 2H_2$$

*Ammonia*
nitrogen + hydrogen = ammonia
$$N_2 + 3H_2 = 2NH_3$$

# Soaps and Detergents

## Soap

Broadly speaking, detergents are substances which help to remove dirt. Soap, then, is a detergent, but we tend to use the name 'detergent' for synthetic products which are used for washing clothes, dishes, and so on.

In very early times people rubbed their bodies with oils, plant juices, ashes, and even a kind of absorbent clay (called Fuller's Earth) to cleanse themselves. Then they gradually realised that it would be easier to cleanse themselves with water and so, eventually, soap was discovered. The Romans, certainly, knew how to make soap.

The basic chemistry of the subject was revealed by the French chemist Michel Eugène Chevreul (1786–1889) in 1823. He showed that natural oils and fats are compounds of fatty acids with glycerine. When they are reacted with caustic soda, something like this occurs:

glyceryl palmitate + caustic soda
   (oil or fat)      (alkali)
      = sodium palmitate + glycerine
              (soap)

Such a process is called *saponification*.

Generally, soap is made from a mixture of fats or oils to give acceptable properties. Thus, coconut, palm, cottonseed, olive, and soya-bean oils or beef tallow, mutton tallow, or hardened whale oil may be used.

## Detergents

Shortages of these natural fats and oils, limitations in certain properties of the finished soaps, and simply pure economics caused chemists to search for other ways of making soaps. This is how the modern detergents industry was born, which relies on petroleum (crude oil), coal, limestone, and air for its raw materials. It was soon found that these detergents, whilst not necessarily replacing soap for ordinary bathing or washing, could do many other jobs more efficiently and more cheaply.

Detergents are able to do their job of cleaning because their molecules have two different ends—one water-liking (*hydrophilic*) and the other water-hating (*hydrophobic*), or oil-liking. When a dirty, greasy object is placed in water containing detergent, the oil-liking end joins up with the grease. The grease is then effectively coated with the water-liking ends and therefore dissolves in the water. Some dirt is not greasy—mud, for example. In this case the detergent makes the water 'wet' the fabric better, thereby tending to 'push' the dirt off. Agitation helps this operation, and that is why washing machines have an impeller to agitate the clothes.

Laundry compounds are much more complex than ordinary soap. They usually contain a detergent, a water-softener, dirt absorbents, and an optical 'brightener'.

The increasing use of detergents at first created pollution problems at sewage farms. Processes of treatment were upset and excessive frothing occurred in the water.

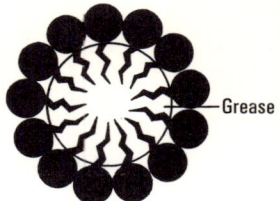
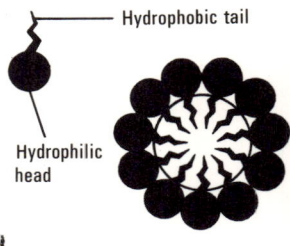
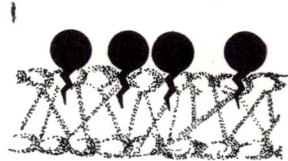

Detergent action. The hydrophobic "tails" of the detergent molecules attach themselves to greasy particles, effectively coating them with hydrophilic "heads". This allows the grease particles to be floated away.

Below: A flow diagram illustrating detergent manufacture. The most important ingredient of a detergent is the wetting agent, or surfactant. This is produced in the sulphonator cooler. Phosphates are added to help loosen dirt the sulphate and silicates ensure that the finished powder will be free-flowing; sodium carboxyl methyl cellulose helps to suspend the dirt in the water.

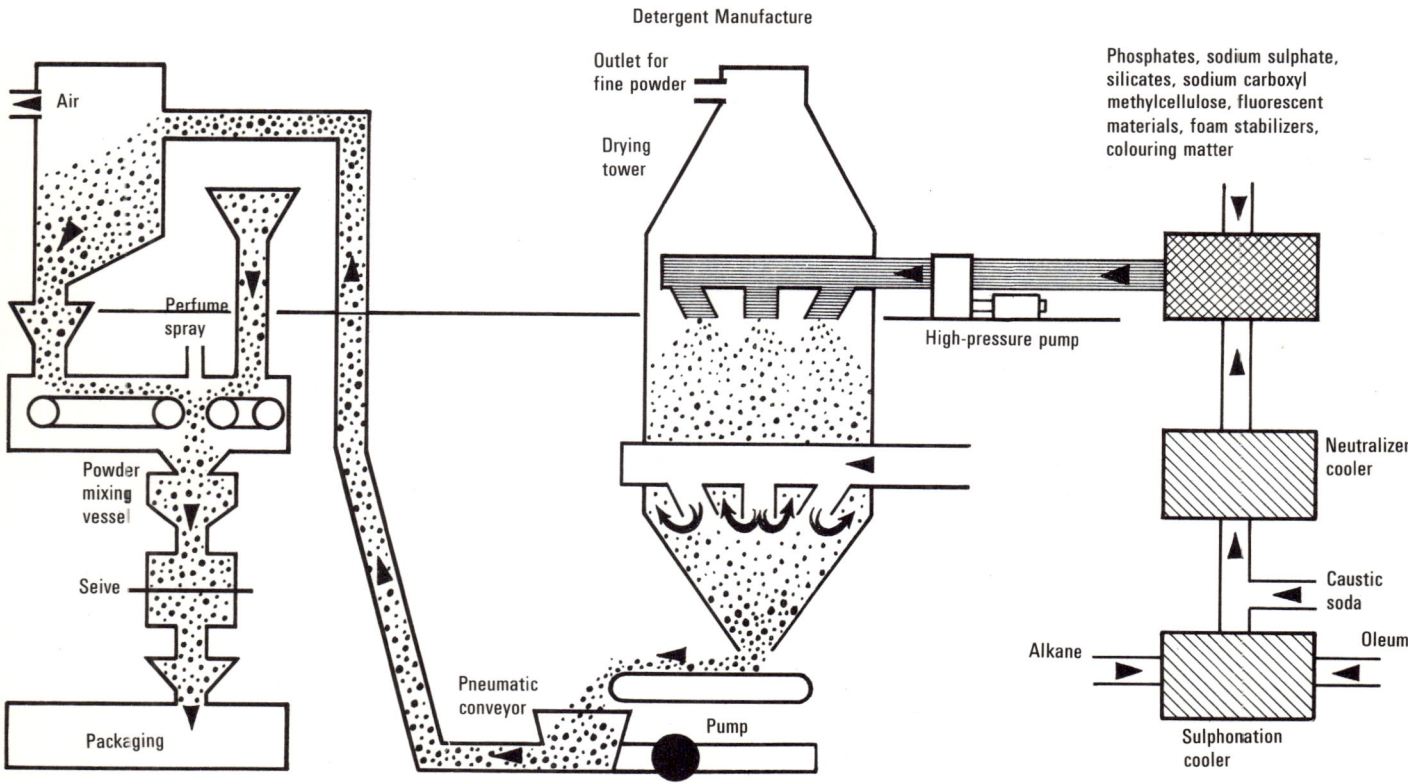

# Explosives

Although most people would describe an explosion as 'a loud bang', probably few realize that it is really an extremely violent flash of flame. The bang is caused by the gases released from such a flash shooting out in all directions and pushing air and solids out of their way. Substances which decompose violently when fired, liberating gases, are called *explosives*. There are two main types: very rapid burning, in which fire spreads from one particle to the next, and detonating, in which all particles 'go up' together.

Missiles such as shells and bullets are charged with a combination of each type. A small amount of detonator is exploded mechanically (for example, by the hammer of a gun) and sets off the main charge. Since the charge has been packed into a small space (a cartridge or a shell), the energy released by gas formation causes the container to shoot forward violently.

*Gunpowder* is probably the oldest known explosive. The Chinese had knowledge of gunpowder thousands of years ago, but the first known military use of something like gunpowder was in A.D. 673, when the defenders of Constantinople used 'Greek fire' against the attacking Saracens. Roger Bacon in England (in about 1294) was among the first people to write a scientific account of gunpowder.

Modern 'black powder', as it is called, is virtually the same product described by Friar Roger Bacon, but it has now been largely superseded by smokeless powder. Smokeless powder is really *nitrocellulose* formed into flakes, grains, or cylinders. It is the basic propellant material used in pistols, shotguns, and so on, where too violent an explosive would be dangerous to anyone setting it off. *Cordite* and *ballistite*, which are nitrocellulose mixed with a little *nitroglycerin*, are much more powerful.

Nitrogen is present in some form or another in all non-nuclear explosives. Oxygen is another essential element which is usually present as part of the molecular structure of the basic explosive. *Liquid oxygen* by itself can be used as a tremendously powerful explosive: some material soaked in liquid oxygen is set off by a detonator.

*Mercury fulminate* is a detonating explosive: it immediately breaks down when struck into nitrogen, carbon monoxide, and mercury vapour, with a tremendous release of energy. *Nitrogen iodide* is another incredibly sensitive detonator. A fly walking on a particle of nitrogen iodide will blow itself up! Nitrogen and phenol give *picric acid,* an explosive used in shells. One of the most powerful explosives is *T.N.T.* (trinitro-toluene), which is nitrogen reacted with toluene (a liquid obtained from petroleum or coal tar). Nitrogen and glycerin give nitroglycerin. Any absorbent substance soaked in nitroglycerin is known as *dynamite*. *Amatol* is T.N.T. mixed with ammonium nitrate. *Ammonal* is a mixture of T.N.T., powdered aluminium, charcoal, and ammonium nitrate.

One major use for explosives is in quarrying. Dynamite or gelignite—a mixture of nitroglycerin, nitrocellulose, ammonium nitrate, and woodpulp—are the commonest explosives used.

Demolition of such a thing as an old bridge is made simple by using explosives. Note how the bridge collapsed. Being an arch bridge it was in compression. The charge was exploded at the center of the span. When the center was destroyed, the two sides just toppled over.

# Paint

One of the earliest recorded uses of paint can still be seen in caves in Spain, where Capsians used a mixture of red iron oxide and clay to paint animals and hunting scenes on the walls. Paint of a different type (color mixed with fat) has been used for thousands of years for decorating priests', warriors', and witch-doctors' bodies to frighten enemies and keep away evil spirits, and perhaps to boost the morale of the wearer. Modern greasepaint is somewhat like this. It is used by actors to counteract the brilliant lighting used on stages and on film sets where ordinary skin looks unhealthy.

The two main functions of a paint are to *decorate* and to *protect*. Such a vast range of paints are available today that you can find one suitable for any decorative or protective job you can think of.

Paints are produced in certain basic stock colors. But in the majority of cases the basic colors can be blended together to give a wide variety of different shades. Paint tinting is usually carried out in batches, as shown here.

Below: Industrially, spraying or dipping, as in many auto manufacturing plants, are the commonest methods of applying paint. The result is a good surface finish. Hand spraying is being increasingly replaced by automatic spraying by multi-nozzle machines.

## Paint Recipes

Paint consists of four main ingredients: pigment, varnish, thinner, and driers. The *pigment* gives the paint colour, opacity and protectiveness. 'Opacity' (from opaque) is the ability of a paint to cover a surface. The *varnish* holds the pigment on the surface painted and helps to seal that surface against attack. It is often called the *vehicle*. The *thinner* is a solvent which dissolves the paint and brings it to the right consistency for whatever method of application being

### Special Paints

The effective protection of bridges, gas tanks, chemical plant and other large metal structures raises special problems for the paint-maker. Acid and other deposits fall on to such surfaces from the air, and rain, wind and sun add their quota of destructive effect. The paint has to stand up to some pretty severe conditions. Special brushing air-drying paints are made for this type of work in addition to the red lead paint mentioned above. They often contain lead, zinc, aluminum or stainless steel powder to help combat corrosion. Zinc paints, for example, are widely used by motorists to touch up rust patches on the bodywork of their cars.

Milking parlours, dairies, slaughterhouses, cold stores and hospitals each need a *fungicidal* paint, which resists bacterial and fungal growth.

Railway carriages, wooden partitions in large buildings and stage scenery require *fireproof* paint to help reduce flame spread if they catch light. Boilers, engines, exhaust systems, aircraft, spacecraft, and so on are painted with *heat-resistant* paints. They, too, often contain finely divided metal.

There are paints for many other special uses, too. *Marine* paints protect ships from salt-water attack and barnacles. *Road-lining* paints often contain thousands of tiny glass beads (ballotine) which glisten in headlight beams at night. *Fluorescent* paints glow vividly and are often used on warning signs and even on the clothes of people who must be seen clearly, for example, maintenance workers on highways or builders on a busy construction site.

used. This may be brushing, spraying, dipping, or roller-coating. *Driers* are compounds added to accelerate the drying of the pigment/varnish mixture and cause it to set hard.

The actual 'recipes' for paint-making vary enormously, depending on what kind of paint is needed and what job it has to do. One of the older types of paint still successfully used is *lead paint*. The pigment is either red lead (a lead oxide) or white lead, (lead carbonate). The varnish is linseed oil, which is often boiled to improve its wearing properties. The thinner is white spirit, a liquid somewhat like paraffin. White spirit is often called 'turps' substitute because it replaced turpentine (distilled from pine-tree sap) as a paint thinner. The driers are usually compounds of the metals cobalt, manganese, and lead. Red lead paint is still widely used for painting bridges and other outdoor metal structures.

One of the problems met when linseed oil is used as the varnish is its long drying time, sometimes days. This is one area where science has helped by providing new varnish materials. Their general name is *synthetic resins,* and they are invariably used instead of plain oil. They include alkyd, phenolic, acrylic and polyurethane resins. Some of the other products used as varnishes in paint are cellulose nitrate (known as pyroxylin in the United States), cellulose acetate, rubber and rubber compounds, polystyrene, polyvinyl chloride, silicones, cement, and even glue. You will notice that many of these are familiar to us as plastics (see page 36). That is not really surprising because a layer of paint is more or less a plastic film.

## Domestic and Industrial Paints

House paints are used to decorate and protect the inside and outside of buildings. Apart from different colors, these may be matte, semi-matte, or glossy and may be supplied as priming (first coat), undercoating, or finishing paint. Walls, ceilings and ceiling tiles are usually painted with so-called emulsion paint. Cement paint is often used for outside walls. In both instances, the thinner is water. Water-based undercoat and gloss paints are also now available. The great advantages of these paints for the home decorator is that they are much cleaner to use, for splashes can be wiped away and brushes cleaned simply with water. House paints are often termed 'air-drying' paints.

Industrial paints also include air-drying paints, but they are usually faster drying than house paints. The other two main kinds of industrial paints are stoving and catalyst paints. Articles painted with *stoving* paints

Rigorous testing of paint films is essential to development work. Here testing is being carried out in a chamber that duplicates but accelerates attack by the weather.

Below: the essential stages of common paint manufacture. Resin and drying oil are cooked to form the vehicle. Then pigment is added and ground fine in a ball mill. Then the other ingredients are added and mixed thoroughly.

are cooked in hot ovens, and this yields a very hard and durable finish. Automobiles, refrigerators, and cookers are examples of everyday articles which are 'stove-enameled' in this way. Ceramic paints, or *vitreous enamels* are a form of stoving enamel cooked at high temperature, like potters' colours. *Catalyst* paints are used where stoving an article is impractical. After application of the paint, a chemical reaction takes place in the film and makes it very hard and durable.

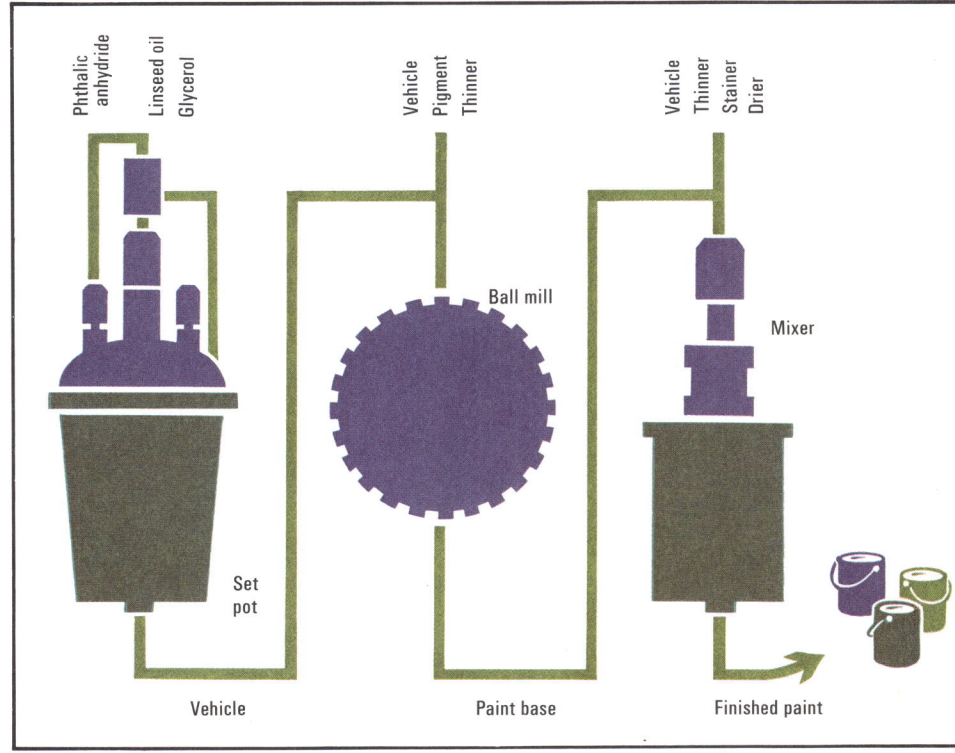

# Pharmaceuticals

The pharmaceutical industry is really a section of the chemical manufacturing and processing industry concerned with making drugs and medicines. Many of the operations performed are similar to those mentioned in the articles on the general chemical and heavy chemicals industries.

The really important difference is the trouble which has to be taken during processing to ensure that whatever is being made is pure. This means that continual chemical analysis and control is exercised during all stages of manufacture. And stringent efforts are made to eliminate what are known as 'trace contaminants'. Frequently, the presence of as little as one part in a million of an unwanted ingredient in a product could cause it to be rejected for human consumption. The other main difference between pharmaceutical and normal chemical manufacture is generally in the quantities of product involved. Drug tonnages are much smaller, but they are worth a lot more money.

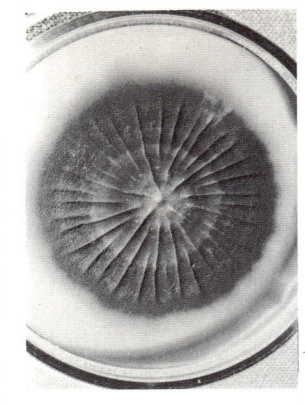

A Penicillium mould, which produces the powerful germicide we know as penicillin. Penicillin, discovered by Alexander Fleming in 1928, brought about a revolution in medicine.

Below: Research and manufacture of penicillin and drugs like it must take place in scrupulously clean conditions, hence the mummy-like appearance of these workers.

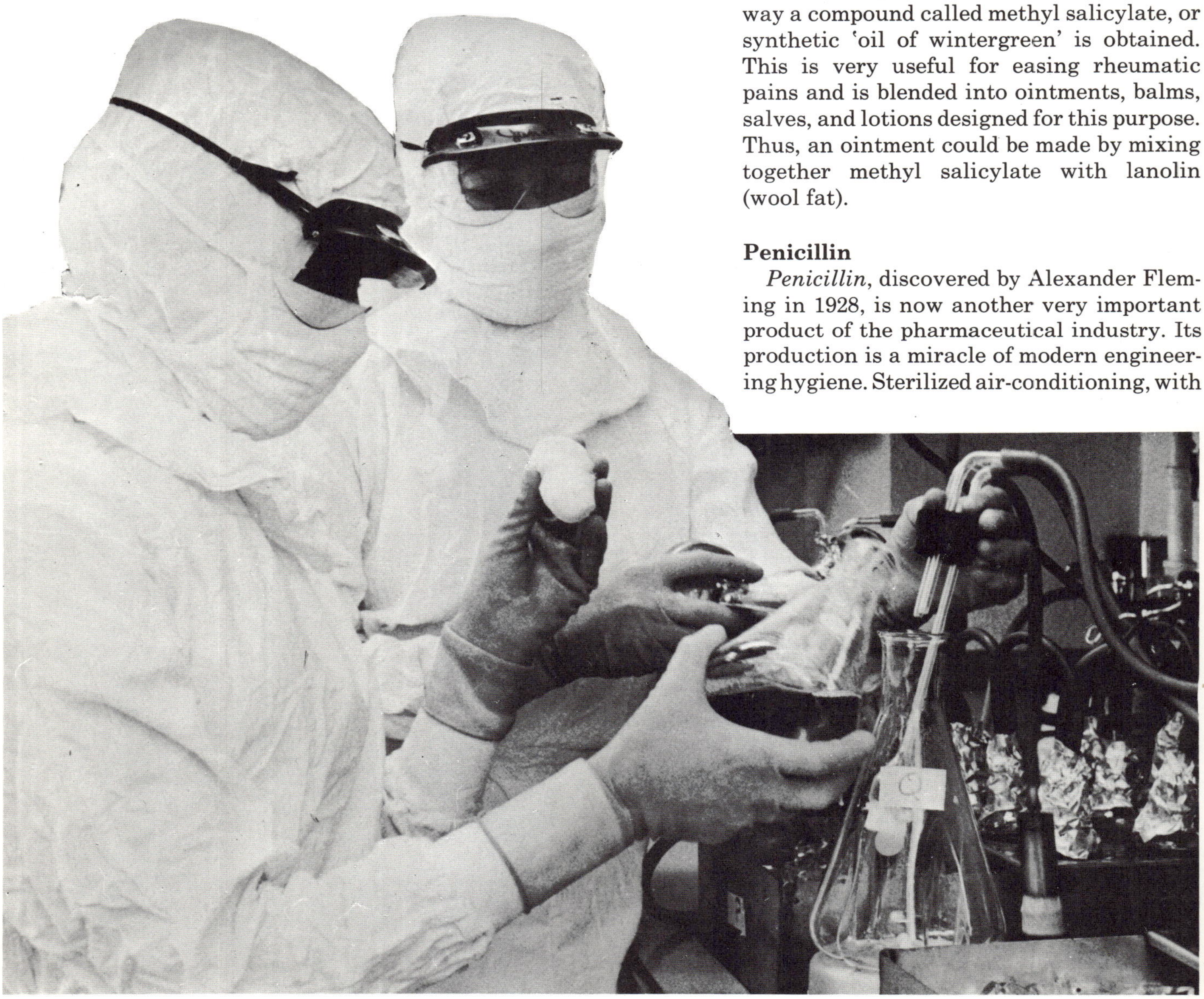

## Aspirin

One of the best-known products of this section of the chemical industry is *aspirin*. First discovered by Charles Gerhardt, a French chemist in the mid-1800s, its use today is phenomenal. It is prepared basically from phenol (a product from coal tar, sometimes known as carbolic acid). Phenol is heated with caustic soda to give sodium phenate. Reacting sodium phenate with carbon dioxide gas under high pressure yields salicylic acid. In turn salicylic acid is treated with acetic anhydride to yield acetyl salicylic acid, which is aspirin powder. This powder is blended with neutral substances called *extenders* to render it easily acceptable by the stomach, and the mixture is then tableted and packaged ready for sale.

If the salicylic acid is treated in another way a compound called methyl salicylate, or synthetic 'oil of wintergreen' is obtained. This is very useful for easing rheumatic pains and is blended into ointments, balms, salves, and lotions designed for this purpose. Thus, an ointment could be made by mixing together methyl salicylate with lanolin (wool fat).

## Penicillin

*Penicillin*, discovered by Alexander Fleming in 1928, is now another very important product of the pharmaceutical industry. Its production is a miracle of modern engineering hygiene. Sterilized air-conditioning, with pressurized entrance and exit seals, is used to ensure that no unwanted bacteria can enter the penicillin via the air (which would certainly happen under ordinary working conditions). The workers wear masks over their faces, and their clothing is all especially sterilized.

## Sulpha Drugs

*Sulphonamide* drugs, or *sulpha* drugs for short, are another example of highly specialized manufacture. The original sulpha drug, *sulphanilamide,* was discovered in 1908 by P. Gelmo, a German chemist. But it was not until the 1930s that a German biochemist, Gerhard Domagk, discovered that such a compound would overcome streptococcal blood poisoning. From then on new derivatives of the original discovery were prepared, and today their value in preventive medicine cannot be assessed.

## Saccharin

Another quite different product of this fascinating industry is *saccharin*. It is a white powder that is sweeter than sugar by about 550 times. It is prepared from a product derived from coal-tar or petroleum called toluene. Saccharin was discovered by accident in 1879 by a German chemist, Constantin Fahlberg. Although saccharin is very sweet, it has no food value. And people who are trying to slim often use saccharin as a sweetener. It is also widely used by sufferers from diabetes who must not eat sugar. Other compounds have since been discovered that are even sweeter than saccharin – *sucramine,* for example.

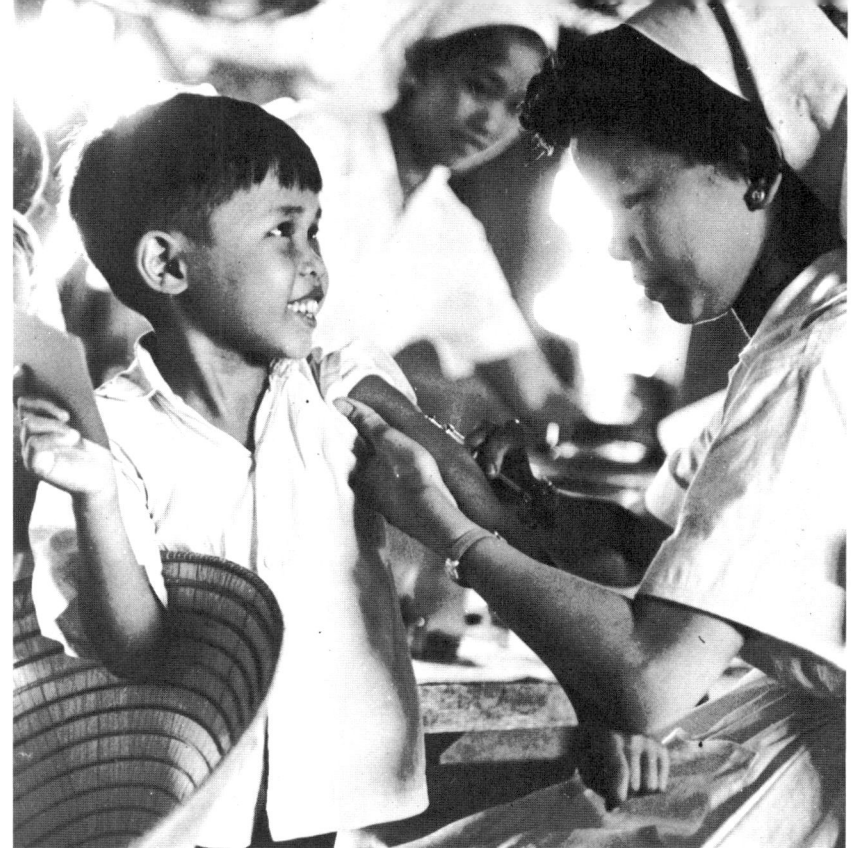

Above: Inoculation of children against common childhood diseases is now widespread. Inoculation is the process of injecting dead or weakened germs into the body to produce antibodies to combat the disease.

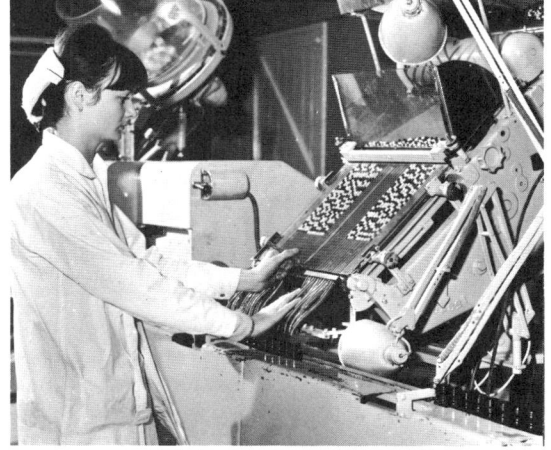

Left: A great many drugs are now administered in capsule form, the capsule body being made of gelatine. The picture shows a bottle-filling machine.

---

**Common Types of Drugs**

*Anaesthetics* are drugs administered during surgical operations of all kinds to help prevent the patient feeling pain. *General* anaesthetics cause complete loss of consciousness, and are used in major operations, such as appendicectomy and tonsilectomy. They include laughing gas (nitrous oxide), chloroform, ether, and sodium pentothal. *Local* anaesthetics deaden the nerves in one part of the body only. Dentists use them, for example, when drilling or extracting teeth. Local anaesthetics include lignocaine and procaine.

*Analgesics* are substances which prevent or relieve pain, but the person taking them retains consciousness. They include simple drugs like aspirin and bromides and powerful ones like morphine and heroin.

*Antibiotics* are drugs that have the power to destroy body germs quickly and to prevent germs growing. They are compounds that are produced by bacteria and moulds, tiny plant organisms (micro-organisms). Antibiotics are thus probably the biggest life-savers today. They combat pneumonia, typhoid, syphilis, scarlet fever, and many other serious illnesses. Penicillin, streptomycin, chloromycetin, and terramycin are among the most important.

*Antihistamines* relieve the symptoms of asthma, hay fever, and other allergies. They counter excess production in the body of substances called *histamines* which causes sneezing and a streaming nose.

*Hormones* are used as drugs when the body has a hormone deficiency that causes disease. Insulin is probably the best-known hormone drug. It is given to sufferers of diabetes, who have an insulin deficiency. The insulin injected is obtained from the pancreas of sheep, cattle, and pigs. Adrenalin, cortisone and ACTH are other important hormone drugs.

*Narcotics* deaden the whole nervous system and prevent a person feeling pain. They may make him sleep or go into a coma. Opium and the drugs derived from it – codeine, heroin and morphine – are the most widely used narcotics.

*Sedatives* are soothing drugs that generally send a person to sleep. Common sedatives are barbiturates and bromides. Narcotics and anaesthetics have sedative effects, too.

*Tranquillizers* are taken to calm the nerves and prevent worry. They do not, like sedatives, dull the nervous system, or slow down mental or physical activity.

*Vaccines* help the body to develop resistance to disease. They consist of dead or weakened germs, and are introduced into the body by means of an injection, a scratch, or by mouth (orally). The word *vaccine* comes from the Latin word meaning 'cow'. The original vaccine was of cowpox virus. It was used by Edward Jenner in 1796 to combat smallpox. Today vaccines have been developed for poliomyelitis, diptheria, whooping cough, tetanus, and many other dangerous illnesses.

# Astronomy

Astronomy, the scientific study of the heavens, is one of the oldest of sciences. And anyone who has gazed at the stars on a really clear night in the country will readily understand why. The never-changing patterns of stars, the constellations, wheel slowly around the dome of the sky. The so-called 'wandering' stars, the planets, move gradually across the starry background. The silvery Moon rises and sets, its shape changing from night to night. A few 'stars' appear to fall right out of the sky or shoot upwards. By day the light from the Sun blots out the stars, and usually the Moon, as it brings us warmth.

Small wonder, then, that the heavens were regarded with awe in ancient times, or that the Sun was worshipped as a god (the Greeks and Romans called him *Apollo*) and the Moon as a goddess (the Greeks called her *Artemis*; the Romans, *Diana*).

Naturally enough the ancient peoples tried to explain what they saw happen in the heavens and also made up stories about how the heavens came about. Before men did much travelling, they thought that the world is flat and if you go too far you will fall over the edge! But the Greeks believed that the Earth is a sphere and that it is the fixed center of the universe. All the other heavenly bodies revolve around it. The stars are fixed in a great rotating celestial sphere with the Earth at the center. The great Greek astronomer Ptolemy of Alexandria (A.D. 100s) held this view, and Ptolemy's picture of the universe was accepted for fourteen hundred years.

In 1543 the Polish astronomer Nicolaus Copernicus (1473-1543) challenged this view in his famous book *Concerning the Revolutions of Celestial Spheres*. He believed that the Earth is just a planet, and that the planets circle around the Sun. This view of a Sun-centered, or *solar* system was supported by the Italian genius Galileo (1564-1642), who made the first telescopic observations of the heavens and found that the planet Jupiter has a miniature 'solar system' of its own.

The matter was finally settled by Johannes Kepler (1571-1630). Kepler, assuming a solar system, found that the movements of the planets can be accurately described if the planets move not in circles but in *ellipses* around the Sun. He formulated his celebrated laws of planetary motion which showed exactly how the solar system behaves. He explained 'how' but not 'why'.

Sir Isaac Newton (1642-1727) was the first to explain 'why'. He demonstrated with his *law of universal gravitation* how the universe is held together (see page 20). His law of gravitation explains not only why objects fall to the Earth when dropped, but why the Moon keeps in its *orbit* (path) around the Earth, and the planets in their orbits around the Sun. Newton's and Kepler's laws formed the basis from which modern astronomy developed.

Among Newton's many other achievements was his invention of the *reflecting telescope*, which uses a mirror rather than a lens to gather and focus light rays. It is free from the colour distortion of refracting (the lens-type) telescopes. All of the biggest optical telescopes today, such as the 200-inch Hale Telescope at Mount Palomar Observatory in California, are of the reflecting type. In recent years radio telescopes have opened up a whole new 'window' on the universe, and their use has resulted in the discovery of such phenomenal bodies as the quasars and the pulsars.

Two of the greatest astronomers of all time; left, Galileo, and right, Sir Isaac Newton.

For many years the largest telescope in the world, the 200-inch reflector at Mount Palomar Observatory in California, USA. It has now been eclipsed in size by a 236-inch reflector sited in the Caucasus Mountains in Russia. This telescope can locate objects of the 25th magnitude. This is equivalent to being able to detect the light of a candle at a distance of 15,000 miles!

# The Earth

The Earth is a great sphere of rock surrounded by a relatively thin layer of air called the *atmosphere,* which is kept there by gravity. The Earth is not exactly a sphere, though. It bulges slightly at the Equator. The diameter of the Earth at the poles is about 27 miles less than its diameter at the Equator (7,927 miles).

The upper crust of the Earth is made up of rock. These are the deep-seated *igneous rocks,* the melted and recrystallized rocks like granite; *sedimentary rocks* like sandstone, made up of compressed layers of fragments of preexisting rocks; and *metamorphic rocks* like marble, formed when igneous or sedimentary rocks are transformed by heat and pressure. Study of the rocks forms the branch of science called *geology.*

Beneath the crust is the relatively soft *mantle,* and beneath that an outer and inner *core.* The outer core appears to be fluid. The solid inner core may be of iron or nickel. The presence of iron and nickel, which are magnetic, could explain the magnetic field which surrounds the Earth.

The atmosphere has no definite limits. It is densest near the ground and then gradually gets thinner and thinner, until at almost 100 miles there is scarcely any at all. Then we are on the fringes of space. Nevertheless, air molecules are still present even more than 300 miles up. Satellites in orbit at that altitude will still experience slight drag.

Like all the planets, the Earth has two distinct movements in space. It rotates on its own axis, like a top. And it travels in orbit around the Sun. It travels once on its axis in 24 hours. For part of the time (day) Earth is illuminated by the Sun; for the remainder, it is dark (night). It takes $365\frac{1}{4}$ days, or one year to travel once in its orbit around the Sun. As you can see, this 'year' is $\frac{1}{4}$ day longer than our calendar year. To make up for this discrepency we add an extra day to the calendar year every four years to make a *leap year.* (An extra day is added to February.)

The axis on which the Earth spins is tilted slightly with respect to the plane of the Earth's orbit. This means that at any time some parts of the Earth are closer to the Sun than others, and this position changes constantly as the Earth circles the Sun. The tilt thus gives rise to the *seasons.* When the northern hemisphere is tilted towards the Sun, it is summer there but winter in the southern hemisphere. Six 'months' later, when the Earth is on the opposite side of its orbit, the situation is reversed. The northern hemisphere is tilted directly at the Sun on about June 21 each year (the longest day). This is the *summer solstice.* It is tilted directly away on about December 21, the shortest day. This is the *winter solstice.* Midway between these dates are the *spring,* or *vernal equinox* (March 21) and the *autumnal equinox* (September 23), when the Sun lies directly over the Equator 'travelling' northwards (vernal) or southwards (autumnal).

Igneous and sedimentary rocks, and sandstone in the making. At left is dark volcanic, that is, igneous rock. At right is bedded chalk, a sedimentary rock. The sand could in time be compacted by material above it into solid rock. That is how our sedimentary rocks were formed millions of years ago.

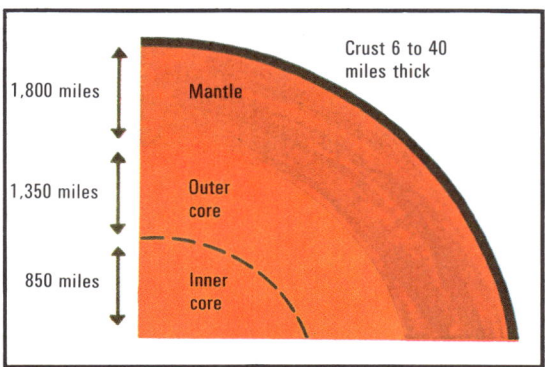

Right: A cross-section of the Earth, with approximate dimensions. The core is thought to consist of iron and nickel, which could explain the Earth's magnetism.

Below: The Earth as seen from space. This picture was taken by the Apollo 11 astronauts on their epoch-making journey to the Moon in July 1969. They were 106,000 miles out from the Earth at the time.

# The Moon and The Sun

**The Moon**

We know more about the Moon than we do about any other heavenly body because it is our nearest neighbour in space, a mere 239,000 miles away. And man has actually travelled to the Moon and landed on it (see page 24). He has seen for himself just what it is really like.

The Moon is Earth's only natural satellite. Some of the other planets have many more satellites (see page 182). The Moon is almost a quarter of the size of the Earth and naturally its mass and therefore its gravity is much less. The Moon's gravity is only one-sixth that of the Earth. Astronauts exploring the surface therefore feel extremely light on their feet and have some difficulty in walking. They almost have to hop around like a kangaroo! But their feet sink slightly into the lunar soil, causing distinct footprints. The soil is very crumbly, like newly ploughed soil on Earth.

The overall picture of the surface is one of extreme desolation. Gravity is so low on the Moon that there is no atmosphere, which means no life, no weather, no sound. Everywhere there are craters, mostly quite small, but some so large you can practically see them with the naked eye back here on Earth! Strewn around the craters are jagged lumps of rock. Everything is sharp and jagged because there can be no 'weathering' to smooth the surfaces as there is on Earth.

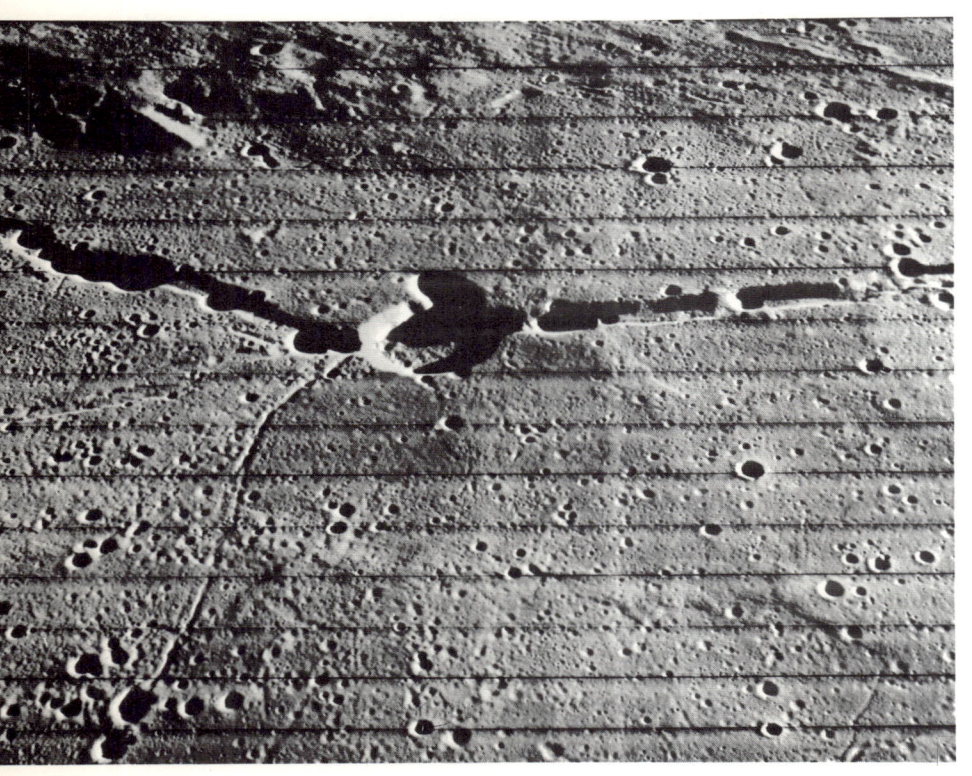

Both pictures on this page have been built up from photographs taken by American Lunar Orbiter craft in orbit around the Moon. The one above shows part of the Moon's hidden side. That below shows a prominent linear feature known as a rill, about 2 miles wide.

Most of the craters are probably caused by the impact of the lumps of rock from space called *meteorites* (see page 184). Only some of the big craters may well be volcanic in origin.

There are mountain ranges on the Moon, too, and also 'seas', called *maria*. They are not really seas, of course, because there is no water on the Moon. They are, in fact, great plains.

The temperatures on the Moon are extreme. In the lunar day, they reach more than the boiling point of water (100°C or 212°F). In the lunar night, they drop to $-150°C$, or about $-240°F$.

The Moon shines by reflecting the sunlight that falls on it. It moves around the Earth about once a month ('months' comes from 'moon'). And each time it orbits the Earth it turns once on its *axis*. This means that we always see the same side. As the Moon moves around the Earth, the area illuminated by the Sun changes, giving rise to what are called the Moon's *phases*. The *new moon* is when the Sun is illuminating the back of the Moon, and we can scarcely see the Moon at all. A week later half of the Moon's disc is lit up ('first quarter'), and in another week, the whole disc ('full moon' or 'second quarter'). Then the area lit up gets smaller to the 'third quarter' and then disappears again at another new moon.

The Moon's presence so close to the Earth gives rise to the *tides*. The Moon's gravity

Left: The large crater is Copernicus, the smaller one Eratosthenes. Note the typical mountain peaks at the center of the craters. (See page 17 for a close-up picture of Copernicus.)

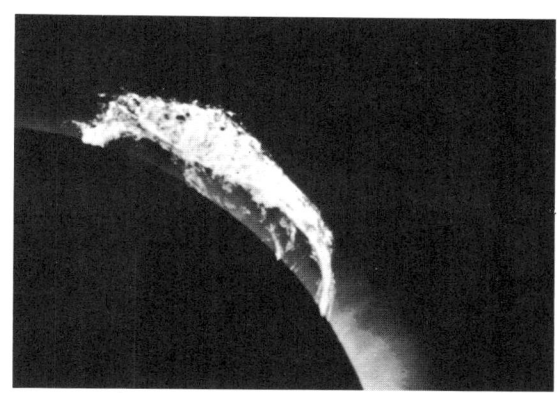

Right: A solar prominence visible on the "rim" of the Sun during a solar eclipse, when the photosphere is masked.

tries to 'pull' the water in the oceans away from the Earth. And to a certain extent it succeeds. In some areas of the world the water level drops as much as 50 feet from high to low tides!

## The Sun

Ninety-three million miles away from the Earth in space is the Sun, an intensely hot ball of glowing gas that is the centre of the solar system. The Sun is nothing more nor less than a star. It has very much the same size and brightness as millions of other stars in the universe. It appears to be so big only because it is so close to us. Ninety-three million miles may not seem very close, but it is in terms of distances in space (see page 186).

The bright disc of the Sun is called the *photosphere*, or 'lightsphere'. Often the photosphere has dark spots showing on it, called *sunspots*. They appear to be whirling gases, slightly cooler than the surrounding photosphere. Sunspot activity is at its maximum about every 11 years or so. At such times sunspots thousands of miles across appear on the surface of the Sun and they may last for months. We sometimes feel the effects on Earth. Somehow the sunspots set up magnetic forces that reach us here and disrupt long-distance radio communications. Compasses and other magnetic and electrical instruments, may be affected, too.

Surrounding the photosphere is a kind of thin 'atmosphere' of gases. It becomes visible only at times of a total solar eclipse, when the Moon's disc obscures the dazzling photosphere. During an eclipse the lower part of the Sun's 'atmosphere' shows up as a red fringe. It is called the *chromosphere,* or 'colour-sphere'. Up through the chromosphere shoot great flaming 'fountains', or *prominences.* The outer atmosphere, called the *corona,* or crown, shows as a great halo, millions of miles in extent.

How does the Sun continue to pour out heat and light, day in day out, without get-

Below: Sunspot activity on the solar disc. The sunspot regions are darker and appear to be cooler than their surroundings.

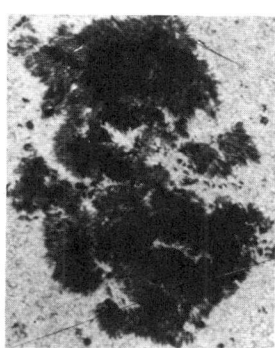

Below: The solar corona photographed during an eclipse. The corona consists of very tenuous gases and extends for millions of miles from the Sun.

ting cooler or dimmer? The answer is that it burns, not an ordinary fuel like coal or oil, but nuclear fuel. It acts like a mammoth nuclear reactor, changing one kind of atom into another and producing incredible amounts of heat and light in the process. The central core of the Sun is so hot that atoms of hydrogen combine, or fuse together to form atoms of helium. This process of fusion is the one that is imitated in the immensely destructive hydrogen bomb, and you know what incredible amounts of energy are released then (see page 33). Most of the other stars in the universe produce energy in the same way. Hydrogen, which is the simplest element in the universe, is therefore probably the most common element, too.

The Sun, like the Earth, has two distinct motion. It rotates on its own axis and travels in an orbit of its own around the far-off centre of the great star system, or *galaxy* (the Milky Way) to which it belongs (see page 187).

# The Planets

There are nine planets, including the Earth, in the solar system. Going outwards from the Sun, they are Mercury, Venus, (Earth), Mars, Jupiter, Saturn, Uranus, Neptune, and Pluto. They are very different in size and composition. Mercury, the planet closest to the Sun, is a scorching ball of rock not much bigger than the Moon. Jupiter, the biggest planet, is a gigantic ball of cold gas thirteen hundred times bigger than the Earth. And, of course, there are great differences in the planets' 'years' (the time they take to circle the Sun) because of their different distances from the Sun. Funnily enough, the planets' 'days' (the time they take to turn on their axis) are shortest for the biggest planets.

One thing the planets do have in common, however, is that they orbit the Sun in more or less the same plane. Only Pluto's orbit is appreciably inclined (20°). Also, the planets are all travelling the same way around the Sun. If you could look down on the solar system from a point near the North Star, say, you would see the planets moving anti-clockwise around the Sun. Most of the planets spin on their axes in an anticlockwise direction, too. But it appears that Venus has a clockwise, or *retrograde* motion.

Most of the planets have moons, or *satellites* revolving round them like a solar system in miniature. Mercury, Venus, and Pluto are the exceptions. The Earth, of course, has 1 – the Moon. Mars has 2, Jupiter 12, Saturn 10, Uranus 5, and Neptune 2. Most of these satellites revolve around their respective planets in an anti-clockwise direction, although a few do have a retrograde motion.

The planets fall roughly into two groups, the *minor,* or *terrestrial* planets – that is, those that more or less resemble Earth – and the *major* planets, so-named because of their size. The terrestrial planets are Mercury, Venus, Mars, and Pluto. They have a hard, rocky core, like that of Earth. The major planets Jupiter, Saturn, Uranus, and Neptune are primarily gaseous. Other ways of grouping the planets are as *inner* and *outer* planets. The inner planets from Mercury to Mars form a natural unit in the solar system as far as distance is concerned. Then there is a comparatively large 'gap' before we reach Jupiter and the outer planets. In this gap there is a great 'ring' made up of lumps of rock and known as the *asteroid belt* (see page 185).

## Comparison of the Planets

| Planet | Distance from Sun (million miles) | Equatorial diameter (miles) | Circles the Sun in* | Turns on axis in† (hours:mins) |
|---|---|---|---|---|
| **Mercury** | 36 | 3,010 | 88 days | 59 days |
| **Venus** | 67 | 7,650 | 225 days | 247 days§ |
| **Earth** | 93 | 7,927 | 365 days | 23:56 |
| **Mars** | 142 | 4,220 | 687 days | 24:37 |
| **Jupiter** | 483 | 88,760 | 11¾ yrs | 9:50 |
| **Saturn** | 887 | 74,160 | 29½ yrs | 10:14 |
| **Uranus** | 1,783 | 29,300 | 84 yrs | 10:49 |
| **Neptune** | 2,794 | 27,800 | 165 yrs | 15:48 |
| **Pluto** | 3,674 | 3,700 | 248 yrs | 153 hrs |

\* i.e. its 'year'.  
† i.e. its 'day'.  
§ retrograde

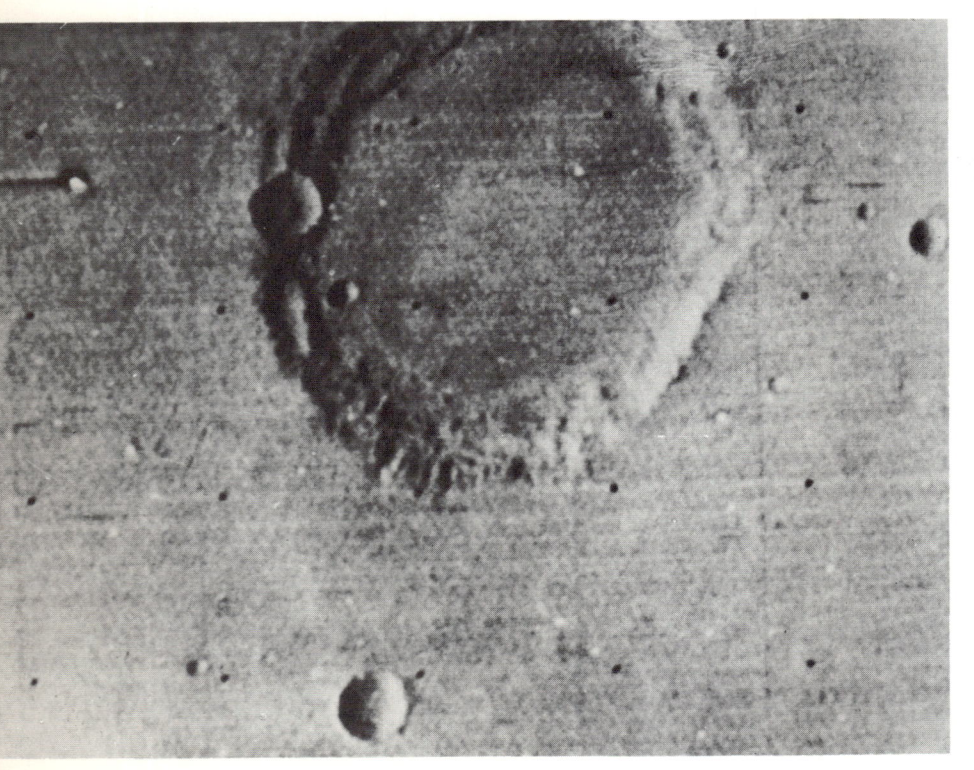

Mars was the first of the planets to be photographed from close to, by American Mariner space probes. This photo, taken from about 2,300 miles up shows a large crater some 24 miles across. Mars, like the Moon, is heavily cratered. But on Mars there is much evidence of erosion having taken place.

## The Inner Planets

*Mercury* lies so close to the Sun that the sunlit side is scorching hot. Being small and having therefore weak gravity, Mercury has no atmosphere to equalize the temperatures at all. Mercury is so close to the Sun that it can hardly ever be seen from the Earth. Once every while, however, it can be seen crossing the bright face of the Sun. Such events are called *transits*.

*Venus* is very similar to Earth in respect of size, and it has an appreciable atmosphere. But the atmosphere is composed of the heavy gas carbon dioxide, not air. Dense clouds in the atmosphere permanently obscure the surface features of the planet. These clouds reflect sunlight brilliantly. This, in conjunction with its closeness to Earth, makes Venus the brightest body in the heavens after the Sun and Moon. Sometimes it rises before the Sun and is known as the *morning star;* sometimes it sets after the Sun and is called the *evening star*. It is the easiest of all the planets to identify. And,

Above: Mars as seen from Earth. Dark regions and the polar ice cap can be clearly distinguished. The next planet beyond Mars is the giant planet Jupiter. It is pictured below compared in size with the Earth. Farther out still is the second largest planet Saturn, notable because of its fascinating ring system (right).

like the Moon, Venus shows distinct phases.

*Mars* has been given the name of the *red planet* because it appears distinctly red to the eye. Most of its surface appears to be a reddish colour. But at the poles there are great white 'caps' which grow larger and then smaller at different times of the Martian year. Also dark patches appear, grow bigger, and then disappear again.

Mars has definite seasons, like Earth, because its axis is tilted to the plane of its orbit. And the coming and going of these surface features correspond with changes in the Martian 'seasons'. It has therefore been suggested that the "caps" are ice and that the dark patches are areas of primitive vegetation. But available evidence tends to rule out these theories. For one thing the atmosphere of Mars is very thin and there appears to be no oxygen present to support life and little, if any, water vapour. There is, however, a little carbon dioxide, so the caps could be dry ice, or frozen carbon dioxide.

American *Mariner* space probes have taken some remarkable photographs of Mars in recent years. The pictures show the surface to be very much like that of the Moon, covered with hundreds of craters. The surface features are not as rough, nor as jagged as those of the Moon, though.

**The Outer Planets**

*Jupiter* is made up chiefly of hydrogen, which means that, although it is very big, it has a comparatively small mass. It probably has a central core of frozen hydrogen, and its atmosphere is made up of methane and ammonia. Swirling currents in the atmosphere give rise to dark-coloured bands in it parallel to the planet's equator. And a mysterious Red Spot can usually be seen, too, changing its position all the time!

*Saturn* is similar in composition to Jupiter, and has similar bands in its atmosphere, too. But its most striking feature is its *ring system*. Around its equator there are three flat rings, one inside the other and in the same plane. They are believed to be made up of either pieces of rock or frozen gas.

*Uranus* and *Neptune* are probably similar to Jupiter and Saturn in composition, but a lot colder. *Pluto* is so far away that we know little about it. Even its size is not accurately known. But we do know that it must be very, very cold.

**Asteroids**

As we mentioned earlier, there is an apparent 'gap' in the solar system between Mars and Jupiter, where we would expect to find a planet. But we do not find a planet, but a great belt of bodies called asteroids, or planetoids. It is possible that they are indeed the remains of another planet, but evidence points against it. Most of the asteroids are too small and too far away to be detected with the instruments we have on Earth. But over 1,500 have been discovered.

**Meteors**

If you remain star-gazing for any length of time on a clear night, you are almost bound to see a falling, or shooting 'star'. It is not a star, of course, but the bright streak of a

Pluto, remotest planet in the solar system, can only just be distinguished from Earth as a faint 'star' (arrowed above). Even the biggest telescopes are not powerful enough to reveal any detail.

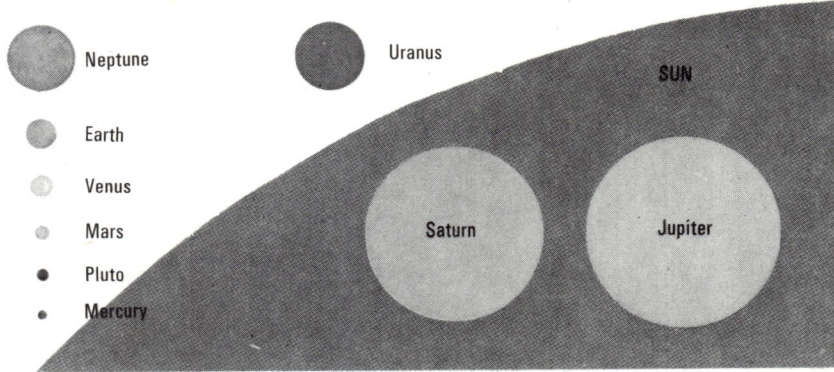

Comparative sizes of the planets. Note the vast difference in size between the four biggest planets Jupiter, Saturn, Neptune and Uranus (the so-called major planets) and the others, which are termed minor planets. But note also how even the biggest planets are dwarfed by the Sun, whose diameter is 865,000 miles.

Relative positions of the planets in the solar system. Earth, as you can see, is comparatively close to the Sun, being on average about 93 million miles away. Compare this with the 3,674 million miles between Pluto and the Sun.

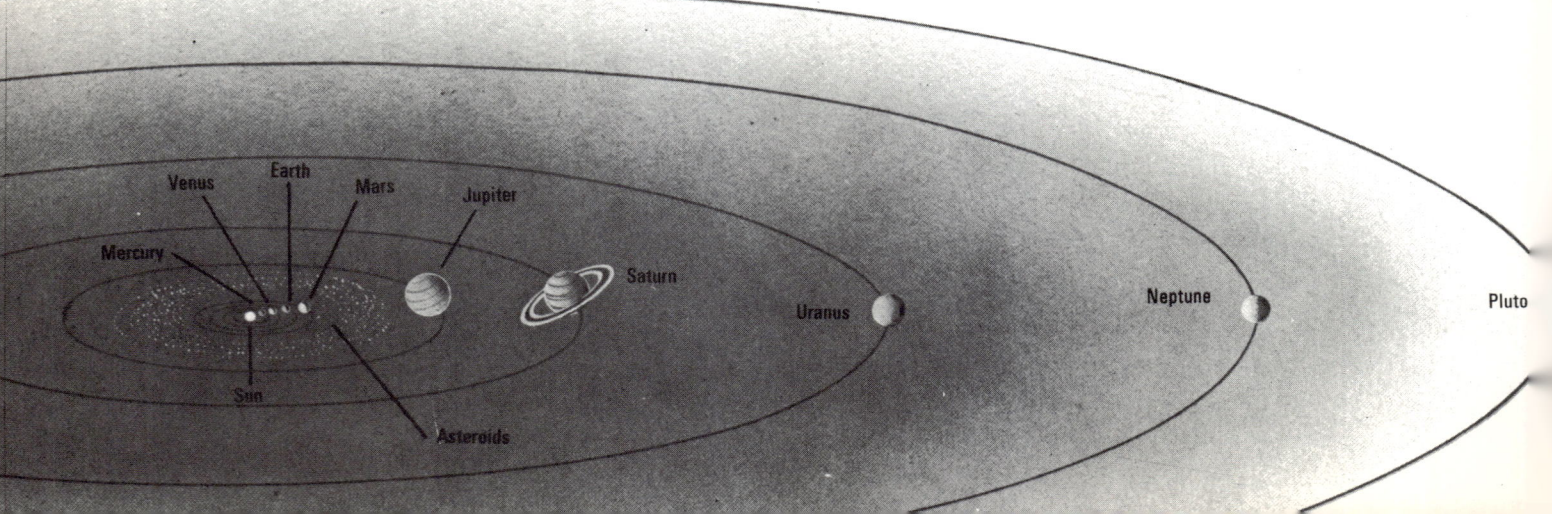

*meteor*. The streak is caused by the passage of lumps of rocks from outer space through the Earth's atmosphere. The lumps are travelling very fast and friction with the upper air, thin though it is, heats them up until they glow, first red, and then white hot. As the heat becomes fiercer and fiercer, they disintegrate, and usually burn up to nothing.

Sometimes, however, the lump of rock is so big that parts of it reach the ground intact. We call these pieces *meteorites*. Some of these meteorites are not rock-like at all, but metallic. They are made up chiefly of iron and nickel.

A few meteorites have been really gigantic. They made great craters as they struck the Earth with tremendous force. The largest known crater is the one in Arizona, which is nearly a mile wide, and several hundred feet deep. Scientists guess that the meteorite weighed hundreds of thousands of tons and that it plummetted to Earth maybe 50,000 years ago. More recent falls of large meteorites occurred in Siberia in 1908 and 1947.

## Comets

Other members of the solar system which we can occasionally see in the heavens are the *comets*. Comets can be the most spectacular of all the heavenly bodies. They are like great glowing balls moving against the background of stars. Some of them have what looks like a tail fanning out from them. But the tail does not stretch out behind them as they move across the sky. It points away from the Sun.

The head, or *nucleus* of the comet is probably made up of a mass of dust and gas, frozen solid. The tail is presumably composed of minute specks of dust from the head. The 'pressure' of streams of particles from the Sun pushes the specks away from the head to form the tail.

Comets, too, have definite orbits and reappear regularly, although some break up after a while. The most famous comet of all is Halley's Comet, last seen in 1910 and due to return in either 1986 or 1987.

Above: The most famous of the terrestrial meteor craters, the Arizona meteor crater near Canyon Diablo in north-east Arizona, USA. 570 feet deep, it has a diameter of 4,200 feet.
Below: The most famous of all comets, Halley's comet, named after second Astronomer Royal Edmond Halley. This comet has been observed at 29 returns, the earliest in 240 BC. The picture was taken in 1910 when it was last seen.

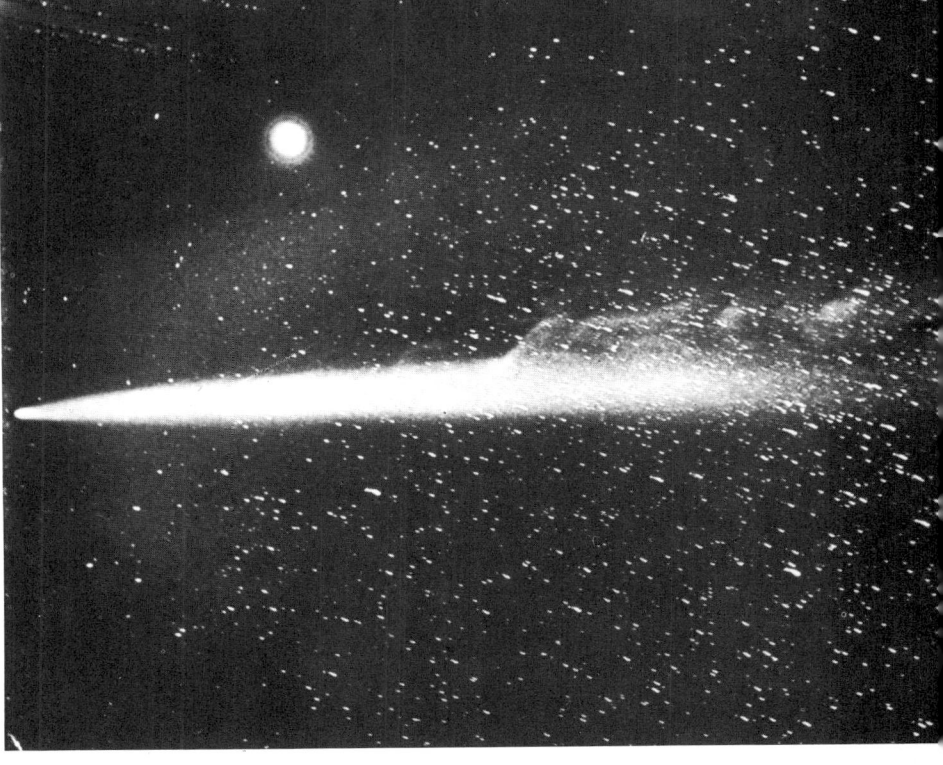

# The Stars

## Measuring Distances in Space

The Sun is our star, 93 million miles away. The next nearest star lies about 25 million million miles away! Light travelling at 186,000 miles every second takes over four years to reach us from that star (Proxima Centauri). The mile as a unit of measurement for such distances is hopelessly small. It is far easier to express the distance by saying that light takes over four years to reach us, or that the star is over 4 *light-years* away. (1 light-year is equivalent to about 6 million million miles.) There are objects in the sky that astronomers estimate to be over 10,000 million light-years away!

Astronomers measure distances in space in a variety of ways. Where they can, they like to estimate them in more than one way as a check. The method they use for the nearest stars is called the *parallax* method.

Stars cluster together in space in great star islands, or galaxies. A great number of the galaxies are spiral in form, like that above. Our own Milky Way galaxy is spiral. Many others are elliptical, such as that below.

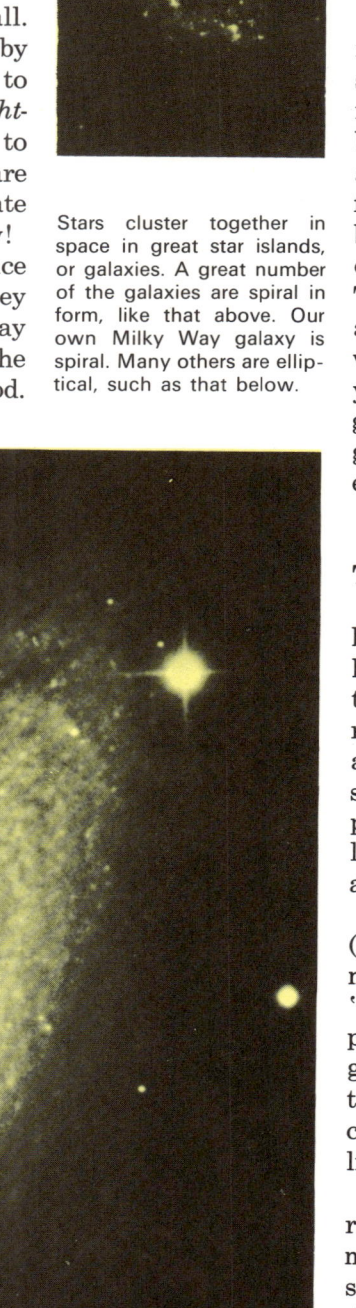

It is based on the principle that the position of an object relative to its background appears to change if it is viewed from different points. Notice how your finger appears to move when you look at it first with one eye and then the other. The farther the viewing points are apart, the greater will be the apparent displacement. That is why astronomers view the stars from opposite ends of the Earth's orbit. Even then they find that only a few show a slight displacement in position.

The parallax method takes the astronomer as far as about 100 light-years. To 'measure' still farther, he looks for certain stars in the sky called *variable stars*, whose brightness increases and decreases regularly. He knows that the *period* (the time from bright to bright) is related to the true brightness of the star. And he knows by observation what the apparent brightness is. Therefore he can calculate roughly how far away the variable star, and stars associated with it, must be. After about 5 million light-years, variable stars cannot be distinguished. The astronomers can then only guess at the distances involved, using their experience of nearer stars.

## The Spectroscope

Stars are made up of large amounts of hydrogen, which they 'burn' to produce heat and light (see page 181). They also contain helium, calcium, sodium, and other relatively light elements. How do we know all this? The answer lies in the spectroscope, which is a combination telescope and prism (see page 47). It splits up the white light from a star into its component colours as a *spectrum*, like the colours in a rainbow.

Crossing the spectrum are dark lines (*Fraunhofer* lines), which identify the elements present in the star's cooler, outer 'atmosphere'. The density of this atmosphere can also be estimated, because the greater the density the wider the lines. Even the temperature can be guessed, too, because a hot atmosphere gives rise to fewer lines than a cool one.

The spectrum of a star has even more to reveal. It can tell us how fast the star is moving! We find that the spectral lines of a star moving towards or away from us are shifted from where we would expect to find them if the star were stationary. The amount of the shift indicates the star's speed. The direction of the shift tells us if the star is approaching or receding from us. A star approaching us has a *blue shift*, or a shift towards the blue end of the spectrum. A receding star has a *red shift*.

Astronomers find that most stars have red shifts and are receding. The farthest stars seem to be travelling fastest. This leads us to believe that the whole universe is *expanding,* as though from a gigantic cosmic explosion millions of years ago (see page 189).

## Constellations

When you look up into the night sky, you can see recognisable patterns of stars there. With a little imagination, you might say that certain patterns remind you of, say, a flying swan, a scorpion, or a bear. There are, in fact, patterns that do resemble these things. They are some of the so-called *constellations* which man named thousands of years ago. But these star groups are groups only insofar as the human eye is concerned. Few bear any relation to the actual distribution of the stars in space.

## Galaxies

But stars are associated together in definite systems, called *galaxies.* At least a thousand million galaxies have been discovered. The Sun and all the bright stars in the sky form part of the *Milky Way* galaxy. The galaxy is thought to look something like a rotating Catherine wheel – a disc with vast spiral arms. In its centre is a dense mass of stars. Not all galaxies are spiral galaxies like our own. There are elliptical galaxies, for example, which have no 'arms'.

The Sun lies quite a way from the centre of our galaxy. When you look at the fuzzy band of stars in the sky we call the Milky Way, you are actually looking edge-on at the disc of the galaxy. You cannot see the great mass of stars at the centre of the galaxy because of great 'clouds' of gas and dust in between. These obscuring 'clouds' are called *nebulae.* They may be dark or bright, depending on whether they have any stars within them.

The centre of the galaxy is thought to be surrounded by vast closely knit groups of stars called *globular clusters.* We can see only a few of them because they are mostly obscured by the nebulae, but those we can see (with a telescope only) are astonishing. They contain thousands upon thousands of stars. Elsewhere in the sky there are more loosely knit groups of stars called *open clusters.* The Pleiades, or Seven Sisters is the best-known and most easily visible open cluster.

## Binary and Pulsating Stars

Most stars shine steadily year in, year out. (They twinkle, of course, but that is caused by dust and air currents in the atmosphere.) Some, however, vary in brightness in a regular way. If we examine some of these so-

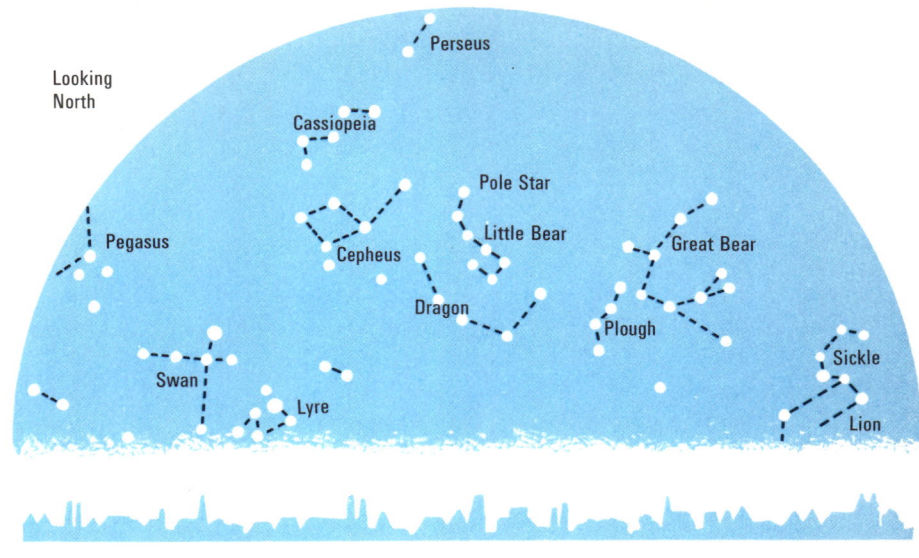

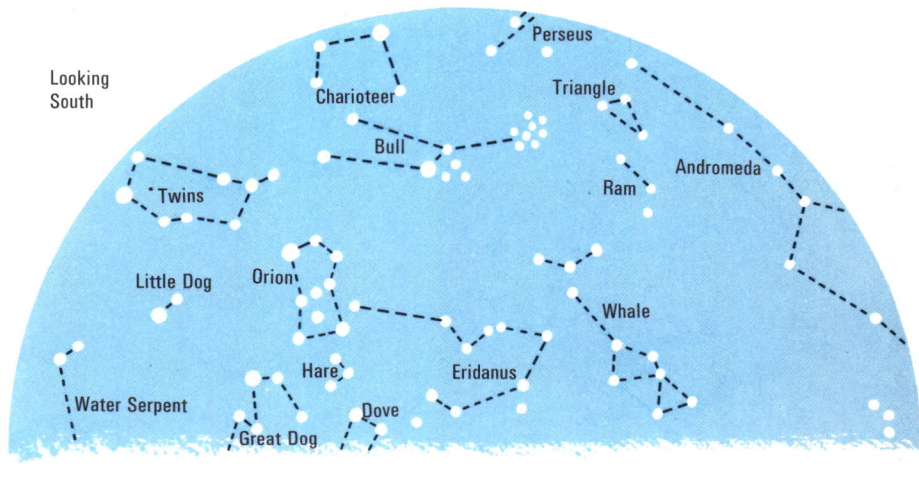

Above: Constellations that can be seen in the night sky in winter, looking north and south. The Pole Star is situated almost directly above the Earth's geographical North Pole. During the night all the stars appear to move in circles around the Pole Star because of the Earth's rotation on its axis.

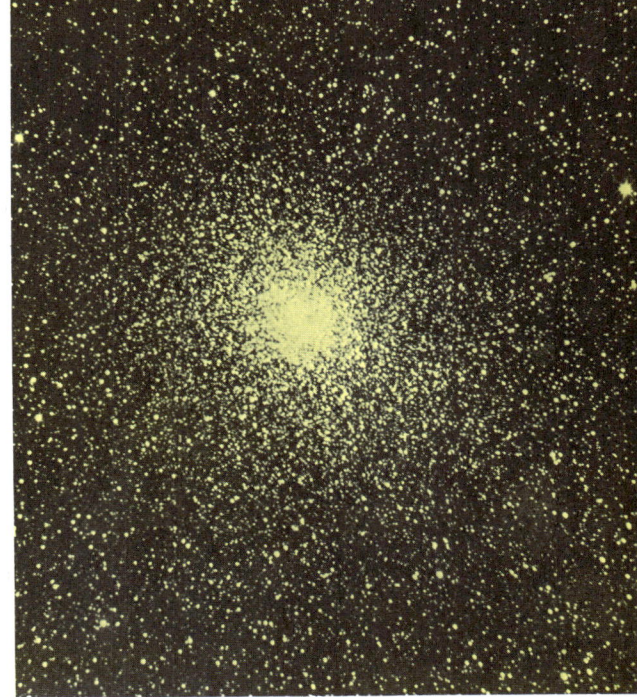

Right: Tens of thousands of stars grouped together in what is termed a globular cluster. Such tightly knit clusters of stars seem to be situated around the centre of the galaxies.

187

called *variable stars* carefully, we find that they are made up of not one but two stars. These two stars rotate around each other in such a way that regularly one of them passes in front of the other and obscures its light. Therefore the total brightness of the system decreases until the stars are clear of each other again.

Such a star system is called an *eclipsing binary*. There are many other binary systems in which the stars do not cross our line of sight, and therefore they maintain their brightness as far as we are concerned. A few of these binaries can be distinguished with the naked eye. But many are so close together that they can be distinguished only by using a spectroscope.

The true variable stars are totally different. Their variation in brightness comes in some way from within. They are called *pulsating stars* because they are thought to get alternately bigger and smaller. But it is not at all certain what they do. Some vary as regularly as clockwork and it is these which we use to estimate distances (see page 186). The best-known variables of this type are called the Cepheids.

### Novae and Supernovae

Pulsating stars seem to be inherently unstable, as though they are liable to explode at any time. Some stars, however, become so unstable that they do, in fact, explode. They suddenly flare up for a short while and then fade just as suddenly. Before it flared up, the star may not have been noticeable. And it would appear when it 'explodes' that a new star has appeared. That is why astronomers call such an event a *nova*.

Occasionally the explosion is so catastrophic that it literally blows the whole star apart. Then it is called a *supernova*. There were supernovae recorded in 1054, 1572, and 1604, but none have been recorded since.

### Radio Sources

The Sun emits radio waves, and so presumably do the other stars but they are too weak to reach us.

However, very intense radio waves have been detected coming from the universe. Astronomers use special, *radio telescopes* which have great metal troughs or dishes to gather and focus the radio waves onto a receiving aerial. Many of the so-called radio sources are invisible, but some seem to be associated with visible objects. Some galaxies emit radio waves; so do the remains of supernovae. And so do some mysterious bodies called *quasars*, or *quasi-stellar radio sources*.

Quasars appear to be simply small stars from their photographs. But they seem to be travelling incredibly fast and be thousands of millions of light-years away. For this distance they have a fantastic brightness. They are thought to be 100 times brighter than the brightest galaxy, yet thousands of times smaller! Yet even quasars seem ordinary when compared with the *pulsars*, which are radio sources that pulsate rapidly and regularly. They release bursts of energy equal to the output of tens of thousands of stars. The pulses may last only for a ten-thousandth of a second, and may be repeated 30 times a second. Yet it appears that pulsars can only be about 20 miles across! X-ray and optical pulsars have also been found.

### The Origin of the Universe

Controversy rages over how the universe came into being. It is unlikely that anyone will ever know for certain. But there are two main theories. One is that the universe came into being at a definite time and evolved into what it is today. The other is

Two of the most famous nebulae are associated with the constellation Orion. Left is the Great Nebula in Orion, which is a bright nebula visible to the naked eye. Below is the aptly named Horse's Head Nebula, which is a dark nebula seen against a bright nebula and giving the impression of a horse's head.

that the universe has always been in existence and is basically the same now as it was millions of years ago and will be millions of years from now. You could hardly conceive two more divergent theories!

The first theory, the so-called *evolutionary*, or *big-bang* theory, supposes that all the matter in the universe was once concentrated together in one place. Between ten and fifteen thousand million years ago, this concentrated 'lump' exploded. Gas and 'dust' were thrown out in all directions. Eventually, some of the cosmic material 'condensed' into denser masses which evolved into the stars and the galaxies, always expanding.

The second theory, the so-called *steady-state* theory, supposes that there was no beginning, and there will be no end to the universe. Matter is being continuously created. In any part of the universe galaxies form and fly off into space; new galaxies form from the new matter to keep the overall state of the universe steady.

So far it seems that the big-bang theory accounts closer for the observed state of the universe than the steady-state theory.

Some of the most spectacular advances in astronomy in recent years have been in the field of radio astronomy —that is, 'viewing' the universe at radio rather than light wavelengths. Incredible bodies such as quasars and pulsars have been revealed, whose nature and methods of energy production continue to baffle astronomers. The picture below shows the antenna of a powerful radio telescope. Signals from such an instrument are recorded, right.

# Textile Fibers

Practically all our cloth and other textiles are made from fine fibers. By twisting the short, weak fibers together, a continuous strong yarn can be made. This is the basis of *spinning*. By interlacing the yarn in various ways, many kinds of fabrics can be made. *Weaving* and *knitting* are two common ways of doing this. Most of the fibers we use come from natural sources – plants, animals and even minerals (asbestos). But more and more artificial, or man-made fibers are being used.

## Plant Fibers

The most important fibers obtained from plants are cotton and flax, from which linen is made. Like all plant fibers, they are made up of almost pure cellulose. Cotton fibers come from the seed pod, or *boll* of the cotton plant, a shrub which thrives in warm, moist tropical and sub-tropical regions. When ripe, the bolls burst to expose the seeds, which are covered with a fluffy mass of delicate white fibers. On the vast plantations in the southern states of America, which produce more than a third of the world's cotton, cotton picking is done by machine. But elsewhere much is still picked by hand. The picked cotton goes to the ginnery, where the cotton gin separates the long, spinnable fibers, called *lint*, from the seeds.

Flax for making linen thrives in the cool, moist climate of northern Europe. It is grown in much the same way as cereals such as wheat and oats. At harvest time the flax is pulled from the ground by hand or by machine so as not to damage the stalks, which contain the fibers. The first stage in processing the flax is *rippling,* which means removing the seed heads from the stalks. Afterwards, the stalks are laid out in the fields in shallow pools of water or in tanks, where they are allowed to rot slightly. This

Cotton is the world's most widely used fiber. These days the ripe bolls are often harvested by machines such as those shown below. But a great deal of cotton is still picked by hand.

process, called *retting*, loosens the fibers within the stalks. After drying, the fibers are separated from the stalks in a *scutching* machine.

**Animal Fibers**

The most valuable animal fibers are wool and silk, which are made up of complicated substances called proteins. Wool comes from the fleece of sheep. It is really a kind of hair that is naturally curly, or crimped. The finest wool comes from the Merino breed of sheep. Australia is by far the world's largest wool-producer, with fifteen times as many sheep as people. The sheep are generally sheared in the late spring. Nowadays most shearing is done with power shears, which are somewhat like a barber's electric clippers.

Silk, the most prized of our textile materials, comes from the cocoon spun round itself by the silkworm, which feeds on mulberry leaves. The 'worm' is actually the larva, or caterpillar, of a large, cream-coloured moth. The cocoon is made up of an extremely fine,

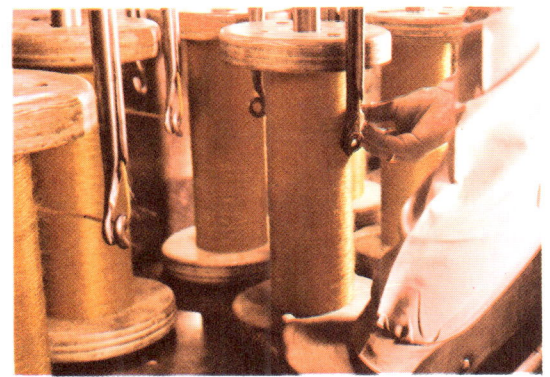

**Other Plant Fibers**

There are several other plants which, like flax, contain fibers that can be processed into textile materials. Although the fibers are too coarse to make fine cloth, they nevertheless have plenty of uses. *Hemp* is widely cultivated in the warm temperate regions of the world for its long, coarse stalk fibers. Rope, twine, canvas, and sailcloth are some of the products made from them. *Jute* is grown in tropical countries, especially in India and Pakistan, also for its stalk fibers. Jute fibers find a vast outlet in the form of sacking, hessian, and the backing of carpets. Some kinds of tropical plants called agaves contain tough, white fibers in their long, swordlike leaves. The fibers, which are called *sisal*, are made into ropes, sacking, and mats. Even the hair on coconuts, called *coir*, is a useful fiber. It is made into coarse floor coverings, the so-called coconut matting. In some countries banana leaves are shredded into fiber and made into ropes, mats, and occasionally into paper.

**Other Animal Fibers**

Other animals besides sheep produce useful fibers. The hair of goats, for example, is widely used in many parts of the world to make carpets and fabrics. The hair of most common goats is too coarse to be suitable for anything but the coarsest fabrics. But two goats, the Cashmere and Angora goats, have fleeces of fine, silky hair which is much more lustrous than wool because it is not so crinkled. The softest of all fabrics are made from the downy underhair of the Cashmere goat, which lives in the mountains of Tibet and northern India. The smooth, long hair of the Angora goat, a native of Turkey,

is better known to us as *mohair*. It is a popular material for sweaters, lightweight suiting, and upholstery. It is very hardwearing.

Several members of the camel family grow fleeces which yield valuable fibers. The two-humped Bactrian camel, from the desert regions of central Asia, grows a thick, shaggy coat to protect itself from the severe winters. In summer it moults, or sheds its coat. The famous camel-hair cloth, used for high-quality suiting and overcoats, is made from the moulted fleece. The camel-like llama, alpaca, and vicuna, which live high in the Andes mountains of South America, also yield valuable fleeces.

Above: Viscose rayon emerging from the acid bath where it has been regenerated in fine filament form. After cotton, viscose rayon is the most widely used fiber.

Above left: Acrylic yarn being wound onto bobbins after manufacture. Acrylic fibers are widely used for making carpets and blankets.

Left: Wild agaves of the kind that yield tough leaf fibers useful for ropemaking.

continuous thread, which the silkworm secretes from a spinning gland, or spinneret, in its head. Unreeling the fine thread from the cocoon is a very delicate job, usually done by women. They throw several cocoons into hot water to soften the gum binding the threads together. Then they catch the thread ends and draw them onto a rotating reel. More than a thousand yards of the fine thread may be unwound from a single cocoon.

Of all the natural fibers, only silk is produced in the form of a continuous thread,

An expert sheep-shearer at work. The fastest shearers can shear up to 500 sheep a day! Merino sheep yield the heaviest and finest quality wool. As much as 28 pounds may be taken from a full-grown ram.

or *filament*. The other natural sources yield short, *staple* fibers.

Each kind of fiber has its own particular properties and advantages. Cotton is strong and hard-wearing. Linen is crisp and shiny. Wool is warm and resists creasing. Silk is soft and lustrous. They all absorb water readily, which makes them suitable for clothing next to the skin, which comes into contact with perspiration. They can be dyed and finished easily. But they rot and may be attacked by insects, such as moths.

The range of natural fibers suitable for making textiles is rather limited. These days an increasing variety of *man-made fibers* are being used as well. These are fibers which do not occur in Nature but are fashioned in some way by man, either from natural materials or from chemicals.

### Regenerated Fibers

The original man-made fiber, and still the one most widely used, is *rayon*. In its early days, it used to be known as 'artificial silk' because of its appearance and 'feel'. Rayon is made from a natural material – the cellulose in plants. It is known as a *regenerated* fiber because in the manufacturing process, the cellulose is dissolved and then reformed, or *regenerated* in a different form. Other regenerated fibers are made from the proteins in milk (casein), maize (zein) and ground nuts. But they are not much used.

There are several different kinds of rayon, each with different properties. The most common kinds are viscose, acetate, and triacetate. Triacetate rayon is better known under such trade names as Arnel (United States) and Tricel (Britain). The starting point for all three kinds is either purified wood-pulp or cotton. Both are forms of pure cellulose. The cotton used is the linters, or short fibers that are unsuitable for normal spinning processes.

To make *viscose* rayon, the cellulose is treated with chemicals to form a syrupy solution. This solution is then pumped through tiny holes in a spinning cap, or *spinneret,* and into an acid bath. The acid changes the fine streams of cellulose solution emerging into fine threads, or *filaments* of pure cellulose. This method of forming fibers is known as *wet spinning*. The filaments are gathered into a single thread, which may then be wound on high-speed reels to form what is called continuous-filament yarn. Alternatively, the thread may be chopped up into short lengths to form staple fibers, which can then be spun in the same way as cotton and wool.

*Acetate* and *triacetate* fibers are produced in a slightly different way. They are not pure

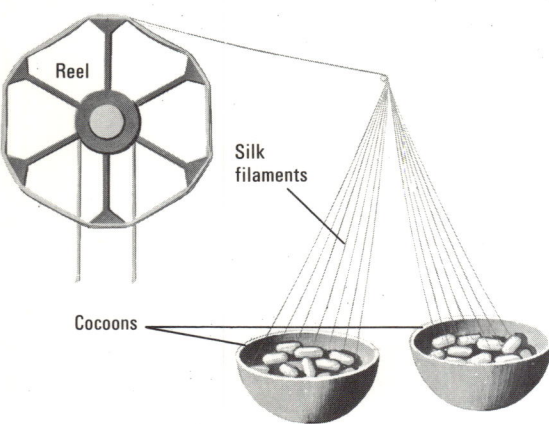

Preparing silk for spinning is a comparatively simple process. The cocoons are thrown into hot water to release the gummed-up filaments, which are gathered into a loose rope and wound onto a reel.

cellulose but compounds of cellulose. The fibers are formed differently, too, by a process called *dry spinning*. The cellulose compounds are dissolved in a solvent that evaporates in the air. The fibers are formed when the solution is forced through a spinneret into a stream of warm air.

After cotton, rayon is the most widely used of any natural or man-made fibers. Vast quantities are used in continuous-filament form to make the corded fabric in motor tires. In staple form it is spun into yarn for use in all kinds of clothing fabrics. One of the biggest uses of staple viscose is in carpets. Acetate fabrics are valuable because of their great similarity to silk. Both acetate and triacetate fabrics keep their shape well. They can be pressed into permanent pleats which do not come out during washing. Also, they need little or no ironing.

## Synthetic Fibers

The other class of man-made fibers includes the *synthetics*. Synthetic fibers are produced, not from any kind of natural materials, but entirely from chemicals. The original and best-known synthetic fiber is *nylon*, which is used in all kinds of textiles from underwear and overcoats to carpets and parachutes. Nylon and all the synthetics are kinds of plastics which can be drawn out into long fibers. They are often made by heating together certain chemicals in a kind of pressure cooker called an *autoclave*. The small, individual molecules of the chemicals link up to form the long-chain molecules typical of plastics. This process is called polymerization, and the product is called a polymer (see page 38).

Nylon is the name by which we know the *polyamide* fibers. The other leading synthetics include the *polyesters* and the *acrylics,* which are better known by their trade names. Terylene, Dacron, and Fortrel are polyesters. Acrilan, Courtelle, Dynel, and Orlon are acrylics.

Fibers are made from the polymers in a variety of ways. Nylon and the polyester fibers are made by *melt spinning*. The polymer is melted and then forced through the tiny holes of a spinneret. The long filaments formed quickly harden in a stream of air. They are collected into a single yarn and wound on bobbins. Later the yarn is drawn, or stretched, either cold (for nylon) or hot (for the polyesters) to give it greater strength.

The acrylics cannot be made into fibers by melt spinning because they break down when they are heated. They first have to be dissolved and forced through the spinneret as a solution. The fibers are formed either by dry spinning into the air or wet spinning

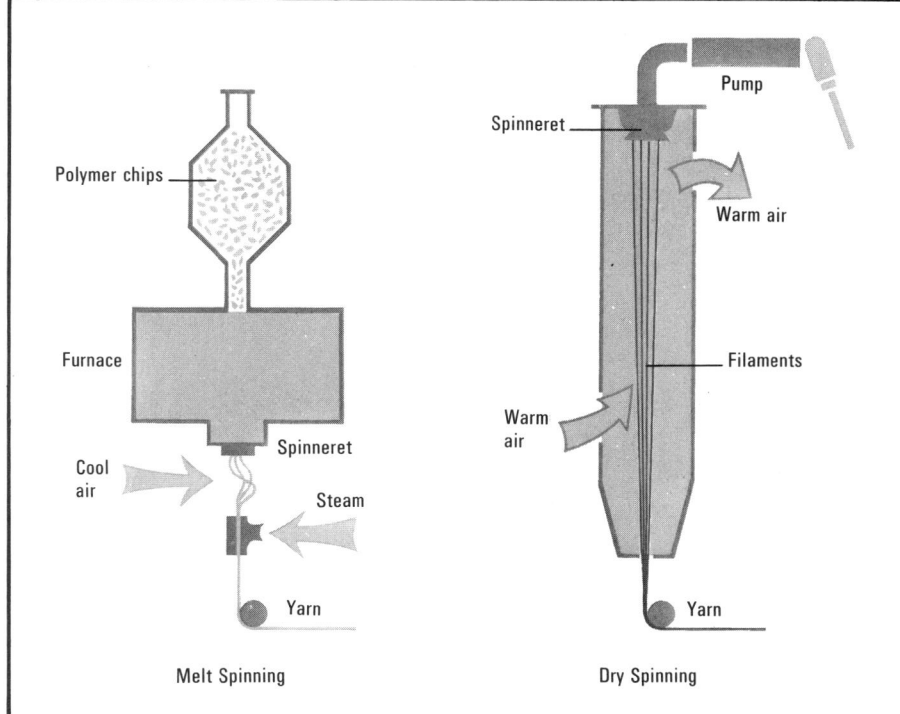

Above: Two common ways of making synthetic fibers. Melt spinning involves pumping molten plastic through the tiny holes of a spinneret and cooling the emerging filaments. Dry spinning involves pumping a solution of the plastic in a volatile solvent through the spinneret and then evaporating the solvent.

Right: Calendering cotton cloth. This process is used to impart a lustre to the fabric.

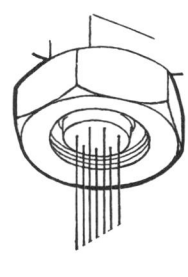

A Spinneret

into a solution (as for the rayons). It all depends on the particular solvent used to dissolve it in the first place.

The continuous-filament yarns of synthetic fibers may be used directly for making textiles. But often they are chopped up into short lengths to make staple fibers, which are then spun in the normal manner. In this staple form they can be blended with other synthetic or with natural fibers.

Synthetic fibers are much stronger than natural fibers and rayon. They are resistant to insect attack and do not rot. Also, they do not absorb water. This makes them suitable for use in outer garments and easy-to-wash, 'drip-dry' fabrics. Synthetic fibers are naturally crease-resistant, too.

Synthetic fibers are in widespread use in a variety of textiles. In their pure state they look much alike, but there are significant differences between them. Nylon, for example, is elastic and is therefore ideal for close-fitting clothing like stockings and underwear. Polyesters on the other hand resist stretching. Fabrics made from them therefore keep their shape better.

# Spinning and Weaving

All natural textile materials, except silk, are made up of fairly short, fine fibers. Cotton, for example, has fibers up to 2 inches long. Wool fibers are a little longer. Individual fibers are too short, too fine, and too weak to be woven into cloth. They must first be gathered together, drawn out into a long 'rope,' and then twisted tightly to make yarn. This is the basis of *spinning*. The great bulk of the world's yarn is spun by high-speed automatic machines.

It was the introduction of the forerunners of these machines which sparked off in the 1700s the period of industrial expansion known as the Industrial Revolution.

## Preparing the Fibers

The 'raw' fibers of cotton, wool, or flax from the plantation or farm are by no means in a fit state for spinning. A variety of preliminary operations must be carried out on them first. Bales of raw cotton, for example, arrive at the spinning mill in a tight, tangled mass, full of dirt, leaves, and bits of stalk. There, they are fed in turn into a series of machines called the *bale-breaker,* the *opener,* and the *scutcher*. These machines break up the bales, beat and suck out the dirt and other bits and pieces, and open out the fibers into their original fluffy state. The cotton fibers emerge from the scutcher as a continuous fleecy sheet, or *lap*. The lap then goes

Above: Threading the warp yarn onto the warp beam prior to weaving on the loom. Below left: A traditional spinning wheel, the first mechanical aid to spinning, which was originally introduced in China in the 1500s. Below right: A modern knitting machine being used to knit a circular, seamless fabric. The latest machines can produce millions of stitches a minute.

into the *carding machine,* or *engine,* which straightens out the fibers from their natural tangled state. The engine also removes short fibers which would otherwise make the resulting yarn weak. The carding engine consists of a number of rotating rollers covered with wire 'teeth'. The rotating teeth open out and disentangle the fibers. For fine-quality cottons, the fibers are straightened further by *combing.* In the combing machine, the fibers are held and combed even straighter by needle-like teeth.

The fibers leave the machines in the form of a filmy web, which is then split into sections and gathered into loose ropes called *slivers.* Several slivers are combined, drawn out through rotating rollers, and slightly twisted. This process produces an intermediate yarn called *roving.* And it is in this form that the fibers go to the spinning machines.

Raw wool, too, is very dirty when it arrives at the spinning mill. It is greasy and contains bits of straw, soil, and dried sweat. In fact, dirt and grease make up half of the weight of the raw wool. The grease is removed by *scouring,* or washing the wool with soap, detergents, or with chemicals. Vegetable impurities are removed by a process called *carbonization* in which the wool is treated with acid and heated.

After cleaning, the tangled mass of wool fibers is treated with olive oil to make it supple and to prevent it tangling more in later processes. These processes differ according to whether worsted or woollen yarn is required. *Worsted* yarn is made from only the long wool fibers. The wool undergoes the same carding and combing operations as outlined above for high-quality cottons. *Woollen* yarn is made from both the long and the short fibres. The wool goes into a kind of carding machine which reduces it to a web in which the fibres lie in all directions.

The wool coming from the combing or carding machine is gathered into sliver form, washed, and then made into roving for spinning as before.

Flax fibers, too, undergo preliminary cleaning and carding. The machine which does it is called a *hackling* machine. Again, the emergent slivers are made into roving.

### The Spinning Machines

We now have our fibers in a suitable state for spinning by one or other of the spinning machines. The machines differ mainly in the way in which the yarn is twisted and wound. The two most important spinning machines are the *ring-spinning frame* and the *spinning mule.* Variations of the ring frame are the

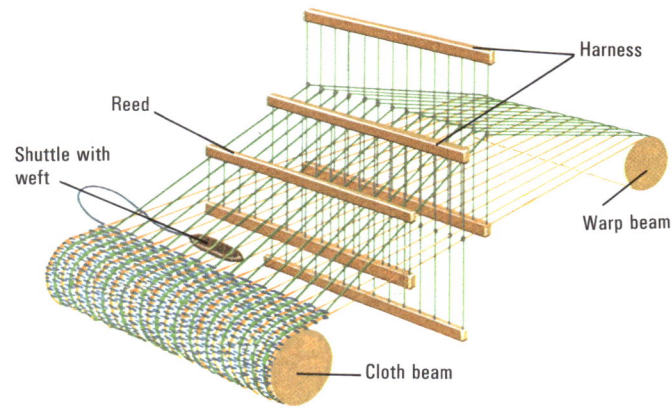

The main elements of a simple loom. Movement of the harness up or down parts the warp threads, and the shuttle shoots through the gap (shed) created. The reed moves in and presses, or 'beats' the new line of weave hard against the previous line.

Below left: A carding engine. A tangled 'lap' of fibers enters the machine and is transformed into loose ropes of parallel fibers called slivers. The slivers are drawn out into finer rope called roving which is then ready to go to the spinning machines. On the ring-spinning frame below right the roving can be seen entering at the top. The bobbins of fine, spun yarn are being removed from the front of the machine.

*cap frame* and the *flyer frame*. Cotton is usually spun on the ring frame or the mule. Wool can be spun by any kind of machine. Linen is always spun on the flyer.

On the spinning machines the prepared roving passes through several pairs of rollers, each pair revolving at a slightly higher speed than the previous one. This draws out the roving into finer and finer strands of yarn.

On the ring frame, the yarn is then guided onto rapidly rotating bobbins by metal clips called *travellers*. The travellers move around the bobbins and twist the yarn as they do so. On the mule, the yarn from the rollers is wound onto spindles which move backwards and forwards on a carriage. As the carriage moves forward, it draws out more yarn. The yarn slips off the spindles and is twisted. The twisted yarn is wound onto the spindles as the carriage moves back.

## Silk Throwing

Raw silk does not have to go through these kinds of spinning processes because it is naturally produced in the form of a long filament. All that is needed is for the filaments to be simply doubled and twisted together into a yarn. This process is known as *throwing*. A certain amount of silk 'waste' is left behind after the cocoon has been unravelled. It consists of a number of short, broken lengths which are not worth joining together to make a long filament. The waste is broken down into short fibers, combed and spun. This produces *spun silk*.

Like silk, man-made fibers are produced in the form of long filaments, which merely have to be doubled and slightly twisted to make yarn. No spinning as such is necessary, although such names as melt spinning, wet spinning and dry spinning are given to the various processes of making filaments. However, these filaments may also be chopped up into short, staple fibers for spinning in the usual way.

## Weaving

Most of our fabrics are made by the relatively simple process of *weaving*. One set of threads is interlaced with another. Curtains and handkerchiefs are typical woven products. If you look closely at them, you will see how the threads criss-cross one another to form a firm cloth. The threads in the woven cloth are so closely packed that it is at first difficult to believe that the cloth actually consists of thousands of separate threads. This is very different from knitted fabrics, such as sweaters where the separate loops of wool can be clearly seen.

The method of weaving every kind of yarn, whether it is natural or man-made, is more

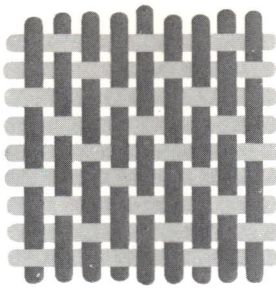
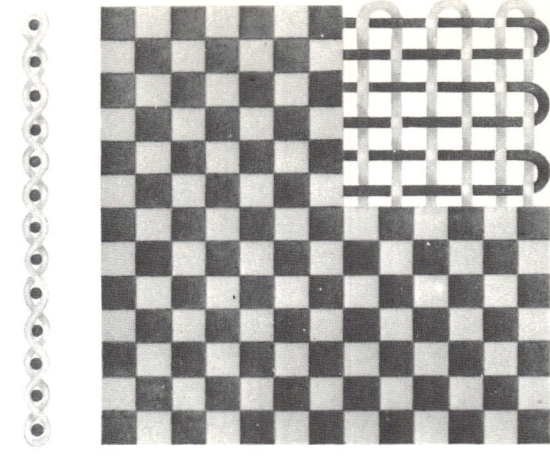

Above right: Plain weave is the simplest basic weave. Above: Interlacing the threads in this way produces the typical zigzag pattern of twill weave (see below).

Plain weave

Twill weave

Intricate woven designs are produced on the Jacquard loom. On this loom each warp thread is moved individually by a separate cord. A mechanism controlled by perforated cards moves each cord up or down according to the pattern. The perforated cards can be seen at the upper left of the picture.

or less the same. One set of threads is passed crosswise under and over a set of lengthwise threads. The crosswise threads are called the *weft,* and the lengthwise threads, the *warp.* Weaving in this up-and-down way would be very slow. It is far easier to raise and lower alternate warp threads and pass the weft thread between them. This is what happens on the weaving machine, the *loom.* At the end of a weft line, the positions of the up-and-down warp threads are reversed and the weft thread is again passed through to make another line. This process is continued until the cloth is finished.

Many of our fabrics are woven in a *plain* weave. The weft thread goes over one warp thread, under another, then over again, and so on. *Twill* and *satin* are other common types of weave in which the weft threads cross the warp in different ways. Cloth with a twill weave, such as drill and gabardine, has a series of diagonal raised lines across its face. A satin weave produces a cloth with a smooth face. *Velvet* is a wool weave with an extra weft thread which is left standing up in little loops. Cutting the loops gives the cloth a 'furry' finish. *Damask,* so called because it was made originally in Damascus, is a woven silk cloth with a design. *Brocade* is similar but the designs are raised.

### Knitting

After weaving, *knitting* is the most important way of making fabrics. It differs from weaving in that only one yarn is used. This yarn is formed into rows of loops by means of knitting needles in such a way that each row hangs from the row above. Knitted fabrics, because of their structure, 'give' much more than woven fabrics. This makes them suitable for garments such as vests and pants, which have to 'give'.

The great disadvantage of knitwear is that if one loop breaks, the loop hanging on it in the row below drops. This in turn causes the next row to collapse, and what is known as a run, or 'ladder' results. This happens particularly in women's nylon stockings, which are made up of fine yarn.

A great many women knit their own woollens by hand. But most knitwear is made by machine. A woman may use up to four needles when knitting. Knitting machines have hundreds of needles, arranged either in a row or in a circle.

### Non-woven Fabrics

*Lace,* one of the most delicate and beautiful of fabrics, is made not simply by interlacing threads, as in weaving, but by twisting them round each other. The interlaced threads form a diagonal, open-mesh pattern. There are a large number of different kinds of lace, each one differing in the way in which the threads are twisted together. Lace-making by hand is a traditional craft in many countries, but most lace is now made by machine.

*Felt* is another kind of non-woven fabric. It is made from wool. Loose wool fibers are first made into a lap, or web. Several layers of lap are piled on one another and pressed and rubbed together in a steamy atmosphere. Then this so-called *bat* is soaked in a soapy solution and pounded until the fibers become firmly matted. The result is felt. One of the main uses for felt is as underlay for carpets. It is also used for making hats and slippers.

Scientists have developed a method of making cloth from synthetic fibers without weaving. The technique involves pressing the fibers together lightly under heat. This causes them to bond together into a fabric.

Fine designs can be applied to fabrics by embroidery. Here embroidery is being done using sewing machines. But the finest and most intricate work is done by hand. The embroiderer uses a variety of stitches to achieve the desired effect. Appliqué is a form of embroidery in which the embroiderer stitches pieces of different material on the background fabric.

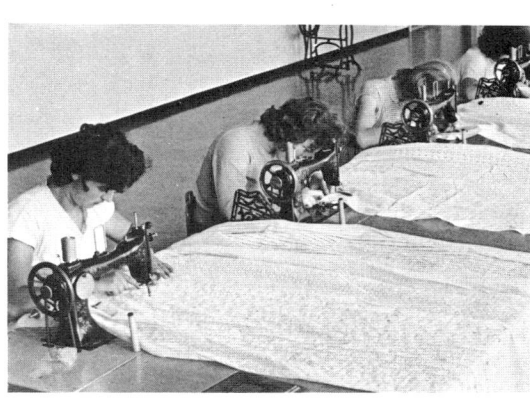

### Making Carpets

Most carpets are made up of a number of tufts of material which have been woven, threaded, or knotted into a firm backing. The tufts form what is called the *pile* of the carpet. The backing, which is usually made of cotton or jute, is usually woven at the same time as the carpet. Wool is the ideal fiber for the pile because it is naturally 'springy'. Nylon is also widely used.

Most carpets are woven on high-speed power looms much like those used for weaving cloth, but the looms have various arrangements for forming the pile. There are two main kinds of woven carpets, which vary in the way in which the pile is formed. In a *Wilton* carpet, the pile is formed from warp yarn. This pile yarn is formed into loops and woven into the backing. Then the loops are cut to make the tufts of the pile. In a *cord* carpet, the loops remain uncut. In an *Axminster* carpet tufts for the pile are cut first and then woven into the backing.

*Tufted* carpets are not woven. The pile yarn is simply needled into the backing in a series of loops, which are then anchored there by a rubber or plastic compound. The loops are cut to form the pile.

A cross-section of a Wilton carpet, showing how it is made up. Extra warp yarn is woven in to form the pile yarn. It is looped and then cut by a blade to form the pile tufts. The stuffers are generally of jute thread, while the fine chains are of cotton.

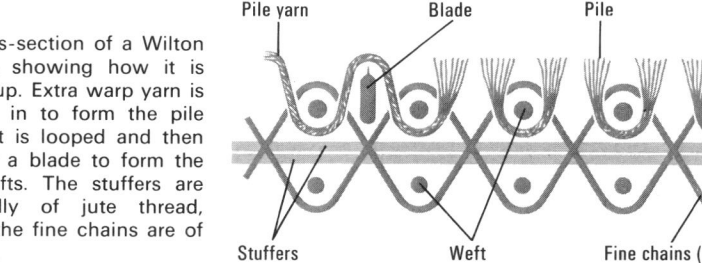

# Dyeing and Finishing

We should not be very happy if we had to use cloth in the state in which it leaves the loom. It is generally dull and dirty, harsh to handle, out of shape, and contains a variety of impurities picked up during spinning and weaving. Before cloth is used, therefore, it undergoes a variety of operations aimed at making it more attractive from all points of view. It may be bleached white or coloured with dyes, and then treated so that, for instance, it dries quickly after washing or resists shrinking and creasing.

Before any of these treatments take place, the cloth is thoroughly washed in a hot or boiling soap solution. This removes most of the dirty impurities. In the case of silk, it removes the natural silk gum around the filaments. This makes the silk a lot lighter. The weight may subsequently be restored by a treatment called *weighting,* in which certain chemicals are deposited within the filaments.

Even when it is clean, however, the fabric is not white. The fibers themselves contain colouring matter of some sort. This is removed by *bleaching.* Different kinds of bleaches and different methods of bleaching must be used for different fibers. Otherwise the fabric will be damaged. Synthetic fibers are more difficult to bleach than natural fibers. They always tend to be rather yellow. But their whiteness can be improved by adding to them a blue whitener, which makes the fabric look whiter.

## Dyeing

Bleaching is essential for another reason, too. It removes natural fats and waxes from the fibers. In this state the fibers can absorb colored dyes readily. The coloring matter from certain plants, such as madder and indigo, and from certain insects, such as cochineal, have long been used for dyeing textiles. But the range of such natural dyes is limited. Most dyes used these days are made synthetically. A wide variety, in literally every color of the rainbow, is obtained from *aniline,* a substance in coal tar. The synthetic dyes are cheaper and more brilliant than the natural ones. And they are *fast*, or resistant to sunlight, washing, and perspiration.

Every kind of fiber requires a different dyeing treatment. A dye that is suitable for cotton is not necessarily suitable for wool. Sometimes a substance called a *mordant* may have to be added before the dye will act. Natural fibers are dyed through the water that they absorb. Synthetic fibers are difficult to dye because they do not absorb water. Sometimes synthetics are colored early on while they are still in their solid, plastic-like state. The filaments are therefore already colored when they are formed. This process is known as *spin-dyeing*.

Dyeing a fabric colors it all the way through the same color. There are several methods of dyeing. One of the commonest is *jig-dyeing*, in which cloth is drawn back and forth through the hot dye solution or

Above: Fabric emerging from the dyeing vat. Dyeing may take at this stage, when the fabric has been woven, but often the yarn itself is dyed prior to weaving. Multi-color designs can then be woven.

Left: The search for new brilliant and color-fast dyes is never ending. Here a number of different dyes are being tested at a research laboratory.

'liquor'. In another method the fabric is dipped in the liquor and then steamed to 'fix' the dye. Dyeing does not always have to wait until after the fabric has been made. The raw fibers or the spun yarn may be dyed. This is usually done by pumping the dye liquor through them. Sometimes fibers of different colors are spun together to form the multicolored yarn.

### Printing

Fabrics can be given beautiful colored designs by *printing* processes. The designs are produced not by dyeing the fabric right through but by coating the surface with the dye. If you look at the 'wrong' side of a printed material, you will see that the colors are much weaker than those on the 'right' side. The dyes used for printing are the same as those used for dyeing, but they are applied to the fabric in the form of a paste, not a solution. The two main methods of printing are roller printing and silk-screen printing.

In *roller printing*, the fabric passes over a number of rollers. Each roller prints on the fabric a particular part of the design in a different color. In *silk-screen printing*, the colored dye pastes are applied to the fabric through pieces of silk. The silk is treated so that it will only let the dye through in certain places, which correspond to the parts of the pattern to be printed in a certain color. After printing by either method, the fabric is steamed to fix the dye.

### Finishing Processes

After bleaching and coloring, the fabrics may go through a variety of other processes to give them desirable properties which they do not naturally have. For example, cotton is not naturally lustrous, but it can be made so by *calendering*. In calendering, the cotton fabric is passed between heavy, rotating rollers. This treatment flattens and 'polishes' the fibers. The luster is only temporary, however, and washes out after a time. A more permanent luster can be obtained by treating the stretched yarn with caustic soda. This process is called *mercerizing*. Sewing cotton, for example, is invariably mercerized to give it a luster.

Many fabrics, especially cottons and rayons, tend to shrink when they are washed. To reduce or prevent shrinking, the fabrics are given a so-called *shrink-resistant* finish, such as Rigmel or Sanforize. The object of the process is to close up the threads in the fabric so that it could not get smaller even if it wanted to.

Cotton, rayon, and linen by themselves crease badly when they are crumpled. This can be reduced by impregnating, or soaking the fibers with a kind of resin. Cotton can also be given a 'drip-dry' finish by a similar treatment.

Synthetic fibers require no such treatment. They drip-dry and resist creasing naturally. And most of them have the valuable property that they can be set into permanent pleats by hot pressing after manufacture. For the natural fibers, the pleats must be ironed in after each washing.

Printing on cloth is very much the same as printing on paper. The sheet to be printed travels round a large cylinder. For each color to be printed there is a printing roller, holding the design. As the roller rotates a dye paste adheres to the design and is then transferred to the cloth.

# Pioneers of the Textile Industry

### John Kay (1704?–1764?)

John Kay's invention of the *flying shuttle* in 1733 was the first of a series of inventions that revolutionized the textile industry and played a great part in England's Industrial Revolution.

Until that time, weavers had to pass the shuttle containing the weft thread by hand through the parted warp threads. For widths of cloth of more than a yard, two men were required, one each side, to pass the shuttle back and forth. Kay's flying shuttle changed all this. A simple, hand-operated device drove the shuttle through the warp threads automatically. Using a flying shuttle, a weaver could almost double his output and greatly increase the width of cloth he wove. However, the flying shuttle did not come into widespread use for several years.

As many of his fellow inventors were to find, Kay was violently opposed by textile workers, who considered his revolutionary invention a threat to their livelihood. He was working in a textile mill at Colchester in Essex when he made his invention, but soon he was forced by hostile workers to leave. He returned to his home town of Bury, in Lancashire, only to receive the same treatment. After his home had been wrecked and the government had refused to uphold his patent claim, he fled to France. There he died in poverty and obscurity.

### James Hargreaves (1722?–1778)

The invention of the flying shuttle created a demand for more yarn. Until the mid-1760s the yarn was spun either by hand on a distaff or on the single-spindle spinning wheel. Neither method could produce enough thread to satisfy the increased demands of the weavers.

In 1764 a poor weaver named James Hargreaves, who lived at Standhill, in Lancashire, began experimenting with a multi-spindled spinning machine. By 1768 he had built a machine which could spin eight spindles at once. It incorporated a moving carriage for drawing out the roving. The machine was called the *spinning jenny*. It may have been named after Hargreaves's wife or after the local word for engine. The spinning jenny was a great improvement on the spinning wheel and could produce enough thread to keep pace with a fast weaver.

Knowing what had happened to Kay, Hargreaves endeavoured to keep his invention secret. But somehow the word got around. His machine was smashed, and he was forced to leave his home and seek safety in Nottingham, where the people recognized the advantages of mechanizing the textile industry. There he built a new machine with thirty spindles and set up a mill. Hargreaves patented his invention in 1770, but like Kay's it was later declared invalid. And so Hargreaves received little financial reward.

### Sir Richard Arkwright (1732–1792)

At the same time that Hargreaves was perfecting his spinning jenny, his fellow Lancastrian Richard Arkwright was building his *roller-spinning frame*. The roving was draw out into yarn by sets of rollers rotating at different speeds.

Arkwright patented the spinning frame in 1769, by which time he had moved from his home town of Preston to Nottingham. His first model was horse-powered, but his later, much improved models were water-powered. His machine became known as the *water frame*. Arkwright built many of these machines, installed them in special buildings, and employed labour to work them. He can surely be regarded as founder of the modern factory.

Although Arkwright's patent was eventually cancelled, he was too experienced and too shrewd a business man to be materially affected. And he died a wealthy and much respected man.

### Samuel Crompton (1753–1827)

Samuel Crompton was a farmer's son, who lived near Bolton. As well as helping on the farm, he assisted his mother on the loom. He became dissatisfied with the quality of the yarn produced by existing spinning machines which was always breaking. And so, in 1774 he began experiments which culminated in 1779 with his invention of the spinning 'mule'. This machine was so-called because it was really a cross between Hargreaves's spinning jenny and Arkwright's water frame.

The mule combined the best features of both. The yarn was first drawn out by rollers and then drawn out further by a travelling carriage. With this machine Crompton began to produce thread of exceptional fineness, which soon attracted widespread attention.

Fearing the trouble that had plagued Kay and Hargreaves, Crompton made his invention public, having been promised adequate reward for so doing. But in the event he received very little and lived in poverty for most of the remainder of his life. In 1812 he received £5,000 in compensation from the government but soon lost it in a business gamble.

### Edmund Cartwright (1743–1823)

As the newly invented spinning machines came into widespread use, yarn was produced much faster than it could be woven on the hand looms of the day, even when equipped with the flying shuttle. The boot, so to speak, was on the other foot. There arose a desperate need for some kind of mechanical loom.

The man who first built a practical *power loom* was neither a weaver nor an engineer but a clergyman, Dr. Edmund Cartwright, who lived in Nottinghamshire. He patented a crude model in 1785, and improved models in 1786 and 1787. A few years afterwards he opened a factory using his power looms, but it was a failure. He even had to give up his patent rights to his creditors.

As usual, manufacturers and workers violently opposed Cartwright's invention. In 1790 the first large mill in Manchester, in which 400 of Cartwright's looms were installed, was burned to the ground. But everyone soon realized the benefits to be gained from mechanized weaving and Cartwright's power loom and many looms based on it were widely adopted.

Left: Samuel Crompton's spinning mule. The mule is still one of the foremost spinning machines.

Below: Hargreaves's spinning jenny, the machine that heralded the widespread mechanization of the textile industry. This model could spin 16 spindles simultaneously.

# Clocks

Every kind of mechanical clock, whether it is as big as the famous 'Big Ben' on the Houses of Parliament in London or as small as a ladies wrist watch, works on the same lines. There must be a means of powering it, a means of keeping it running at a steady rate, and a means of indicating the time measured.

The *power source* is generally one of three kinds: a falling weight, a coiled spring, or electricity, either from a battery or from a utility outlet. The falling weight, which was used in all of the earliest clocks, has the advantage that the power it supplies to the clock mechanism remains the same. And the steadier the power source, the more accurate the clock. Falling weights are used to power the tall grandfather, or *long-case* clocks which stand in the hall and the smaller grandmother clocks which are fixed to the wall. Most of the huge tower, or turret clocks in churches and other public buildings are weight-driven, too.

A coiled spring is used to power small clocks and almost all watches. The great disadvantage of the spring is that the power supplied to the clock mechanism gradually decreases as the spring uncoils. This means that the clock runs faster when the spring is fully wound than when it is nearly run down. Accurate spring-driven clocks incorporate a device called a *fusee* to keep the power from the spring constant.

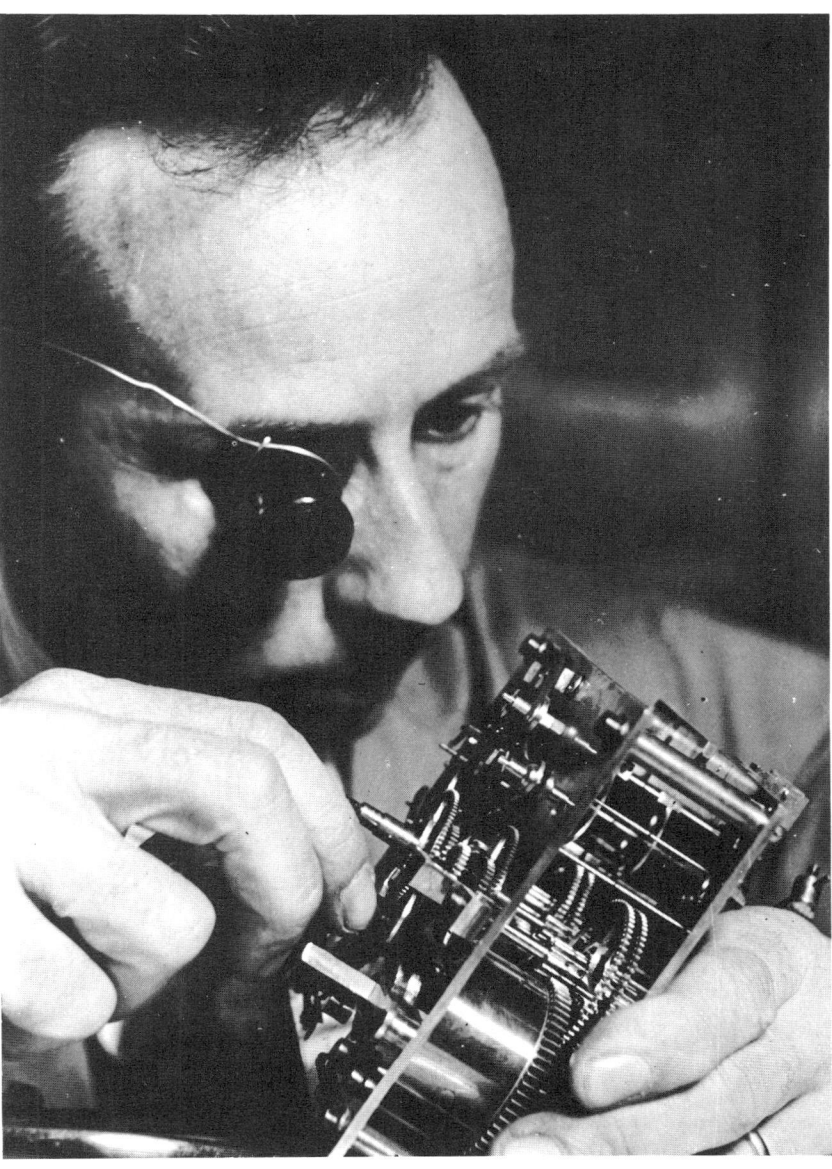

## Escapement

A device called the *escapement* lets power 'escape' to turn the clock mechanism at the rate laid down by the regulator. At the same time it passes on the power to keep the regulator in motion.

One of the commonest and simplest forms of escapement is the *anchor* escapement in a pendulum clock. The power source of the clock moves a set, or *train* of three gear wheels. The third wheel turns a specially shaped toothed wheel called the *escape wheel*. Above this wheel is the anchor-shaped escapement, which is linked to the pendulum. The pendulum swings back and forth and rocks the anchor. The rocking ends of the anchor alternately engage and release the escape wheel one tooth at a time. Thus the clock mechanism goes forward in little 'jumps'. The 'tick-tock' of a clock is the sound of the mechanism jumping forward.

## Regulator

The 'heart' of the clock is the *regulator*, which is a device that has a steady regular

A clockmaker putting the finishing touches to the mechanism. Clock making and mending demand the utmost skill and concentration. The sophisticated, precision modern mechanisms contrast strongly with the crude devices introduced in the thirteenth century.

### Balance Wheel

The regulator used to control most watches and small clocks is the balance wheel. It is a delicate wheel pivoted at the center. It is connected to a very fine spiral spring called a *hairspring*. Whichever way the balance wheel is turned, the hairspring pulls it back. As it comes back, it rocks the escapement and allows the clock mechanism to turn. The escapement nudges it every time to keep it *oscillating*, or moving back and forth. The balance wheel makes a good regulator because the time for each oscillation is the same. It is usually made of two different metals to compensate for expansion or contraction caused by temperature changes, which would otherwise affect its rhythm.

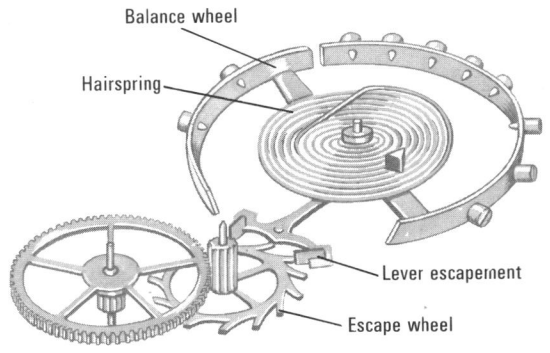

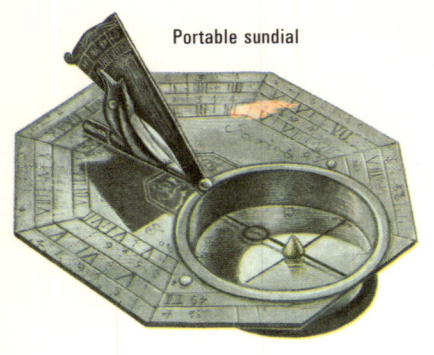
Portable sundial

Wall sundial

Before the advent of the mechanical clock sundials and hour-glasses were widely used for telling the time. Sundials were mounted on pedestals or walls. And there were portable sundials with an inset compass so that they could be correctly orientated every time. The early clocks were simple and crude and kept extremely poor time.

Hour-glass

Death's-head watch, 1600's

Italian clock, 1400's

Escapement / Weight / Pendulum

movement. It is this movement which splits up the time into equal, measurable portions. The two commonest forms of regulator are the *pendulum* and the *balance* wheel. But anything with a regular rhythm could be used. Some watches, for example, use the vibrations from a tuning fork. Very accurate clocks use the vibrations of certain crystals and atoms.

The time is indicated by the *hands* of the clock moving across the *dial*. The minute hand is fitted to the end of the first wheel shaft and turns with it. The hour hand is geared to the minute hand to turn at a slower rate.

**Pendulum**

One of the best simple devices for regulating a clock is the *pendulum*. All it is really is a rod with a weight, or *bob* on the end. The reason it can be used as a regulator is that it swings to and fro at a constant rate.

You can check this for yourself. Tie a weight on a piece of string and let it swing back and forth. You will find that it always takes the weight the same time to swing from one side and back again. It does not matter whether the path travelled by the weight is long or short. This interesting fact was first discovered by the brilliant Italian scientist Galileo in 1584.

Galileo also found that the time of swing of a pendulum depends only on its length. The weight of the bob at the end does not matter. A pendulum about 10 inches long swings back and forth in 1 second. One about 39 inches long takes 2 seconds, that is, each sweep of the pendulum takes 1 second. For this reason it is often called a *seconds* pendulum. The pendulum of 'Big Ben' on the Houses of Parliament in London is 13 feet long. It takes 4 seconds to swing to and fro.

The main elements of a pendulum clock. The anchor escapement attached to the upper end of the pendulum rod rocks back and forth regularly, allowing the clockwork mechanism to move at the required rate.

**Watches**

Most people these days wear a watch, which is really nothing more than a portable clock. There is little difference in the general principles of design between watches and clocks. Everything is just more delicate and on a smaller scale. Most watches are powered by a *mainspring* and regulated by a *balance wheel* fitted with a *hairspring*. The mainspring is a thick steel ribbon, wound inside a barrel called the *going barrel*. The escapement normally used is a small rocking lever. The main moving parts of the watch, the escape wheel, the balance wheel, and so on, turn in bearings made of artificial rubies. Such bearings, called *jewelled bearings,* are so hard that they withstand year after year of constant use without wearing. The lever ends are jewelled, too.

Most modern watches have a sweep second hand in addition to the usual hour and minute hands. They may even show the day and date as well. Some wind themselves automatically. They have a semi-circular disc of heavy metal that swings back and forth with every movement of the hand. Each swing winds up the mainspring of the watch a little more. A simple locking device ensures that the mainspring is never overwound.

Some of the latest watches work electrically. They are powered by tiny, flat mercury batteries.

**Electric Clocks**

Many clocks, too, operate by electricity these days. Some of them merely use electricity as a source of power to drive the clock mechanism. For example, in some clocks a battery operates a solenoid, or electric magnet, which keeps rewinding a short mainspring. In some large public clocks, utility electricity is used to power a motor which winds up a weight driving the clockwork.

The most common electric clocks, however, use electricity not only to power them but also to regulate them. They are called *synchronous* clocks and work on utility electricity. The utility electricity can be used as a means of regulating because it has a built-in, constant 'rhythm'. The electric current flows first one way and then the other so many times every second. That is why we call utility electricity *alternating current* (A.C.). In Britain the utility supply alternates 50 times a second; in the United States, 60 times a second.

In the synchronous electric clock the motor driving the mechanism keeps in step with the rapid alternation of the electric current. *Synchronous* just means 'keeping in step with'. If the electricity alternates a steady 60 times a second, the motor driving the clock mechanism rotates a steady 60 times a second.

Electricity is also used to drive several clock dials. An accurate pendulum clock called the *master* is wired to a number of dials called *slaves*. The swing of the pendulum in the master clock sends out electric impulses along the wires which move the hands on the slave dials. This is a common method used in factories, schools and other large buildings to keep all the clocks at the same time.

### Accurate Timepieces

Modern 'conventional' clocks and watches which we use every day are pretty accurate. The better ones will keep time to within a second or so each day. But this is not accurate enough at sea for purposes of navigation. The *marine chronometer* achieves precision by incorporating a fusee mechanism, and it remains accurate to within a second or two each year. But even this order of accuracy is not sufficient for astronomers and researchers in some branches of science. And so scientists have developed two incredibly precise clocks which do keep time to their extremely exacting standards. They are the *quartz crystal clock* and the *atomic clock*. In massive form quartz is crystalline. Sometimes it occurs clear as rock crystal.

Rock crystal has interesting electrical properties which made it suitable for regulating the mechanism of a clock. When an electric current is passed through it, the crystal vibrates at an almost perfectly constant rate. The rate of vibration depends on the thickness of the crystal. The thinner it is, the faster it vibrates. The faster it can be made to vibrate, the smaller is the time interval that can be measured. In the quartz-crystal clock the crystal is made so thin that it vibrates 100,000 times a second. This means that time intervals of a few thousandths of a second can be measured.

The quartz crystal clock is so accurate that it gains or loses little more than a second or two in 50 years. This is a vast improvement even on the relatively accurate marine chronometer. The quartz crystal clock is a standard feature in major astronomical observatories, where extreme accuracy is required. But the even more

precise atomic clock is used to keep a check on the accuracy of the quartz clock.

The atomic clock can be considered the ultimate in accurate time-keeping. It gains or loses no more than a second in 1,000 years, and it can measure time-intervals of a millionth of a second. What a far cry this is from our ordinary clocks! The atomic clock looks nothing like an ordinary clock either. It is a complicated mass of pipes, pumps, tubes, and electronic gadgets.

The atomic clock is so called because it uses the vibrations of certain atoms, the tiny particles of matter, as a regulating device. The substance used in the clock is a vaporized metal called caesium, whose atoms vibrate very regularly and very rapidly – thousands of millions of times every second.

A far cry from the crude Italian clock pictured opposite, a twentieth century atomic clock at Britain's National Physical Laboratory. This clock, which employs the vibrations of the atoms of vaporized caesium for regulation, has a stability of between 1 in $10^{10}$ and 1 in $10^{11}$, which means in other words that it can keep time to within about a second over a period of 1,000 years or so.

# The Automobile

At least 200 years ago engineers were building kinds of motor vehicles. But these early kinds were driven by steam. They were slow, very noisy, and often dangerous. From the 1830s onwards, electric cars driven by batteries were developed. They were quieter, but could travel only a short distance before the batteries needed recharging. By far the most important development during this time was the invention of the *internal combustion engine*. This kind of engine is the type still used in automobiles today. But the early ones used natural gas, not gasoline as fuel. The French engineer Jean Etienne Lenoir (1822-1900) designed the first one in 1858. In 1876 Nikolaus August Otto (1832-1891), a German engineer, built a natural gas engine in which the fuel was compressed before it was burned. This considerably increased the power output of the internal combustion engine.

In 1885, two German engineers, working independently, produced 'horseless carriages' that can be considered the forerunners of today's automobiles. The engineers, Karl Benz (1844-1929) and Gottlieb Daimler (1834-1900), both used internal combustion engines with petrol as fuel. Their vehicles were slow, open to the weather, and bumpy to ride in because the wheels had solid tyres.

From then on, development was rapid. By 1909, the American manufacturer Henry Ford was mass-producing his famous *Model T*, or 'Tin Lizzie' by the thousands. To produce cars in quantity Ford introduced the modern concept of the moving assembly-line. He used lighter weight alloy steel for the body, which made it much flimsier to look at than its contemporaries, hence the 'tinny' image. No major changes were made to the *Model T* in its 19-year production run.

Today, we can travel quickly, safely, and comfortably in a warm, cushioned automobile. The automobile industry has become one of the biggest in the world. There must be almost 150 million automobiles on the world's roads. Nearly half of them are in the United States, where there is one automobile to about every three people!

## The Gasoline Engine

Most automobiles are powered by an engine burning gasoline as a fuel. The gasoline is burned in the engine *cylinders*. An automobile has a number of cylinders, usually four, six, or eight. They may be arranged in a straight line or in the shape of a 'V'.

Fitting into each cylinder is a *piston*,

Rolls-Royce, 1904 vintage, at speed in the London-to-Brighton Veteran Car Rally, an event commemorating the 1896 run celebrating the repeal of the Red Flag Act, a law requiring that a red flag or lantern be carried ahead of a horseless vehicle.

Our ideas of automobile design have changed radically over the years as tastes and vehicle performance have altered. These two fine specimens, conceived nearly half a century apart, epitomize the change that has taken place. The sleek Italian design below has a vigorous aerodynamic design that matches its superlative performance.

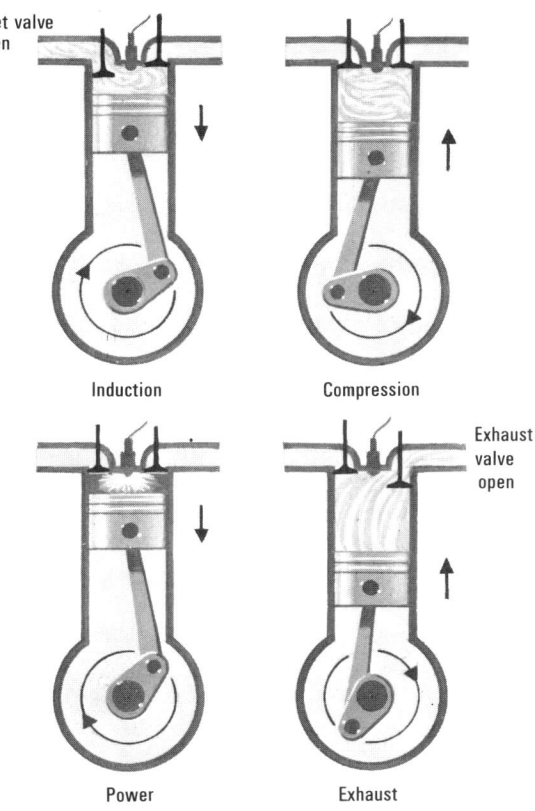

Induction / Compression / Power / Exhaust

Principle of the four-stroke cycle, showing the relative piston and valve movements. On the induction stroke gasoline vapor is drawn into the engine cylinder. On the compression stroke the vapor is compressed. Ignition of the vapor by spark drives the piston down on its power stroke. Then on the final exhaust stroke the spent gases are expelled.

Principle of the two-stroke cycle. 1. Piston moves upwards, compressing the fuel mixture and drawing in fresh fuel in the space below the piston. 2. On ignition the piston is driven downwards. 3. At the end of the downstroke the transfer port is uncovered, allowing fresh fuel into the combustion space. The exhaust port is also uncovered, allowing spent gases to escape.

which is shaped like an upturned barrel. The top of the cylinder is closed, except for two small holes which can be opened or closed by *valves*. An explosive mixture of gasoline and air enters through one of the valves, and is fired, or *ignited* by a spark. The gases produced drive the piston quickly down the cylinder.

The piston is attached by a rod to a shaft called the *crankshaft*, shaped like a complicated winding handle. Because of its shape, the crankshaft changes the downward movement of the piston into a turning movement. When the piston has finished travelling downwards, the crankshaft pushes it back up the cylinder again. From the crankshaft, the turning motion is carried to the driving wheels by the *transmission system*, which includes the clutch and gearbox (see page 211).

**The Four-Stroke Cycle**

Each upward or downward movement of the piston is called a *stroke*. The gasoline engine works on what is called the *four-stroke cycle*. This is so called because each piston has to travel four strokes – two up and two down – before it repeats its sequence, or *cycle* of operations. The four-stroke cycle is also called the *Otto cycle* after Nikolaus Otto, who first used it.

The first stroke of the cycle is the *intake*, or *induction* stroke. The piston is pulled down the cylinder by the crankshaft. At this point, one of the valves (the *inlet valve*) in the top of the cylinder opens. Gasoline and air mixture is sucked through it. The valve closes when the piston starts to travel upwards again.

The second stroke is the *compression* stroke. The piston compresses the gasoline mixture at the top of the cylinder. When the piston is near the top of its stroke, a spark ignites the mixture to produce hot, expanding gases.

As we have already described, the gases force the piston down to produce power. This third stroke is the *power* stroke.

The fourth and final stroke in the cycle is the *exhaust* stroke. As the piston moves upwards, the other valve (the *exhaust valve*) opens. The piston pushes the exhaust gases from the cylinder. On the next downward stroke, more fuel is drawn in, and the cycle is repeated.

Most automobile engines work on this four-stroke cycle. But some, called two-stroke engines, work on a *two-stroke cycle* and produce power every two strokes. Mostly, two-stroke engines are used in motor-cycles, lawnmowers, and the like.

**Two–Stroke Cycle**

A two-stroke engine produces power on every downstroke. Its cylinders are designed differently from those of a four-stroke engine. For one thing, they have openings called *ports*, instead of valves.

The two-stroke engine goes through stages similar to the induction, compression, power, and exhaust stages of the four-stroke engine. But the stages all take place in two strokes of the piston instead of four.

As the piston moves upwards on its compression stroke, it uncovers the *inlet port*. This allows gas-air mixture to enter the space below the piston. So this stroke is also the induction stroke.

After ignition, the piston travels downwards on its power stroke. It uncovers the *exhaust port* to let the exhaust gases escape. It also uncovers the *transfer port*. The fresh gas mixture from below the piston rushes through the transfer port into the cylinder. As it does so, it forces out the remaining exhaust gases. Then the piston travels upwards again, compressing the mixture, and the cycle is repeated.

The top of the piston is not flat, but slightly ridged to deflect the fresh fuel mixture into the top of the cylinder. This is designed to prevent mixing the fuel mixture with outgoing exhaust gases. In practice some fuel may be expelled with the exhaust gases, which wastes fuel, or some of the exhaust gases remain behind, which prevents the fuel mixture from burning properly.

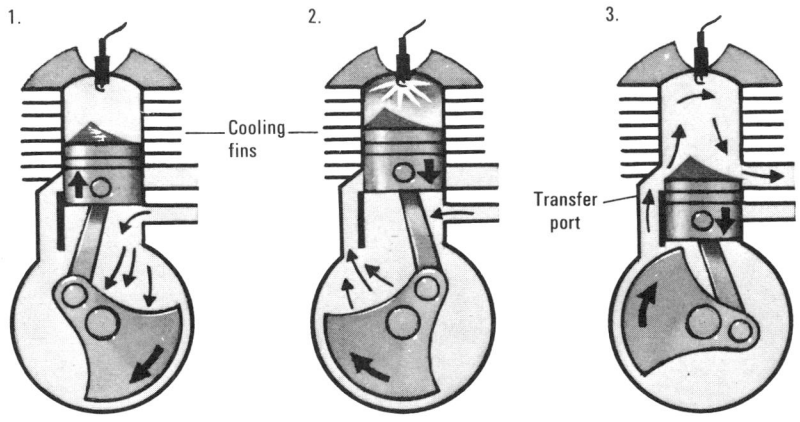

# Parts of the Engine

## Cylinder Block and Cylinder Head

The main metal casting of the engine, which contains the cylinders, is the *cylinder block*. The cylinder head bolts onto the top of it. Most cylinder blocks are made of iron and cast in one piece. Some are cast in aluminum, which is much lighter than iron. Aluminum blocks have steel cylinder liners to withstand wear by the pistons. The pistons naturally have to fit the cylinders tightly to prevent gas escaping past them when the engine is running. They are made gastight by springy bands of metal called *piston rings,* which fit around them.

The *cylinder head* is bolted to the block. There is a thin copper sheet, or *gasket,* sandwiched in between to form a gastight and watertight seal. Bored in the cylinder head are holes for the valves and the sparking plugs, which provide the spark to burn the fuel mixture. Hollows in the block form the *combustion chambers* at the top of the cylinders. In an engine with overhead valves, the cylinder head carries the valves. It also has passages for fuel, water, and the burnt, exhaust gases. The cylinder head is made separately from the cylinder block so that the engine may be cleaned and repaired more easily.

## Valves and Camshaft

The *valves* in the cylinder head are mushroom-shaped. They are carefully ground into their 'seats' so that they form a gastight seal when they are closed. Fuel enters the combustion chamber through the *inlet* valve. The exhaust gases leave it through the *exhaust,* or outlet valve. Each valve opens for only one stroke of the engine cycle. The inlet valve opens only during the induction stroke. The exhaust valve opens only during the exhaust stroke. In most car engines the valves lie above the cylinders. They are called *over-head-valve* engines. Some cars have *side-valve* engines.

The *camshaft* is a rotating shaft inside the engine, driven by gears or a chain from the crankshaft. Raised pieces on the shaft, called *cams,* move linkages called *tappets, push-rods,* and *rocker arms* to operate the valves. The cams are set on the camshaft at

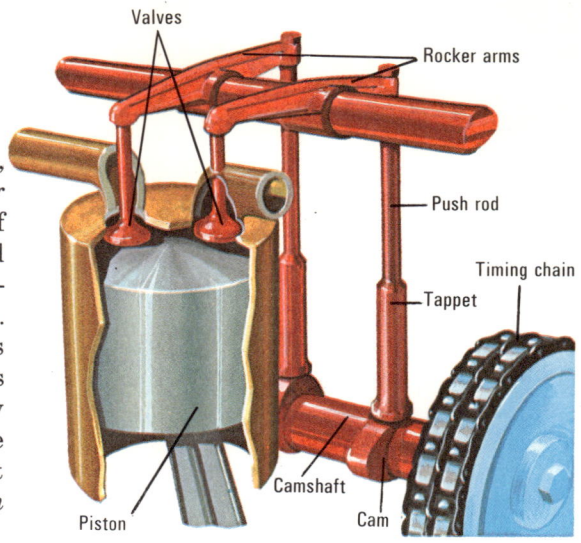

The valve gear of an automobile engine. The camshaft is driven by chain from the crankshaft. The cams open and close the valves at the appropriate times, acting through tappets, push rods and rocker arms. The valve springs, which snap the valves shut, are not shown.

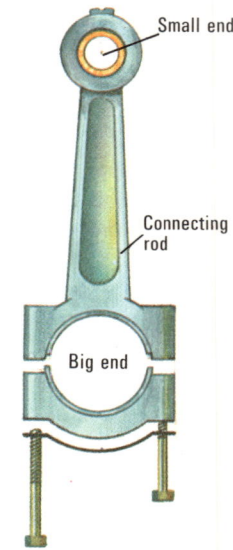

Above: The connecting rod. The small end is attached to the piston by means of a gudgeon pin. The big end connects with the crankshaft.

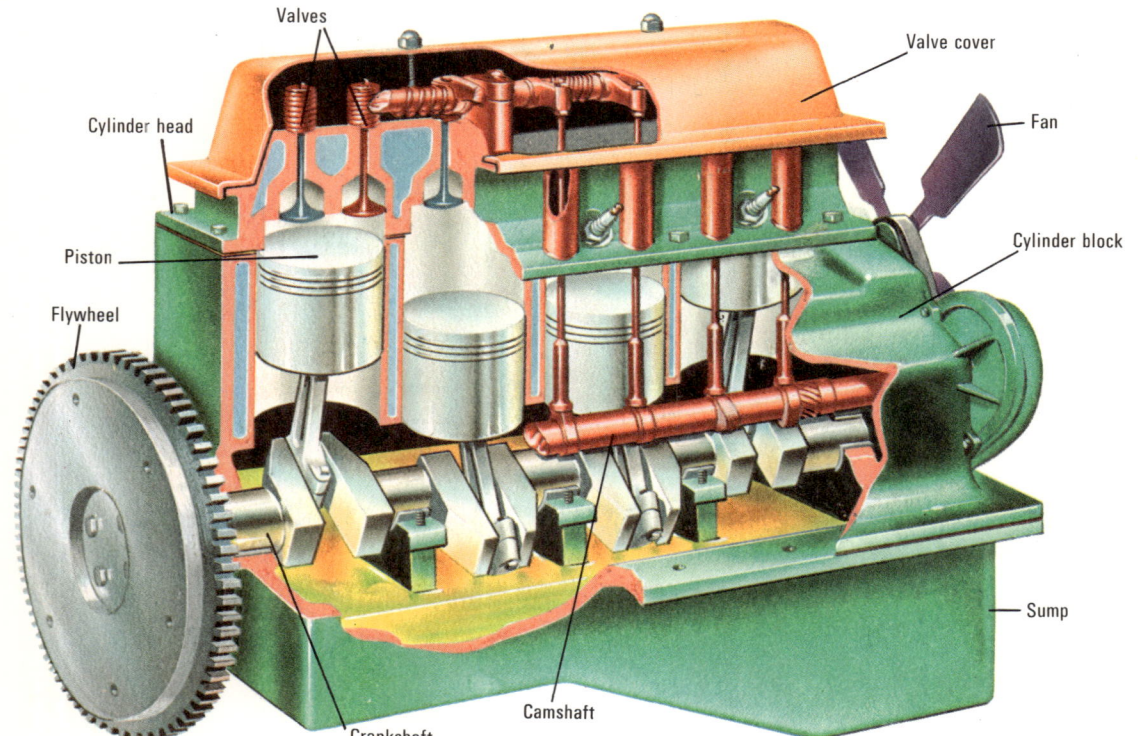

A cut-away view of a typical four-cylinder engine, showing how the main components fit together. The engine is of the normal overhead-valve, single-camshaft type. Note the toothed starter ring around the flywheel which meshes with the pinion of the starter motor to start the engine.

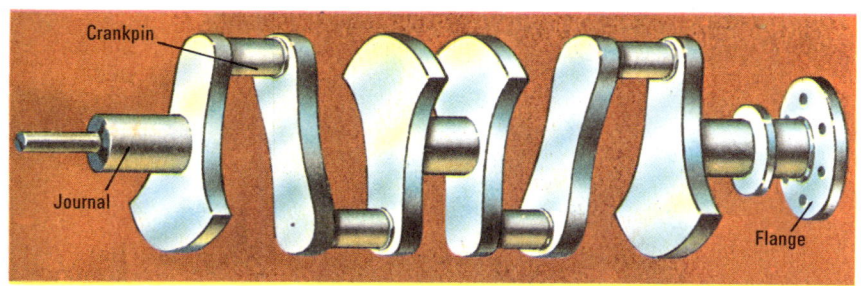

The crankshaft of a four-cylinder engine. The big-ends of the connecting rods fit round the crank pins. The crankshaft rotates in heavy journal bearings. This particular engine has only three main bearings. Five is more common these days, which makes for smoother running and longer life. The flange at the end of the crankshaft is bolted to the flywheel, a heavy wheel that has great inertia and helps to make the engine run smoothly.

different angles so that they open the appropriate valves at the right time. The cams push the tappets up and down. In turn the tappets move the long push-rods resting on them. The push-rods see-saw the rocker arms. As the arms rock forwards, they open the valves by pushing on the valve stems. As they rock back, they release the stems, and the *valve springs* snap the valves tightly closed again.

### Crankcase and Crankshaft

The lower part of the engine unit is the *crankcase*. It houses the crankshaft and its bearings. Generally it is cast with the cylinder block itself. Bolted to the bottom of the crankcase is a trough called the *sump*. It holds the oil to lubricate the engine. A rod called the *dipstick* dips into the sump to register the oil level.

The *crankshaft* is designed to change the reciprocating, or up-and-down motion of the pistons into a rotary motion. It has a number of cranks on it, one for each piston of the engine. They are set at different angles so that the power strokes of the pistons reach the crankshaft at the right time. But the power stroke that drives the crankshaft is only one of the four strokes of the piston. For the other three strokes, the crankshaft drives the piston. To keep the crankshaft turning between power strokes, a heavy disc, or wheel is attached to one end of the crankshaft. This so-called *flywheel* is so heavy that, once it is set in motion, it tends to keep on turning. The motion of the flywheel helps to keep the engine running smoothly.

Around the edge of the flywheel is a toothed ring, called the *starter ring*. When a driver presses the engine self-starter, the starter motor spins a small toothed wheel against the starter ring. This turns the flywheel, which turns the engine over until it fires.

The flywheel is the part of the engine that transmits motion to the transmission system. It is connected to the system by the clutch. In some cars with what is called *automatic transmission* the flywheel and clutch are combined into a unit called the *fluid flywheel*.

The crankshaft is linked to each piston by a *connecting rod,* or 'con' rod. The upper, or *little end* of the con rod is attached to the underside of the piston by a *gudgeon pin*. The lower, or *big end* of the con rod is attached to the crankshaft by a heavy bearing. The big-end bearings take a great deal of 'punishment' and sometimes fail. You have probably heard drivers complaining that their 'big-ends' have gone. The crankshaft itself turns in the heavy *main bearings* of the engine.

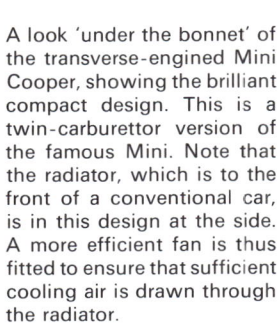

A look 'under the bonnet' of the transverse-engined Mini Cooper, showing the brilliant compact design. This is a twin-carburettor version of the famous Mini. Note that the radiator, which is to the front of a conventional car, is in this design at the side. A more efficient fan is thus fitted to ensure that sufficient cooling air is drawn through the radiator.

# Fuel and Ignition Systems

We often talk of engine *systems*, which supply the important needs of the engine. For example, the fuel system supplies the fuel in a convenient form to be burned in the combustion chambers. The ignition system produces the spark to burn the fuel. The cooling system takes away most of the heat from the burning fuel and prevents the engine overheating. The lubrication system delivers oil to the moving parts and allows them to run smoothly with the minimum of wear.

**The Fuel System**

The fuel system supplies a constant flow of gasoline to the engine. It consists of a storage tank, fuel pump, and carburettor. These are connected by the fuel pipe. The gasoline storage tank is positioned well away from the hot engine to reduce the risk of fire.

The *gasoline pump* pumps the gasoline from the tank to the carburetor. Some cars have an electric pump; others have a mechanical pump driven by the camshaft. The pump draws the gasoline from the tank through a filter made of metal gauze which prevents bits of dirt from choking the carburetor.

In the *carburetor* the gasoline is mixed with exactly the right amount of air it needs to burn rapidly inside the engine cylinders. Some automobiles, especially high-performance ones, have more than one carburettor to increase the flow of fuel into the engine.

The action of a carburetor is much like that of a scent spray. On the induction stroke of each piston, air is sucked through the choke tube of the carburetor. This tube narrows at one point, and it is there that the gasoline enters the airstream. The gas enters through tiny nozzles called *jets* as a fine spray. It then mixes easily with the air.

A steady supply of gas to the jets comes from a small reservoir of gasoline called the *float chamber*. Gasoline pumped from the gas tank enters the float chamber through a needle valve. When the chamber is full, a float on the gas surface keeps the valve closed. When gas is used, the float falls and allows more gas to enter until the chamber is full.

The air enters the carburetor through a *filter*, which prevents dust and dirt from entering the ingoing airstream. Air filters may consist of paper, oiled metal gauze, or an oil bath. The amount of air entering the carburetor is controlled by the *choke valve* in the air-intake tube. The choke is used when the engine is started from cold. At this time, the engine requires a richer mixture (less air, more gas).

A valve called the *throttle valve* controls the amount of gas-air mixture entering the engine and therefore the engine speed. It is connected to the accelerator pedal. When the driver presses the accelerator, the throttle valve opens wider and lets more mixture into the engine.

The fuel-air mixture from the carburetor enters the engine through a system of short

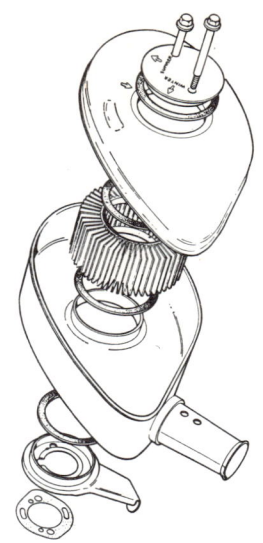

A typical air-cleaner assembly for a carburetor. The 'concertina' paper filter element prevents dust entering and clogging the carburetor jets.

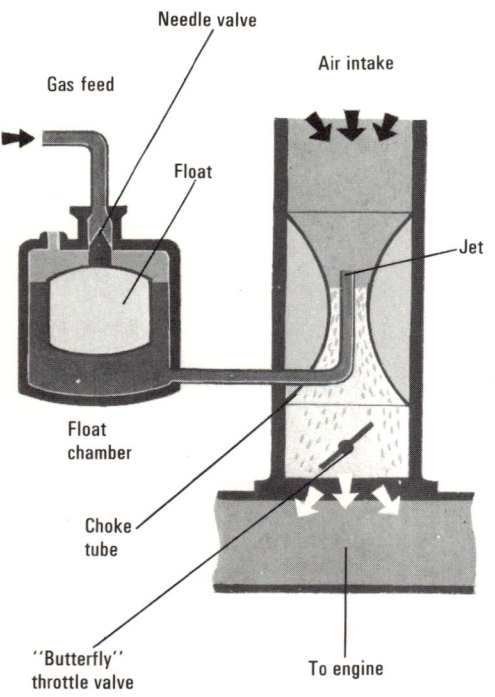

Left: A simple carburetor arrangement. The float chamber supplies a constant 'head' of gas to the carburetor jet. The jet is located at the narrow part of a constricted tube called the venturi or choke tube. Air flows past the jet, entraining gas droplets, and thence into the inlet manifold which leads into the engine cylinders. The quantity of fuel mixture entering the engine is controlled by the 'butterfly' throttle valve.

Right: A widely used type of carburetor, the SU. This operates by means of a suction chamber, connected to engine suction. When the throttle is nearly closed, there is little engine suction, and the piston lifts the tapered needle only a fraction from the jet. This permits only a little fuel to flow out. As the throttle is opened wider the engine suction increases, lowering the pressure in the suction chamber, which causes the piston to lift farther out of the jet, allowing an increased flow of fuel into the engine.

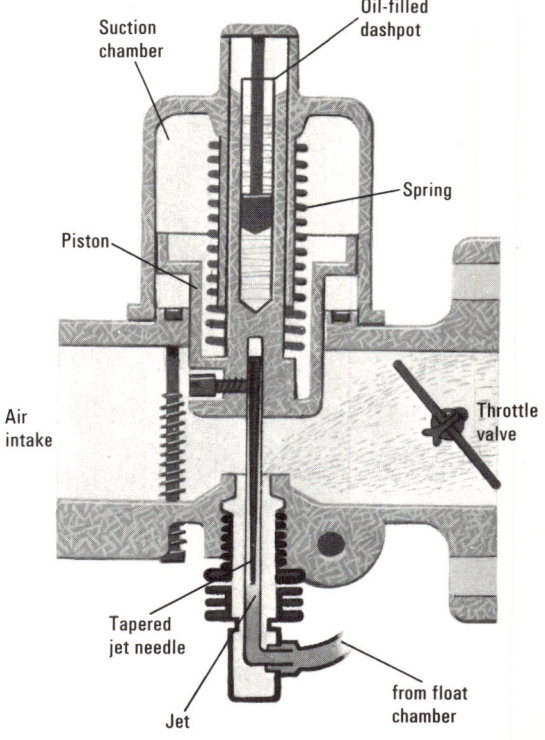

tubes called the *inlet manifold*. The *exhaust manifold* is a similar set of tubes which takes away the exhaust gases from the engine cylinders. The exhaust manifold is the first part of the *exhaust system,* which is designed to lead the high-speed, noisy exhaust gases away from the engine and silence them. The other part of the system includes the *exhaust pipe* and the *silencer*.

**The Ignition System**

The explosive mixture of gas and air entering the engine from the carburetor, is ignited by a powerful electric spark. The ignition system is designed to produce this spark in the various cylinders at exactly the right time. The main parts of the system are the battery, the ignition coil, the distributor, and the spark plugs – one plug in each cylinder.

The *battery* supplies the electricity for the system. It is more correctly called an *accumulator* because it stores electricity produced by the car's dynamo when the engine is running. The most common type of battery is the *lead-acid* battery. It consists of a number of connected cells in a case, often made of a hard rubber. In each cell are two lead plates in a weak sulphuric acid solution. Electric current from the dynamo causes changes in the plates and gives them an electric charge. When needed, current can flow back from the cells to the car's electrical system. Most car batteries have 6 cells, with a total electric charge of 12 volts.

The *dynamo,* or *generator* produces electricity to charge the battery. It is a machine that changes mechanical energy into electrical energy. It is driven by the fan-belt from the crankshaft. The dynamo produces a direct current. Some vehicles have a generator called an *alternator* which produces alternating current initially. An alternator has a higher output than a dynamo, and can produce current when the engine is only just 'ticking over'.

There is between the dynamo and the battery a *voltage regulator,* which ensures that the correct amount of current goes into the battery to keep it fully charged. A switch called the *cut-out* prevents electricity flowing back from the battery to the dynamo when the dynamo is rotating at low speed and producing less electricity than the battery.

The electricity from the battery is not strong enough in itself to produce an electric spark. But the *ignition coil,* in conjunction with the *contact breaker* in the distributor, can magnify the 12 volts of the battery to 30,000 volts! The coil is in effect a kind of transformer. The low-voltage current from the battery circulates in the few windings of the outer primary coil and is rapidly interrupted by means of the contact breaker. As a result of this a very high voltage is induced in the very many windings of the inner secondary coil.

The high voltage is fed back to the *distributor,* which 'distributes' it to the sparking plugs. The contact breaker is a set of points which open and close. It is operated from a rotating spindle driven by the engine camshaft. At the top of this spindle and rotating with it is the *rotor arm,* which receives the electricity from the coil. As the rotor arm rotates, it touches metal contacts in the distributor cap and passes on the high-voltage impulses to each one in turn. The contacts are linked directly to the spark plugs by the plug leads. At the base of the spark plugs, which screw into the top of each cylinder, a spark jumps between the electrodes and ignites the fuel mixture. The normal firing order for a four-cylinder engine is cylinders 1–3–4–2.

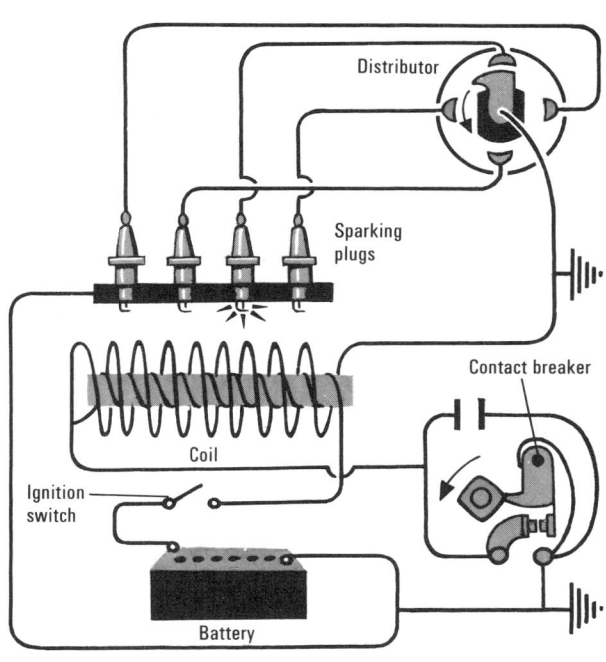

The main elements of the automobile ignition system. Surges of high-voltage current are induced in the induction coil by interrupting the current from the battery. The high-voltage pulses are then fed to the sparking plugs via the distributor.

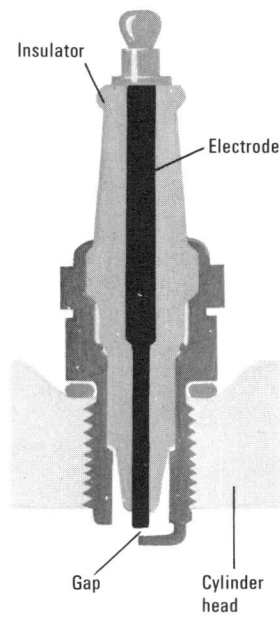

Below: Cross-section of a sparking plug. The high-voltage pulses from the distributor go to the central electrode. The cylinder head, into which the metal body of the plug screws, is earthed. And the electricity 'jumps' from the electrode across the gap, creating a spark to ignite the fuel mixture.

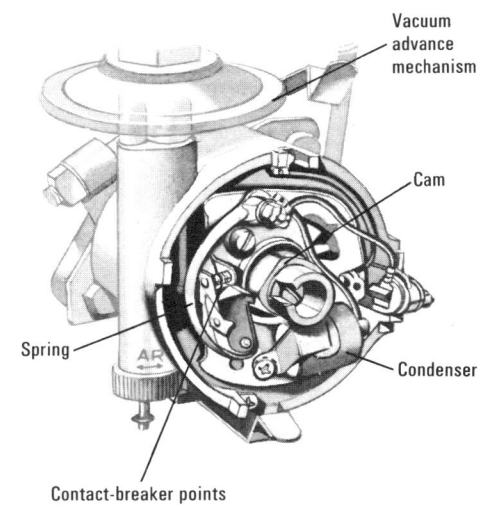

A distributor with the cover removed. Rotation of the cam by the engine periodically forces the points apart. The spring immediately closes them when the cam moves round. The condenser prevents undue arcing. The vacuum advance mechanism, which works by engine suction, automatically advances the point of firing as engine speed progressively increases.

# Lubrication and Cooling

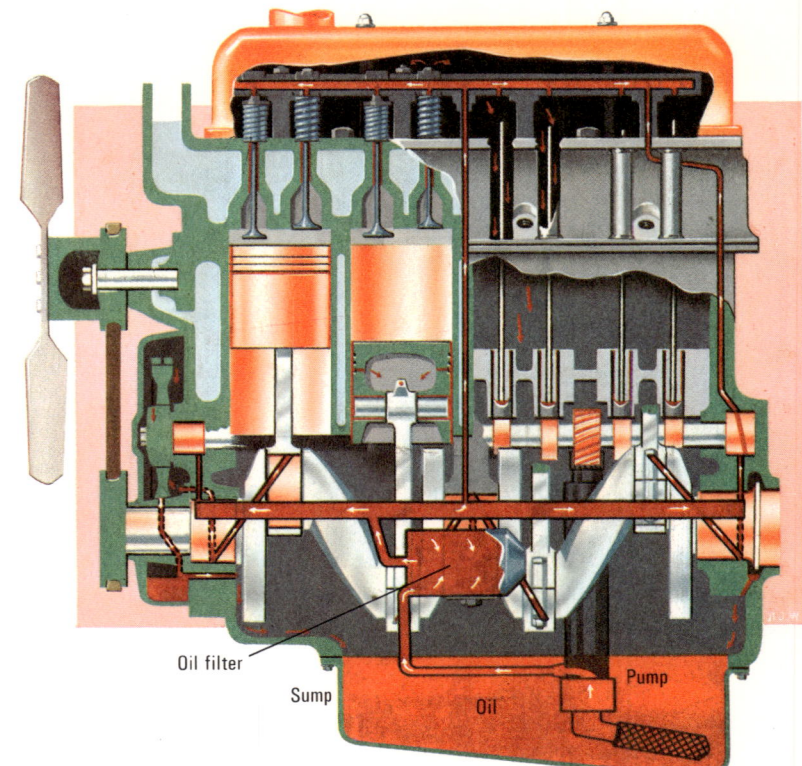

### The Lubrication System

Heat is produced when one piece of metal moves continually over another. The metal will soon wear or even melt, causing the pieces to seize up. A car engine has many metal parts rubbing against one another. These parts must be lubricated all the time to prevent them wearing or seizing up. The oil allows the parts to slide easily over one another with the minimum of wear. The engine lubrication system is designed to keep up a constant oil flow to the moving parts. It also helps to carry away heat.

Most vehicles have a *wet-sump* lubrication system. A reservoir of oil is carried in the sump at the bottom of the crankcase. An *oil pump* in the sump forces oil continually through the engine. It is driven by a gear from the camshaft. First the oil passes through an *oil filter* which removes particles of dirt and metal that might harm the engine. The oil reaches the moving parts in various ways. The cylinders are usually 'splash-lubricated'. The big ends dip into the sump oil as the crankshaft goes round and splash the cylinder walls. Certain parts like the main bearings and the camshaft bearings, are fed by oil under pressure through drilled channels. Other parts receive oil as the oil drips down from the top of the engine. This is known as 'drip-feed'. Eventually all the oil finds its way back to the sump and is again pumped round the system. The oil needs changing about every 5,000 miles, as does the oil filter.

Above: A typical automobile lubrication system. The engine oil is kept in a reservoir, or sump at the base of the crankcase. It is pumped through a filter, and thence through channels to the main bearings. Many parts, however, are indirectly lubricated by drips and splashes.

Below: The majority of automobiles have a water-cooling system like this. Circulating water absorbs heat from inside the engine and then passes into a radiator, where it is cooled by air drawn over the radiator tubes by a fan.

### The Cooling System

The burning of fuel inside the engine cylinders produces tremendous heat. If the temperature were allowed to go too high, the lubricating oil would evaporate, and the moving parts would seize up. The engine's cooling system removes the excess heat from the engine and allows it to work properly. A few cars have an air-cooling system, in which a fan blows air over metal fins fitted around the engine cylinders.

Most vehicles have a *water-cooling* system. Water is pumped through the engine, where it absorbs heat, to the *radiator,* where it is cooled down again. The water is forced through channels in the cylinder block and cylinder head called *water jackets.* It becomes hot and passes to the *header tank* at the top of the radiator. From there it drops down through a network of tiny tubes in the radiator *core.* A *fan,* driven by the fan-belt from the crankshaft, draws cold air through the radiator and over these tubes to cool the water inside. The colder water at the bottom of the radiator is then pumped back through the system. The *water pump* is fixed to the same shaft as the fan and rotates with it.

A valve called a *thermostat* is included in the cooling system to help the engine reach its working temperature quickly when starting from cold. When the water is cold, the thermostat remains closed and prevents the cooling water circulating through the radiator. It opens when the water gets hot.

In winter a compound called *antifreeze* must be added to the water in the radiator. It contains a chemical, such as ethylene glycol, that lowers the freezing point of water.

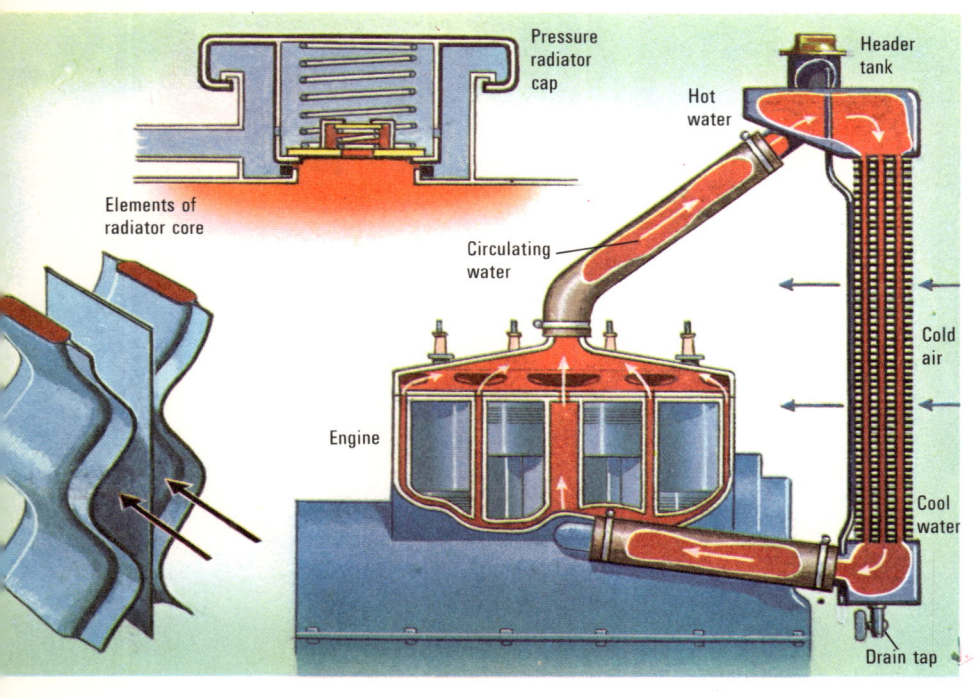

# Transmission

In every motor car, there must be a way of getting the power from the engine to the driving wheels. This is the job of the *transmission system,* so called because it transmits the power. The main parts of the system in an ordinary automobile are the clutch, gearbox, propeller shaft, and final drive.

## Clutch

The *clutch* disconnects the engine from the rest of the transmission system when the driver wishes to engage or change gear. The most common form of clutch is the friction clutch. Most vehicles have a single-plate clutch. The clutch plate is attached to the drive shaft of the gearbox. Normally, when the vehicle is moving, powerful springs press the clutch plate firmly against the flywheel, which transmits the motion of the engine. The plate is lined with a tough material (like a brake lining) which grips the flywheel without slipping.

When the driver wants to change gear, he depresses the clutch pedal. This action forces the clutch plate away from the flywheel and thereby disconnects the gearbox from the engine. The driver can then change gear without the gear wheels 'grinding' together. Then he releases the clutch pedal and allows the clutch plate to take up the drive again. The clutch may operate either mechanically or hydraulically (by liquid pressure).

Automobiles with automatic transmissions have no manual clutch. Their clutch works automatically. In many automatic clutches whirling oil in a kind of turbine transmits the motion from the flywheel to the drive shaft of the gearbox. This arrangement is called a *fluid flywheel.*

## Gearbox

To work well, a modern engine must revolve quickly, even when the road wheels need to turn slowly. The *gearbox* is designed to overcome this problem. Power from the engine goes into the gearbox by the drive shaft. Power to drive the wheels leaves it by the driven shaft. The gearbox changes the speed of the drive shaft into different driven-shaft speeds by bringing together toothed gear wheels of different sizes. A device called a *synchromesh unit* helps to bring the wheels together smoothly. Also, a gearbox can reverse the direction of the drive shaft to make the car go backwards.

Most automobiles have a gearbox worked by hand with three or four forward gears and reverse. Some have a separate higher top

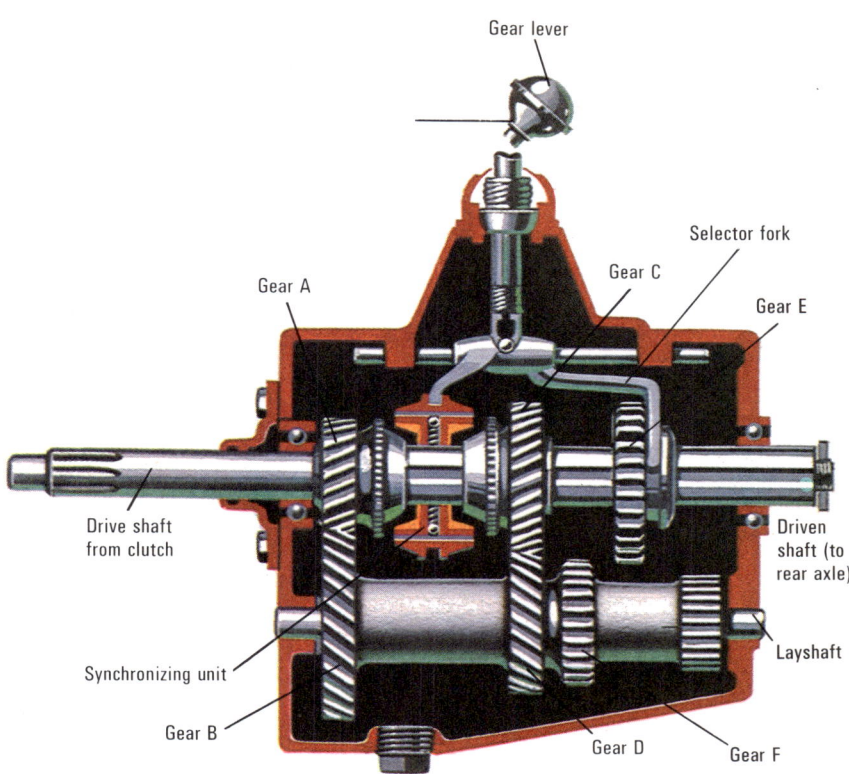

A typical three-speed gearbox, with a synchromesh unit on second and top gears. Gears A and B, and C and D are in constant mesh. The synchromesh unit and gear E rotate with the driven shaft. Gear C is free to rotate around the driven shaft. First gear is selected by moving gear E so that it meshes with gear F on the layshaft, which is driven from the drive shaft via gears A and B. For second gear the synchromesh unit slides against and locks with gear C, so drive is now via gears A, B, D, and C. For top gear the synchromesh unit locks with gear A, giving direct drive.

An ordinary friction clutch. When the vehicle is in motion, the clutch plate is pressed against the engine flywheel and transmits power to the gearbox. When the clutch is depressed, the plate is forced away from the flywheel, thus isolating the gearbox and permitting gear changing.

gear called *overdrive,* which cuts down the engine speed without reducing the road speed. It may be operated manually or automatically.

In normal automobiles the driven shaft from the gearbox is coupled to the *propeller shaft,* which carries the drive to the back axle. At each end of the shaft is a *universal joint.* This is so designed that each end of the shaft can move up and down independently. This prevents the shaft snapping on bumpy surfaces.

## Final Drive

At the rear of the automobile is the *final drive.* The power from the propeller shaft goes to the *differential* in the center of the rear axle. The differential is an ingenious arrangement of meshing wheels and pinions which allows the driven road wheels to travel at different speeds when, for example, the automobile turns a corner. Coming from each side of the differential are the *half shafts* which transmit power to the wheels.

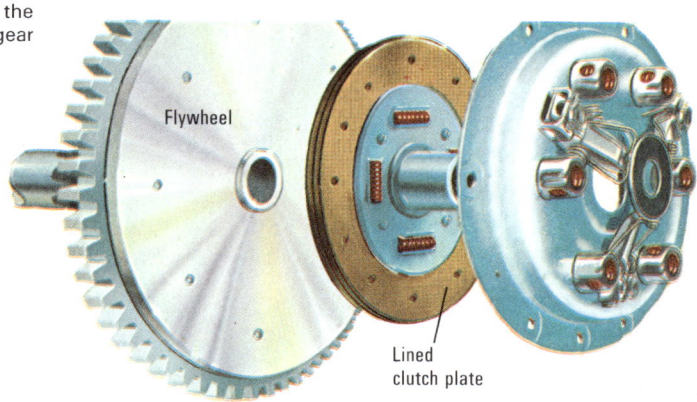

# Steering, Braking and Suspension Systems

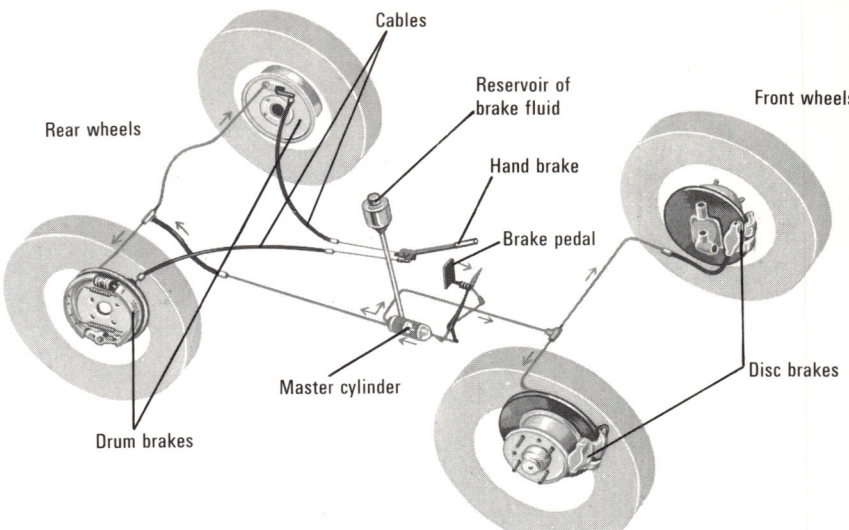

## The Steering System

A vehicle is steered by swivelling the front wheels one way or the other. The driver turns the *steering wheel* when he wants to turn. The steering wheel is attached to the *steering column*. At the base of the column is the *steering unit*, or *steering box*, which transmits the movement of the steering column to the linkages that control the swivelling of the wheels.

Reduction gearing is incorporated in the steering box to make the vehicle easier to handle. A fairly large movement of the steering wheel gives rise to quite a small movement of the road wheels. The linkages between the steering unit and the wheels are so designed that the inner wheel swivels more than the outer one when the vehicle is cornering because it has to turn more sharply.

The steering box varies from car to car. In many designs there is a short worm thread on the steering column which is meshed with a peg or nut on a short shaft. Steering units incorporating this kind of system include *cam-and-peg*, *worm-and-nut*, and *recirculating-ball*. On the short shaft is a 'drop arm' which swings back and forth as the steering column is moved this way and that. The drop arm is connected to the 'drag link', which is itself connected to another arm that makes one front wheel swivel. The swivelling motion of this wheel is transmitted to the other by the 'track rod'. In vehicles with independent front suspension the arrangement of levers and arms is slightly different.

Another widely used kind of steering unit is the *rack-and-pinion*. A pinion on the steering column meshes with a simple horizontal toothed shaft, or rack. Turning the column moves the rack from side to side. This movement is transmitted by 'tie rods' on either side of the rack to steering arms on the wheels.

## The Braking System

Every car has two independent braking systems. The main one is operated by the brake pedal through a hydraulic system.

A typical automobile braking system. It is a dual hydraulic/mechanical system. The mechanical system works only on the rear wheels and is operated by the hand brake. The main hydraulic system works on all four wheels and is operated by the foot pedal. Pressure on the foot pedal forces liquid down the brake pipes to 'slave' cylinders at the wheels which force the brake linings against the drums or discs.

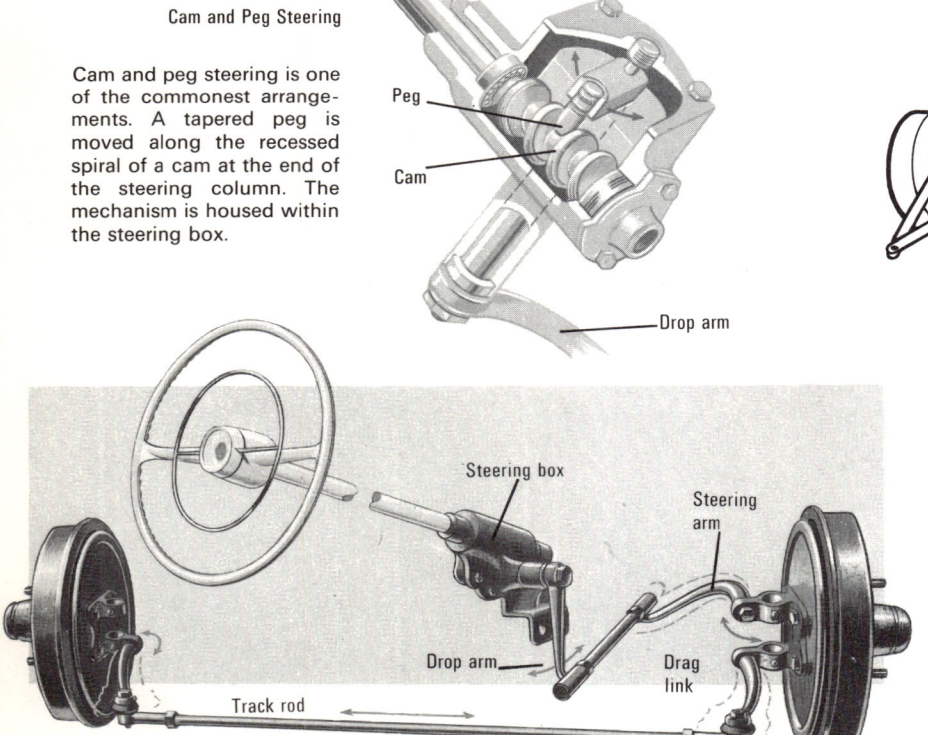

**Cam and Peg Steering**

Cam and peg steering is one of the commonest arrangements. A tapered peg is moved along the recessed spiral of a cam at the end of the steering column. The mechanism is housed within the steering box.

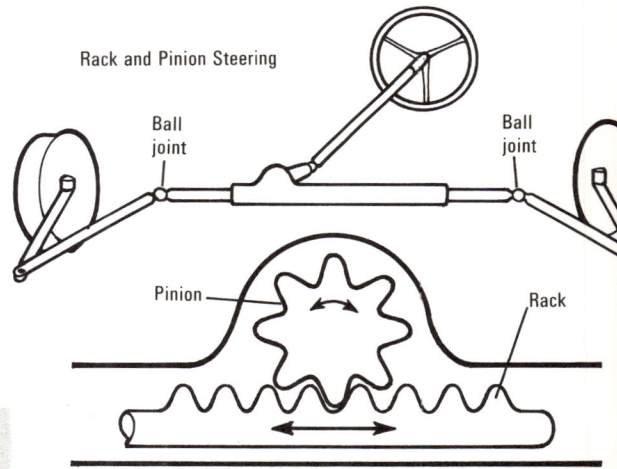

**Rack and Pinion Steering**

Above: Rack and pinion steering, a very simple and very positive steering arrangement. A toothed pinion is connected to the end of the steering column and is turned by it. The pinion meshes with a rack and moves it backwards and forwards. The rack is connected to the wheels via flexible ball joints.

Left: One kind of steering set-up employing a steering box. Movement from the steering wheel is converted into movement of the drop arm, which is communicated to the wheels by linkages.

It acts on all four wheels. The other is operated by the hand-brake lever. It is operated through a mechanical linkage of levers or cables. It acts only on the back wheels. Normally it is used only when the vehicle is stationary, and for this reason it is often called the 'parking' brake.

Two main kinds of brakes are used on automobiles – drum brakes and disc brakes. Drum brakes are generally used on the back wheels and disc brakes at the front. In *drum* brakes, an iron *brake drum* is attached to and revolves with the wheel. Stationary inside the drum are semi-circular *brake shoes* which are faced with a tough, and heat-resistant fractional material called *brake linings*. When the driver applies the brakes, the shoes are forced outwards against the inside of the drum and slow it, and therefore the wheel, down.

In *disc* brakes the wheel has a steel disc revolving with it. When the brakes are applied, two *brake pads*, one on either side, squeeze the disc and slow it down. Disc brakes are very much more efficient than drum brakes. They are also less liable to brake 'fade', which may occur when drum brakes become overheated after continual braking. Disc brakes seldom overheat because most of the disc is open to the air at any one time. Another advantage of disc brakes is that they are self-adjusting. Drum brakes need constant adjustment as the linings wear down.

In high-powered sports cars and heavy vehicles, it requires a great deal of pressure to apply the brakes. Sometimes they may be fitted with what are called *servo-*, or *servo-assisted* brakes. The braking system incorporates a *servo unit* which utilizes the suction produced in the engine's inlet manifold to help apply the brakes.

**The Suspension System**

The suspension system is designed to give passengers as comfortable a ride as possible over bumpy roads. Early automobiles had rigid front and rear axles. When one wheel went over a bump, the whole car tipped up. Springs over each wheel helped to 'cushion' the body from this effect. These days, however, most automobiles have independent suspension, at least on the front wheels. Each wheel can move up and down independently without affecting the other. And rear springing these days is so good that very little of the bumpiness of the road is transmitted to the vehicle body.

Large *coil-springs* form the basis of independent suspension units. The spring is sandwiched between pivoted upper and lower arms that are attached to the short *stub axle* which carries the wheel. The spring absorbs most of the up-and-down movement of the wheels. Incorporated with it in the suspension unit is a *shock absorber*, or *damper*, which 'damps down' the tendency of the spring to oscillate up and down when displaced.

*Leaf-springs* are generally used in non-independent rear-suspension units. They consist of a number of metal strips, or leaves of different lengths bolted together. They are usually arranged lengthways, one over each end of the rear axle.

More elaborate systems of suspension work hydraulically. In the 'Hydrolastic' system, the suspension units on the front and back wheels are linked. The units consist of a liquid-filled chamber and a special rubber spring. Transfer of liquid from the front to the back unit lifts the whole vehicle slightly when the front wheels hit a bump and this reduces pitching, or rocking to and fro.

Another system, called 'Automatic Level Control', maintains a car at a standard curb height regardless of load weight.

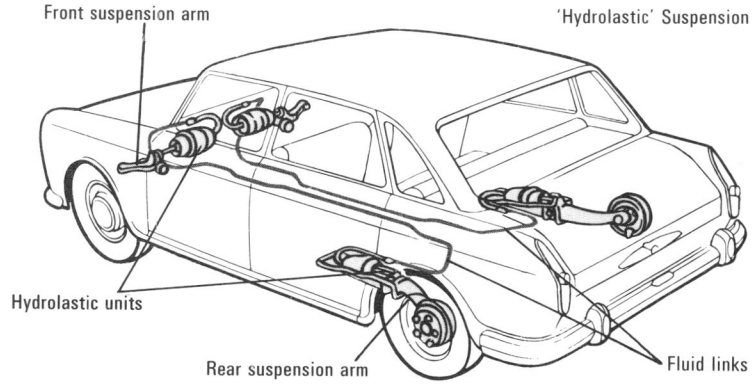

British Leyland's Hydrolastic suspension, as fitted to its 1800 models. The suspension units serve as both springs and dampers. The units on either side of the vehicle are linked by a fluid coupling. When the front wheel rises over a bump, fluid is transmitted from the front to the rear unit, causing the rear to rise as well.

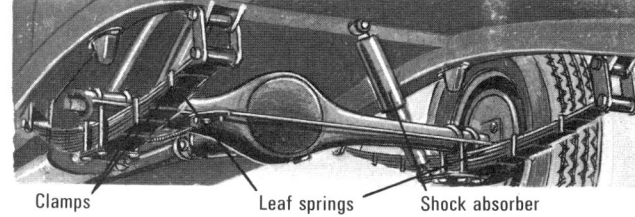

Right: A common type of rear suspension, incorporating leaf springs and shock absorbers. Most ordinary family saloon cars have this simple type.

Below: Independent suspension units are now invariably fitted to the front wheels. Coil springs with inset shock absorbers, as depicted here, are usual.

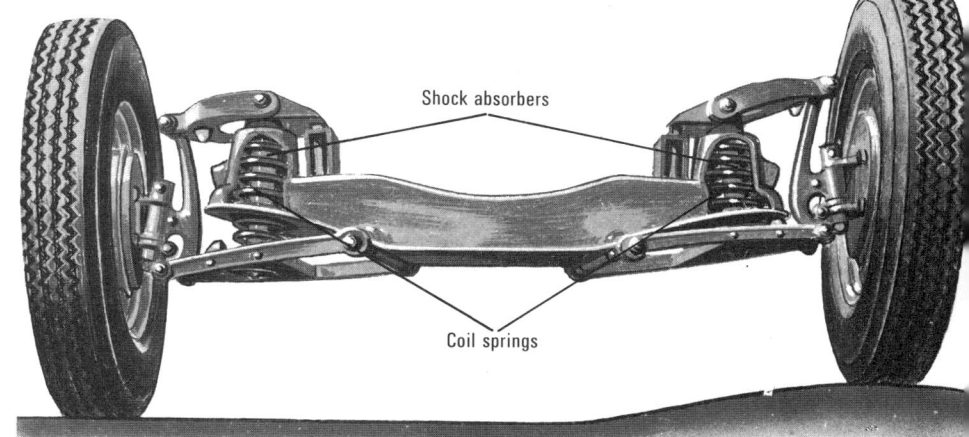

# Mass Production

Modern life offers a vast selection of manufactured goods – refrigerators, washing machines, radios, televisions, automobiles, to mention only a few. People buy goods in ever-increasing quantities and qualities, and so the factories making them are steadily increasing their production to meet the demand. Continuous efforts are made to increase production by such techniques as *simplification, automation,* and *standardization.* Such efforts, are, however, dependent on one major technique – *mass production,* which at its simplest means making 'lots and lots of' something.

To mass-produce an article, the designer breaks it down into all the parts which go into it – nuts, bolts, presswork, mouldings, cut-outs, and so on. When every single *component,* as it is called, has been identified,

The assembly-line system is the key to mass production in the automobile industry. The various parts of the vehicle are assembled from individual components. The workers below are assembling an engine unit, for example. Below right: A view of a production line as the bodies are married to the engine and transmission units.

plans are drawn up as to the best way of making that component. These plans bring to light how much money a part may cost. If the part being considered is something used by other manufacturers, then the designer may well find it is cheaper to buy that particular part ready-made. A simple illustration of this is wood and metal screws.

They are mass-produced by firms who specialize in making screws and nuts and bolts. They make the components for other factories to use. Because they make so many, their production efficiency is high and their costs are low. Therefore it does not pay another manufacturer to instal plant to make just a few screws for his own use.

The designer, having decided which parts to make and which to buy, then plans his production, that is, he works out how many finished articles ready for sale will be made, hourly, daily, or weekly. Each component will have to be made in sufficient numbers to meet this target and will have to be ready at the right place and at the right time in the factory so that the other important process, *component assembly,* can go on. This may mean that the components are placed on moving belts as they are made and carried by this belt to the next operation. For example, in automobile assembly the nearly finished car travels on a moving belt to the wheel fitter; his job is to affix the four wheels (which in turn have arrived to him painted and tired) and to do nothing else.

Each operator, then, is often entirely

dependent on the work done by other operators, and this is one of the weaknesses of integrated mass production. If the flow of one component stops for any reason, everybody involved in work after that component will be held up. The problem can be alleviated to some extent by building up component stocks. But, since storage space is limited, this is a relatively short-term solution. In an automobile assembly factory, for example, there may be sufficient supplies of bodies, say, in stock to cover a temporary supply shortage of a few days. But if the workers in the body factory go out on strike for a week

or so, the automobile-assembly workers will be affected, too.

In essence, then, a mass-production factory can be regarded as comprising many tiny internal factories *(departments)* whose job it is to make components that are simply put together by the big factory. Departments in a factory are often highly specialized in their own work. Integrating their output – that is, fitting all the pieces of the production jigsaw together – to fulfil the objectives of the factory as a whole is called *network planning*.

Inventor of the cotton gin Eli Whitney was

Close-up of a body shell being lowered into position on the assembly line. Note that the cars on this assembly line are mixed, as opposed to those on the assembly line pictured opposite. On a 'mixed' production line very careful phasing of the assembly operation is necessary to bring the correct components together in the right place at the right time. Co-ordinated planning of this sort is now often computer controlled.

one of the first to utilize the principles of mass production. In the 1790s he undertook to supply 10,000 muskets to the government in a year, an impossible task by orthodox gunsmithing. What he did, therefore, was to build machine tools that could produce identical parts. Then all that was necessary was for the various parts to be assembled, which could easily be done by unskilled labour. The modern idea of assembly-line production originated in the early 1900s, when first R. E. Olds and then Henry Ford applied it to car assembly – Henry Ford to the production of his famous 'Tin Lizzie'.

215

# Diesel and Rotary Engines

**The Diesel Engine**

Most automobiles are powered by the four-stroke reciprocating gasoline engine which has been described in detail on the previous pages. But some are powered by other kinds of engines. A few are powered by the *diesel engine,* although the diesel has more familiar uses in heavier means of transport, such as trucks, locomotives, and ships. Locomotive and marine diesel engines are really massive. Diesel engines are also used as stand-by units in electricity generating stations.

A diesel looks very much like a gasoline engine. And, like the gasoline engine, it is a reciprocating piston engine, which may work on the four-stroke or two-stroke cycle. But there are also significant differences.

The diesel engine has neither sparking plugs nor a carburettor, and burns oil not gasoline as a fuel. The oil is not as heavy as engine oil nor as light as kerosene. The fuel is burned by what is known as *compression-ignition.* (This compares with the spark-ignition of the gasoline engine.)

On the induction stroke of the piston, air is drawn into the diesel engine cylinders. On the compression stroke the piston compresses the air so much that its temperature rises to 500°C (about 900°F) or more. Just before the piston reaches the top of its stroke, fuel is injected into the cylinder through a nozzle called an *injector.* The fuel enters as a fine mist and burns immediately as it comes into contact with the very hot air. This drives the piston down on its power stroke. On the fourth stroke, the exhaust gases are ejected. Then the cycle begins again.

The air may be compressed into less than one-fifteenth of its original volume. This is usually expressed in terms of a *compression ratio.* We say that the engine has a compression ratio of 15:1. Higher (up to 20:1) and lower (down to 12:1) compression ratios are used in diesel engines. This is much higher than the compression ratio in gasoline engines.

The diesel engine has a much greater efficiency than the gasoline engine. It can use as much as 40 per cent of the energy in its fuel. A gasoline engine has an efficiency of less than 25 per cent. The diesel has to be built more solidly than a gasoline engine to withstand the higher operating pressures. This means that it is heavier than a gasoline engine of equivalent power and costs more to make. But it lasts longer. It can operate

Marine diesel engines are really massive, as is clearly seen here. Developing thousands of horse-power, they are invariably of two-stroke design. This one has 12 cylinders and is turbo-charged.

better at low speeds than a gasoline engine and can use relatively cheap fuel.

**The Wankel Engine**

Some automobiles have an entirely different kind of engine from normal, called a *rotary* engine. It burns gasoline as fuel like an ordinary gasoline engine, but produces rotary motion. The best-known rotary engine is the Wankel engine. A disadvantage of the normal car engine is that it produces up-and-down motion of the pistons. This has to be changed to rotary motion before it can

Right: The operating cycle of the Wankel engine, showing how each segment between rotor and chamber wall goes through the cycle of induction, compression, ignition/power, and exhaust. The great advantage of the Wankel engine is that it produces rotary power directly, as opposed to the standard piston engine, in which reciprocating motion of the pistons must be converted into rotary motion to drive the wheels.

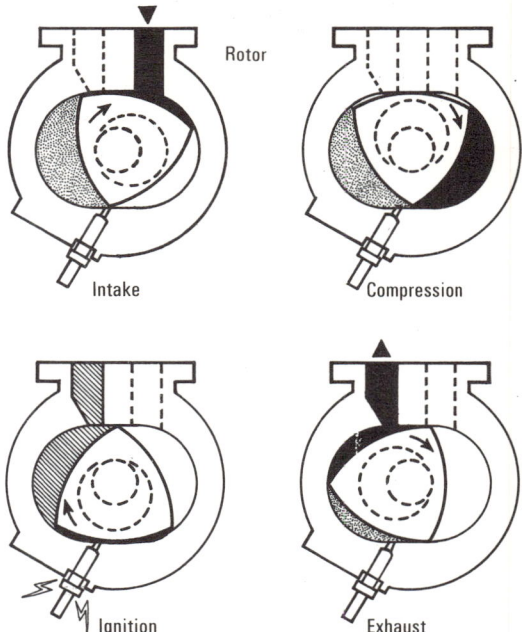

be used to move the car. Also, only one of the four strokes of the piston is a power stroke. The Wankel engine overcomes these problems. It works on the same four-stroke principle as a normal car engine, but produces rotary motion *directly*. It also produces power at every revolution. The Wankel engine has fewer moving parts and is lighter and more powerful than a piston engine of the same size.

The engine casing has a kind of figure-of-eight shape and inside it rotates a shield-shaped *rotor*. The three tips of the rotor are always in contact with the walls, dividing the space inside the casing into three. Each segment goes through stages of induction, compression, power, and exhaust in turn and continuously.

### New Engines

Much research is taking place all over the world into alternative methods of vehicle propulsion. Quite a few manufacturers have produced experimental gas-turbine automobiles using a kind of jet engine as the power plant. The Rover company of Britain produced one as long ago as 1950. One of the main problems with gas-turbines in automobiles is that the gas turbine works best at high speeds of revolution, yet the automobile must run at relatively low speeds.

Great interest is being centered on steam engines once again. Several companies have produced experimental modern steam engines suitable for automobiles. One disadvantage with steam engines for motor vehicles is that of scalding the passengers if the boiler, which holds steam at high pressure, explodes after a collision. One advantage is that steam engines do not pollute the atmosphere to anywhere near the same extent as gasoline- or diesel-burning engines, although they must, of course, burn some fuel to raise steam.

Other experimental vehicles are powered by electricity. Manufacturers are thus returning to an earlier idea. Electric cars were popular in the mid-1800s. Today's models are much better. They have a higher top speed and a greater range, resulting from improvements in design of the batteries. In some models the electricity is produced from fuel cells. Hydrogen and oxygen gas are made to combine under certain conditions in a cell, producing a tiny electric current. This is the same kind of cell that powers the *Apollo* spacecraft on their Moon trips. Another idea borrowed from space is that of solar power. Panels of solar cells on the roof of the vehicle produce electricity to charge a battery. All electric cars have the advantage that they produce no fumes.

Above: The pioneer Rover gas-turbine car, built in 1950. Many manufacturers are again experimenting with this form of propulsion. Below: A modern electric car, the Ford Comuta, designed for city travel. Bottom: A research and development mid-engined coupé built by Mercedes-Benz. Its four-rotor Wankel engine gives an acceleration from 0–60 mph in 4½ seconds and a top speed of nearly 190 mph. Note the upward-opening 'gull-wing' doors.

# How Aircraft Fly

On December 17, 1903, two brothers from Dayton, Ohio, made aviation history on a deserted stretch of sand at Kitty Hawk, in North Carolina. Orville and Wilbur Wright made the first powered flights in a heavier-than-air flying machine. Their flimsy machine was built of wood, cloth, and wire and was powered by a simple automobile engine. It reached only about 30 miles an hour and was airborne for only a few seconds at a time (see page 232).

What a difference in aviation today! Three hundred and fifty ton monster 'jumbo jets' carry almost 500 passengers across the Atlantic in about 8 hours. Supersonic airliners like the *Concorde* travelling at twice the speed of sound (1,400 mph) are designed to more than halve this time. Some jet fighters can travel even faster (more than 2,000 mph).

Aviation has certainly come a long way since 1903. But the basic principles of flight remain the same. The key to heavier-than-air flight lies in the special curved shape of the wings. They have what is called an *aerofoil* cross-section. When the plane travels forwards, the flow of air over the wings produces an upward force on them called *lift*. As the plane goes faster and faster, the lift increases until it is greater than the plane's *weight,* that is, the downward force on the plane. When this happens, the plane will rise into the air and fly.

The force to make the plane travel for-

A Hornet Moth biplane, a direct descendent of the famous Tiger Moth. Biplanes such as these have no place on the commercial aviation scene today, but they are still widely used for private flying.

An English Electric Lightning fighter with its formidable weaponry. This includes a bevy of ground attack rockets, cannon shells and air-to-air guided missiles. Note the huge intake in the nose for its turbojet engine and the thin-section wings.

wards, the *thrust*, is provided by the plane's engines. But, as the plane is thrust forwards, the air pushes against it and tends to slow it down. This air resistance is called *drag*.

In steady flight at constant speed, the four main forces acting on the plane are in balance. Lift equals weight, and thrust equals drag. The study of the forces which act on a body moving through the air forms the science of *aerodynamics*.

## Lift and Weight

Obviously, the lighter an aeroplane is, the less the lift needed to support it and the easier it will take to the air. This is why planes are made with the lightest possible materials, such as aluminium and magnesium alloys (see page 227). The lighter the airframe, the more passengers and freight the plane can carry.

The amount of lift on an aircraft wing is dependent on several things. A thick, very curved aerofoil develops more lift than one of thin section. Increasing the wing area and increasing the speed at which the air flows over the wings also increases the lift.

Low-speed planes therefore tend to have thick and fairly large wings in order to develop enough lift. High-speed planes have wings of thin section and a comparatively small wing area. Wings of thick section would in any case cause too much drag at high speeds.

Another way of increasing lift is to increase the angle at which the wings meet the airstream, the *angle of attack*. A plane is designed to have a slight angle of attack in level flight at cruising speed. The angle is made small in order to minimize drag.

When a plane in level flight slows down slightly, the lift will decrease, and the plane will drop slightly. But the pilot can maintain his height by increasing the angle of attack and thereby increasing lift. He can maintain his height in this way as his speed falls until the angle of attack of the wings is about 15 degrees. If this angle is exceeded, the airflow over the wings breaks down. The abrupt change in flow results in sudden loss of lift, and the plane drops. This condition is known as *stalling,* and the speed at which it occurs is the *stalling speed.* The stalling speed is always the same for the same aircraft.

The particularly dangerous moments for a pilot are on take-off and landing. At these times the pilot wants maximum lift. But his speed is necessarily very low. What he does

Two views of the Anglo-French Concorde supersonic airliner. Able to cruise at twice the speed of sound, Concorde shows typical features of the supersonic aircraft—sharply swept-back wings of thin section, narrow tapering body, and generally clean aerodynamic lines. Note the lack of a tailplane. In the Concorde, rear sections of the extended wing take over the functions of the tailplane, acting as stabilizers and elevators, as well as ailerons.

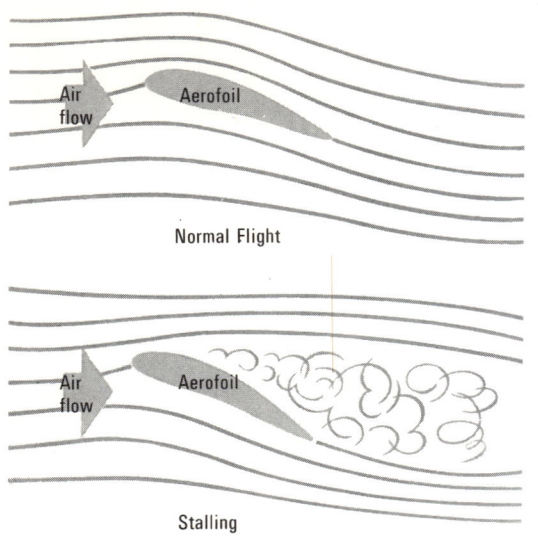

Left: In normal flight the air flow round an aerofoil wing is smooth (top). As the angle of the aerofoil with respect to the incident air flow is increased, a point is reached at which the air flow around the aerofoil breaks down (bottom). This situation, known as stalling, results in an abrupt loss of lift.

## Aerofoils

An aerofoil is curved at the top and fairly flat at the bottom. It has a blunt front, or leading edge and a sharp rear, or trailing edge. Because of this shape air moving over the aerofoil travels faster than air moving under the aerofoil. As air travels faster, its pressure drops. And so the pressure of air above the wings is less than that below it, creating a kind of suction which causes the wing to lift.

You can test that this works for yourself. Hold one end of a sheet of paper up to your lips, and let it droop down in front. Now blow over the paper. You will find that the whole sheet of paper will rise. By blowing over the top surface, you have increased the air speed there and caused a pressure drop.

The propeller of an aeroplane has an aerofoil section, too. The movement of the propeller through the air lowers the pressure in front of it, and the suction produced 'pulls' the propeller along (see page 222). The rotor blades of a helicopter also have an aerofoil section. They provide lift in place of a wing (see page 230).

(The hydrofoils used for high-speed boats are of similar shape and function to the aerofoil, see page 123.)

A transatlantic airliner just after take-off. Note the pronounced 'down' position of the inner and outer flaps at the trailing edge of the wings. This gives the plane increased lift for the take-off, when speed is comparatively low. Note also the undercarriage retracting to its in-flight position.

is to lower *flaps* on the rear, or trailing edge of the wings. This effectively increases the curvature of the wing section, and thereby the lift. Some planes have wings with movable front, or leading edges, called *slats*, which increase lift in a similar way.

## Thrust and Drag

The thrust to propel the plane through the air may be provided by a propeller, a jet engine, or both. The propeller has a special shape that enables it to develop thrust as it spins round. The thrust of a jet engine comes from the reaction to a stream of high-speed gases. (See page 223.)

The drag of the air opposes the motion of the plane. Obviously, the less drag there is, the faster the plane can fly. And so aircraft designers go to a lot of trouble to reduce the drag. They make sure that the area presented to the air stream is as small as possible, because the smaller the area, the less it can be affected. For this reason, the landing gear, or undercarriage, of most aircraft is designed to retract, or fold away inside the body or wings during flight. The actual shape of the body itself is important. A streamlined shape is the ideal one (see page 226). All the surfaces are made as smooth as possible, too, because smooth surfaces cause less drag than rough ones.

## Aircraft Controls

A pilot controls his plane in the air by means of the engine throttle and by moving hinged after-sections of the tail and wings from side to side or up and down. He operates the control surfaces by means of a *control column*, also called a *joystick*, and by foot pedals. (In larger planes the controls are duplicated for the co-pilot.) The control column may be a simple lever or it may take the form of a steering wheel or what resembles the handles of a bicycle. Rods, cables, or, more usually, hydraulic systems connect the column and pedals with the control surfaces.

Pushing and pulling the column operates

the *elevators* at the rear of the tailplane. Pushing the column forwards tilts the elevators and the nose of the plane downwards. Pulling the column back tilts the elevators and the nose upwards. But the action of the elevators alone does not make the plane dive or climb. All it does is to change the attitude of the plane relative to the direction of travel.

Vertical movement of the plane in normal flight is, in fact, controlled simply by opening and closing the engine throttle. All the pilot has to do to climb is to open the throttle and increase his speed. Increasing the speed increases the lift on the wings, and the plane therefore moves upwards. Conversely, decreasing the speed decreases the lift, and the plane drops. A combination of engine and elevator control is generally used for climbing and diving.

Moving the column from side to side operates the *ailerons* on the wings. Moving it to the left causes the left-hand aileron to go up and the right-hand one to go down. This has the effect of making the left wing dip. Moving the column to the right makes the right wing dip. This is called *banking*.

To turn the plane, the pilot operates the *rudder* at the rear of the vertical tail fin. He moves the rudder by depressing the foot pedals. Pressing the left pedal deflects the rudder to the left and swings the nose round to the left. Pressing the right pedal brings the nose round to the right. But the pilot seldom uses the rudder by itself to turn. If he did, the plane would tend to 'skid' or slip sideways, just as a racing car tends to skid on a corner. In fact, the pilot uses the ailerons as well as the rudder so that the plane banks and turns at the same time. In this way the pilot can keep the plane firmly under control without any side-slip.

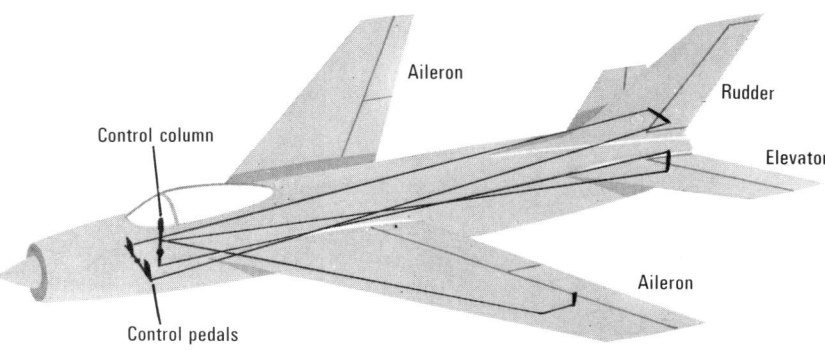

The primary control surfaces on an aeroplane. The ailerons are situated at the end of the trailing edge of the wing and are controlled by moving the control column from side to side. Moving the column back and forth operates the elevators at the rear of the tailplane. The rudder at the rear of the tail fin is operated by control pedals.

### Stability

It would be very irritating if the pilot had continually to be making tiny adjustments to his controls every time, say, a puff of wind hit his plane. And so a plane is designed so that it tends to correct itself. In this state it is called *stable*.

There are three basic movements of a plane in flight which have to be corrected, pitching, yawing, and rolling. *Pitching* is the upward or downward movement of the nose. It is corrected by the tailplane or horizontal stabilizer. *Yawing* is the tendency to turn to the right or left. It is corrected by the tail fin, or vertical stabilizer. You can now understand the importance of a plane's tail. *Rolling* is dipping from side to side. It is corrected by inclining the wings upwards slightly from the fuselage at an angle known as the *dihedral angle*.

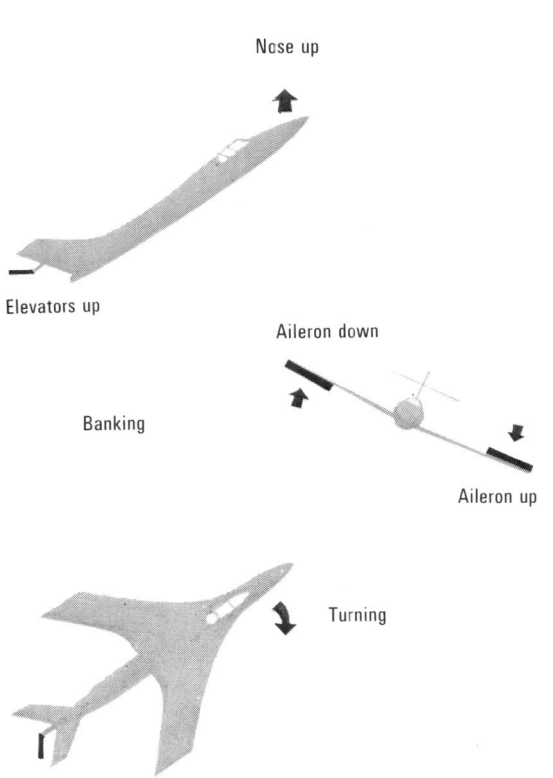

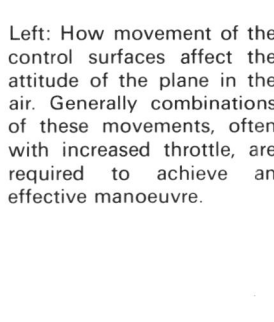

Left: How movement of the control surfaces affect the attitude of the plane in the air. Generally combinations of these movements, often with increased throttle, are required to achieve an effective manoeuvre.

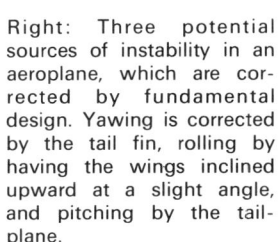

Right: Three potential sources of instability in an aeroplane, which are corrected by fundamental design. Yawing is corrected by the tail fin, rolling by having the wings inclined upward at a slight angle, and pitching by the tailplane.

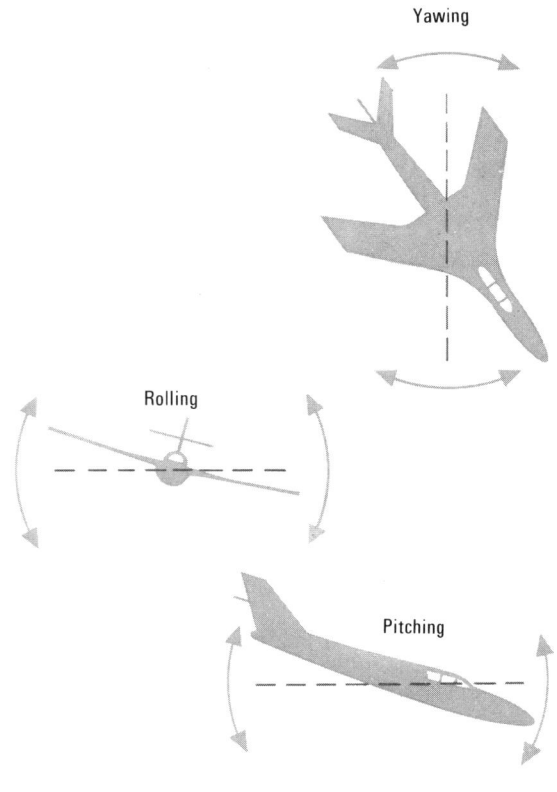

# Aircraft Propulsion

The thrust to drive a plane through the air may be provided either by a rapidly spinning propeller, by the jet of hot gases from a jet engine, or sometimes by both.

A spinning propeller produces thrust because of the shape of its blades. Like the wings, they have the cross-sectional shape of an aerofoil. When they rotate, the pressure drops in front of them, effectively 'pulling' the propeller (and therefore the aircraft) along. The propeller, then, develops thrust in the same kind of way that a wing develops lift.

The propeller blades are angled and curved so that they 'bite' into the air just like a screw bites into wood. This gives the propeller its alternative name – *airscrew*. In most planes the angle, or *pitch* of the propeller blades can be adjusted for different conditions of flight. It can even be reversed so that the propeller is actually 'pushing' against the direction of flight. This 'reverse thrust' is applied immediately after landing to help slow down the plane quickly.

In many aircraft even today the propellers are driven by piston engines. These engines use gasoline for fuel and work on the same principles as the gasoline engines used in automobiles (see page 204). A piston engine consists essentially of a number of closed cylinders in which tightly fitting pistons move up and down. The pistons are connected to the shaft that drives the propeller. Fuel and air are drawn into the cylinders and ignited by a spark. The hot gases produced force the pistons down, which turns the propeller shaft.

Although aircraft engines work on the same principles as automobile engines, they differ from them in several respects. They are relatively much lighter, are generally air-cooled, and have their cylinders arranged in a circle or in two parallel lines. Also, the spark is provided by an electricity generator

The Anglo-French Jaguar is an advanced trainer and tactical support aircraft with twin turbofans, each developing nearly 5,000 pounds thrust. The use of afterburners boosts thrust by some 2,000 pounds. Top speed approaches Mach 2.

called a *magneto* and not by an induction coil. Automobile engines on the other hand are relatively heavy, are generally water-cooled, with their cylinders in a single line or arranged in a 'V' shape, and have a battery to provide the electric power for sparking via the coil.

In most propeller aircraft today, however, the propellers are driven by a kind of gas-turbine, or jet engine called the *turboprop*. This is a development of the simplest and most powerful form of jet engine the *turbojet*, which we shall consider first.

## Turbojets and Turboprops

Basically the turbojet engine is a very simple machine. It burns its fuel, kerosene (paraffin), in compressed air inside a chamber to produce hot gases. As the gases shoot backwards out of the engine, the engine is propelled forwards by reaction.

The turbojet engine is like a huge cylinder, inside which there is a shaft that rotates. Mounted on the front of this shaft are a number of bladed discs (somewhat like a complicated fan) which make up the engine *compressor*. This compresses the air drawn in through the front of the engine and forces it into one or more *combustion chambers*. Then the fuel is injected in a fine spray into the chamber and burned. To start the engine going, the fuel mixture is ignited with a spark. But thereafter it burns continuously as long as fuel is sprayed in. The hot, expanding gases produced by the combustion rush through the open end of the combustion chamber. As they do so they spin the blades of a *turbine* (somewhat like a windmill) mounted on the same shaft as the compressor, which it therefore drives round. The gases then leave through the engine exhaust, or *tail-pipe* as a high-speed jet.

The fuel consumption of a turbojet is enormous. It may be 1,000 gallons an hour or more. To burn this quantity of fuel, the engine must take in almost 200 tons of air every hour. Turbojets operate best at high speeds and high altitudes.

The turboprop, as we mentioned earlier, has a propeller driven by a jet engine. Although the main thrust of a turboprop comes from the propeller, a certain amount also comes from the jet. In the simplest form of turboprop the propeller is mounted on the same shaft as the compressor and, with it, is driven by the turbine. But in most turboprops the propeller is driven by an independent turbine. This kind of engine is known as a *free-turbine turboprop*. Turboprops are more economical to operate than turbojets, but are not suitable for very high-speed flight.

Above: The huge engines currently used in many airliners are often of the turbofan type, which offer greater propulsive efficiency.

Right: The elements of a simple turbojet and a simple turboprop engine. In the turbojet the main propulsive thrust comes from a stream of hot gases produced by burning fuel in compressed air. The exhausting gas stream also spins the turbines that drive the compressors. In the turboprop one of the turbines drives a propeller.

Below: Cutaway picture of a turbofan engine, showing how part of the air compressed by the large front fan bypasses the main part of the engine.

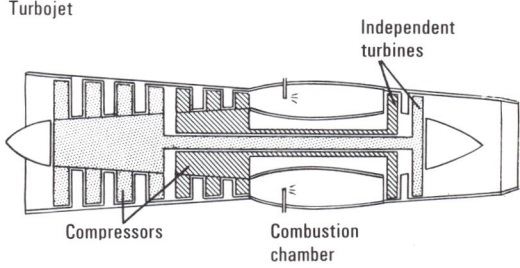
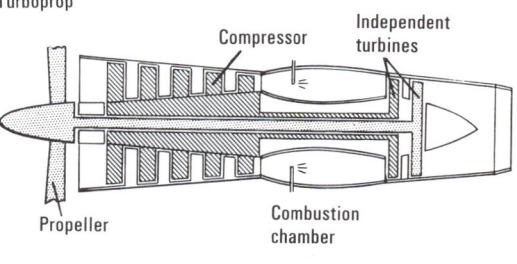
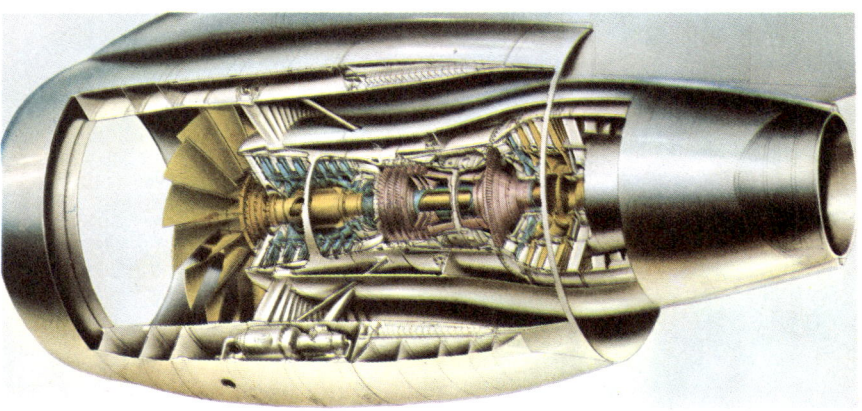

# Aircraft Instruments

As aircraft become bigger and more and more complicated, pilots have to rely increasingly on instruments to fly correctly and safely. With an aircraft costing millions of pounds and carrying hundreds of passengers, nothing can be left to chance. Gone are the days when a pilot flew 'by the seat of his pants' without any instruments and relied on his senses to warn him of trouble.

The cockpit of a modern airliner contains an apparently bewildering variety of a hundred or more instruments, over which the pilot and the other crew members must keep constant watch. Smaller, light planes do not have many more instruments than an expensive, high-powered automobile. However, planes large and small have certain basic instruments that are essential for flight and navigation.

First there is the *air-speed indicator*. This shows the plane's speed through the air, just as a car's speedometer shows its speed over the ground. To find the plane's speed relative to the ground, the pilot must add or subtract the speed of the wind.

The *altimeter* shows how high the plane is above the ground. The commonest form of altimeter works by air pressure, which decreases with increasing height above sea level. Before landing, this altimeter must be 'zeroed' at the air pressure of the airport. The latest planes have a so-called *radio altimeter* which measures the height directly by bouncing radio waves off the ground in much the same way as with radar.

Two of the most important aircraft instruments, the altimeter, above, and the artificial horizon, below. The altimeter shown works by air pressure. The latest types measure altitude directly by bouncing radio waves from the ground.

Below: The complex array of instruments in a prototype Concorde supersonic airliner. In addition to the normal flight and navigation instruments, this particular plane has a host of test instruments to monitor engine, systems and airframe performance.

To find his way through the sky, the pilot uses a *compass*. Some planes have a magnetic compass, just like those scouts and soldiers use. But most now have *gyrocompasses*, which contain a kind of rapidly spinning wheel called a gyroscope. Once set in a particular direction, the gyro is not affected by any kind of movement of the plane. Also, it is not affected by magnetic disturbances in the atmosphere, unlike the magnetic compass.

Three other instruments are essential for flying at night or in bad visibility, when the pilot cannot see the ground. The *artificial horizon*, or *attitude gyro* shows the position of the plane relative to the ground – whether it is level, climbing, diving, or banking to the right or left. The *turn-and-bank indicator* helps the pilot turn the plane properly. The *rate-of-climb indicator* tells the pilot how fast he is ascending or descending.

Other instruments give the pilot information about the engines. The simplest ones are similar to those found in most automobiles including gauges measuring oil pressure, fuel level, and engine temperature.

The latest airliners have panels of warning lights which light up if anything goes wrong in any one of the multitude of systems in the plane. They flash a warning, for example, if the undercarriage fails to lock into position or if the engines become overheated. Many airliners are equipped with electronic equipment to make landing a lot easier or even automatic (see page 235).

# Supersonic Flight

*Sonic* speed, or the speed of sound, is very important in aeronautics, because strange things happen when an aircraft approaches this speed. At speeds below that of sound, that is, *subsonic* speeds, air flows smoothly over the aircraft surfaces. Pressure waves set up by the aircraft pushing against the air travel in front of the aircraft at the same speed as sound.

When the aircraft itself nears sonic speed (or *Mach 1*), the pressure waves cannot, as it were, escape, and shock waves begin to form over the body and wing surfaces. The air flow is no longer smooth but very *turbulent*. This greatly increases the drag on the aircraft and reduces the lift of the wings. In the early days of high-speed flight, aircraft lost control because of the severe buffeting they experienced when they neared sonic speed. It was even considered that the speed of sound constituted a kind of barrier to further progress, and the term 'sound barrier' came into being.

Aircraft designers have found that they can delay the formation of shock waves by building aircraft with sharply sweptback wings of thin section. You have probably noticed that all the supersonic jet fighters and airliners have such wings.

Modern supersonic aircraft are so well designed that they can fly through the 'sound barrier' with little or no noticeable change in handling qualities. Once they are travelling at supersonic speeds, the buffeting ceases and the flight is smooth once more. Shock waves are still formed, but they do not adversely affect the aircraft.

Shock waves do, however, affect people and buildings on the ground. They cause what is known as a *sonic bang*, which sounds exactly like an explosion. And, like the blast from an explosion, it can cause damage. Low-level sonic bangs can shake buildings and shatter windows over quite a wide area.

Another problem becomes increasingly important as speeds rise above Mach 2 (the speed of the Anglo-French supersonic airliner *Concorde*). At these speeds the friction of the air against the metal skin of the aircraft produces quite a lot of heat. Above Mach 2·5 the heat produced is sufficient to weaken seriously the aluminium alloys which are normally used in aircraft construction. This point is often called the 'heat barrier'. It can be overcome by using such metals as stainless steel and titanium, which retain their strength at high temperatures, and by refrigerating the aircraft surfaces.

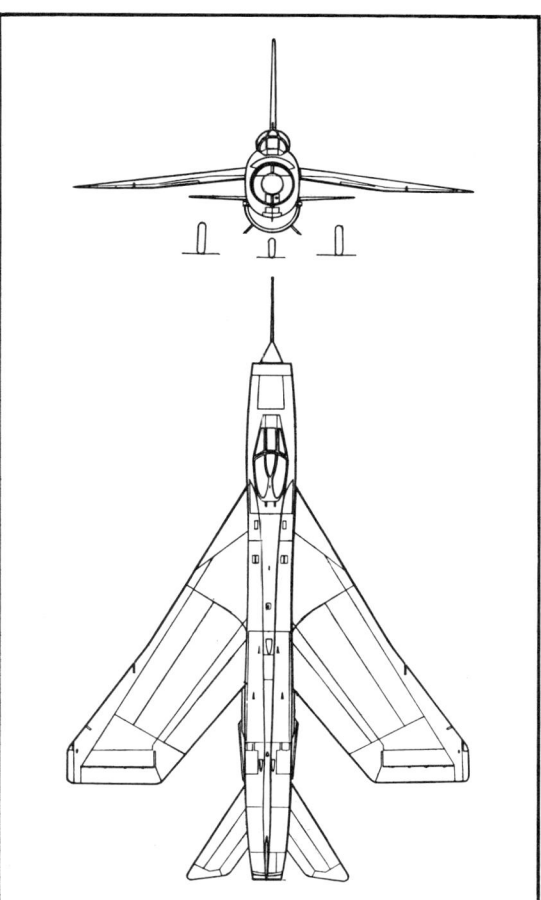

Frontal and plan views of an English Electric Lightning supersonic fighter. We can see clearly here how the fuselage narrows where the wings meet it, exemplifying the so-called wasp-waist design, which has been shown to be effective at high supersonic speeds.

**Mach Number**

The speed of sound is not the same at every altitude. It decreases with increasing height as the air becomes thinner and colder. At sea level it is about 760 miles an hour, but at the cruising height of many jet aircraft (35-40,000 feet) it is only about 660 miles an hour. A pilot therefore wants to know not his actual speed, but his speed relative to the speed of sound. This is called the *Mach number*. Mach 1, therefore, is sonic speed, Mach 2 is twice the speed of sound, and so on.

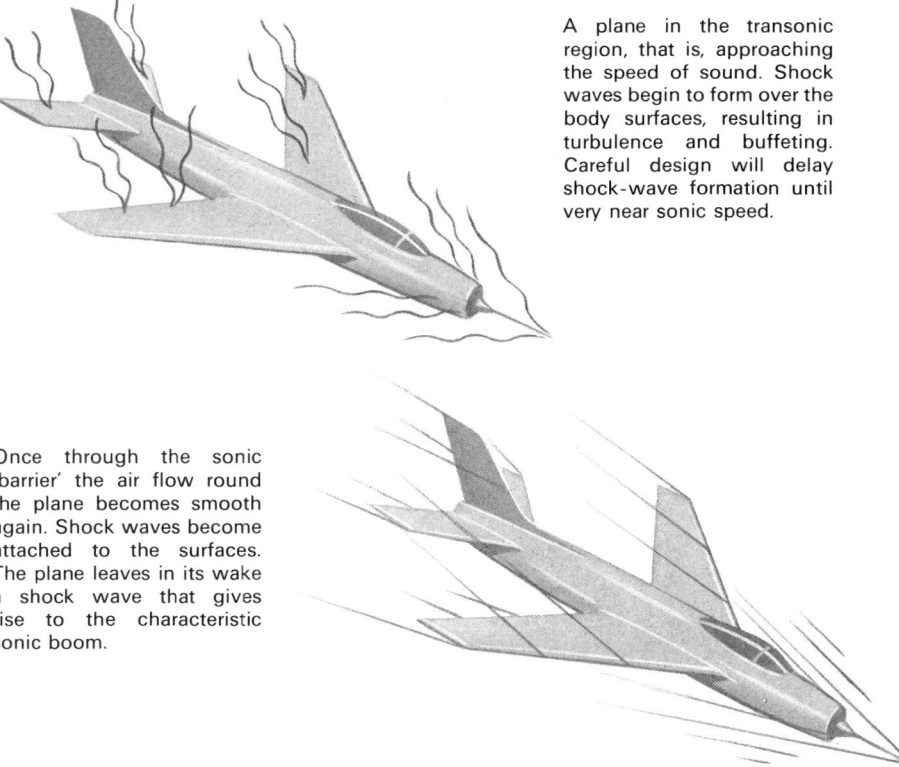

A plane in the transonic region, that is, approaching the speed of sound. Shock waves begin to form over the body surfaces, resulting in turbulence and buffeting. Careful design will delay shock-wave formation until very near sonic speed.

Once through the sonic 'barrier' the air flow round the plane becomes smooth again. Shock waves become attached to the surfaces. The plane leaves in its wake a shock wave that gives rise to the characteristic sonic boom.

# Aircraft Construction

The birth of a new aeroplane takes place in the design office of an aircraft factory. The company may be approached by an airline or by the government to design a plane with a certain seating capacity, range of operation, and cruising speed. Or the company may itself design a plane which they think will be a good seller.

Then they set to and consider all kinds of different designs for their proposed aircraft. The design will be greatly influenced by the desired cruising speed. The shapes of the wings and fuselage of a low-speed plane are very different from those of a plane capable of very high, supersonic speeds (see page 225). For low-speed flight the wings may extend from the fuselage almost at right angles. But for high-speed flight they have to be sharply swept back. The ideal streamlined fuselage shape for a low-speed, subsonic aircraft is that of an elongated raindrop. But this would be no use for supersonic flight. Many supersonic aircraft have what is called a *wasp-waisted*, or *coke-bottle* fuselage which has been shown to be better for very high speed. The 'waist' occurs where the wings join the fuselage.

Other major decisions relate to the number, type, power, and positioning of the engines. They could go on the wings, beneath the wings, or in the rear, leaving the wings 'clean'. Each position has certain advantages and disadvantages.

All these kinds of problems have to be thoroughly investigated before a design can be finalized. Designers test scale models of their new design in wind tunnels to gain an idea how the plane will behave in real flight. They calculate the strengths of the various components to make sure that the plane can withstand the loads and stresses it experiences during flight. They also build a full-size wooden model, or 'mock up' of the plane to help them in detailed design and layout of instruments, fuel lines, electrical systems, hydraulic systems, and so on. Also, potential customers can see exactly what the plane will look like in reality.

When the overall design is completed and the customers are satisfied, work begins on construction. Detailed drawings of the many thousands of components are made; not even the simplest brackets can be omitted. From these drawings the full-size components are manufactured. Many more drawings have to be made to show how the numerous components fit into the overall design.

Stages in the construction of the Concorde. Top, the ribs and struts of the rear fuselage being completed. Middle, work taking place on the engine bays for two of Concorde's four turbojet engines, which each have some 35,000 pounds thrust. The engine bays are fabricated mainly of stainless steel honeycomb and also contain titanium forgings to withstand the intense heat generated by the engines. They are one of the few areas of the airframe where structural materials other than aluminium alloys are employed. The bottom picture shows a general view of the vast Concorde assembly hall.

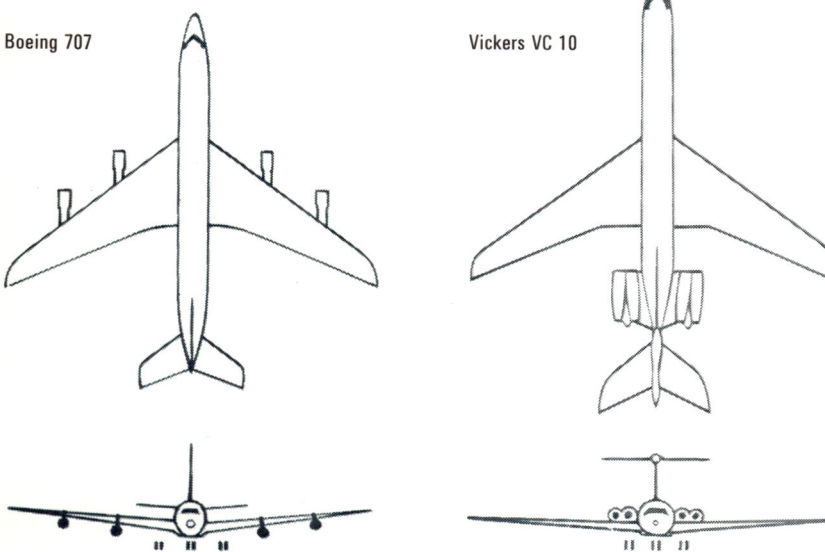

Below: Different designs for aircraft of similar size, range and performance. The VC 10 has its four engines at the rear, leaving the wings 'clean'. It has a high tailplane to avoid the jet exhaust. The Boeing 707 in contrast has its engines slung beneath its wings in 'pods', and has a conventional tail.

## Building

The main part of the body of the plane, the fuselage, is made up of a thin 'skin' of lightweight metal on top of a supporting 'skeleton'. The circular, cross-wise supports are known as *frames* or *formers,* and the lengthwise stiffeners as *stringers*. The wings are constructed in a similar way, with a metal skin covering the framework of ribs and spars. Often they have to be given added reinforcement to take the engines or to accommodate fuel tanks. The metals most widely used in aircraft construction are aluminium alloys, on account of their great lightness and strength, and the ease with which they can be shaped. Magnesium alloys, which are even lighter, are often used on inside structures.

Rigorous testing of all parts of an aircraft must be carried out, and giant test rigs like that above are constructed to test full-scale models of components. This particular rig tests the functioning of Concorde's trimming system, which moves fuel within the airframe structure to alter the centre of gravity in supersonic flight.

In most aircraft the surface 'skin' is made by riveting together large numbers of small sheets. In advanced designs, however, large, complicated panels may be machined from solid metal, using machines controlled by computers. This so-called 'skin milling' gives very much greater strength with less weight. (Riveting requires holes to be drilled in the metal, which naturally weakens it, and adjacent sheets must overlap, which adds to the weight.)

## Testing

As production proceeds, engineers test full-size parts of the plane on test rigs under conditions similar to those found in flight to see whether the parts are strong enough. Many of the parts are twisted and stressed to

Two methods of fabricating wing sections. Top left: Conventional construction with a skin riveted to supporting members. Bottom left: The same result achieved by machining from solid metal. The gain in lightness is considerable without sacrificing anything in strength.

Below: Pressure testing of a fuselage in a water tank in which the fuselage is pumped up with compressed air. This technique duplicates the stresses caused by fuselage pressurization at high altitudes. Pressure testing was pioneered in the 1950s while searching for clues to the Comet disaster.

destruction. One of the great problems to overcome is metal fatigue, which may occur quite suddenly after prolonged stress and vibration. One of the most fascinating tests is the tank test. The whole fuselage is immersed in a water tank and pumped up with air to imitate the pressurization that occurs in actual high-altitude flight (see page 228).

The first planes built, called *prototypes,* are tested thoroughly on the ground and in flight before the plane is ready to go into full production. These days most planes are mass-produced on an assembly line in much the same way as an automobile is.

Aircraft manufacturers need to sell quite a number of a particular model before they can recover the research and development costs and begin to make a profit. Aircraft firms within a country and in different countries are now co-operating to produce new generations of aircraft. The most outstanding example of such co-operation is between British Aircraft Corporation of Bristol and Aérospatiale of Toulouse, who together designed and built the 1,400 mph (Mach 2) *Concorde.*

**Pressurization**

The pilots of high-speed jet aircraft cruise, when they are able, in the region of the atmosphere called the *stratosphere,* at a height of between 30,000 and 40,000 feet. They do this for two reasons. First, at this height they are flying higher than the clouds and 'above the weather'. It is in the lower part of the atmosphere where the clouds form that our weather develops. By flying above it, the pilot has a smoother passage because he avoids rain, hail, snow, thunderstorms, fog, and so on.

The other reason for flying high is that jet engines work more efficiently at high altitudes because the air is thinner there, or in other words, the air pressure is low. This is fine for the jet engine but not for the crew and passengers. Human beings are used to breathing air at ground level, where there is plenty of oxygen, not 6 miles above it, where there is little.

To enable us to breathe at this height, large planes are given an artificial atmosphere corresponding roughly to that which we are used to breathing near the ground. The pressure of air is considerably higher than that outside. This is what is meant by *pressurization.* The fuselage must be made strong enough to resist the extra forces caused by the difference in pressure inside and out.

In an emergency, such as the failure of the pressure system, oxygen masks automatically drop down in front of the passengers. In

An accurately scaled model of Concorde being suspended in a large diameter wind tunnel. Aircraft designers suspend models in an airstream and observe how they perform. By carefully scaling up the observed effects they can predict quite accurately how full-scale aircraft will perform.

this way they can survive until the pilot has dropped to a height at which they can breathe without assistance. It is particularly important, of course, that the pilot and the crew get oxygen immediately. And some airlines insist on at least one member of the crew being 'on oxygen' all of the time the plane is in the air.

Smaller, high-flying planes such as fighters are not generally pressurized. Their pilots wear helmets which supply them with oxygen. Pilots flying up to 60,000 feet and beyond must in addition wear a special pressurized flying suit, which is very much like a spacesuit. This is necessary at such heights because, even if the pilot were supplied with oxygen, there is not enough pressure to force it into his bloodstream. The pressure at 60,000 feet is, in fact, less than one-tenth of the pressure at sea level.

### Wind Tunnel

Aircraft designers have a very good idea of how an aircraft will behave in actual flight a long time before it takes to the air. They gain their information by testing scale models of the aircraft in a *wind tunnel*.

The wind tunnel is specially designed to produce a very smooth, uniform stream of air through the part where the models are suspended. This part is called the *working section*. One side is made transparent for visual observations. A delicate weighing balance in the working section measures the forces produced by the action of the moving air on the models. The models are often suspended upside down, which is done to simplify the resulting calculations.

Electric fans or turbines suck air through the working section over the model, which is, of course, equivalent to the model moving through still air. The forces acting on the model for various air speeds can then be found. By introducing smoke into the tunnel, or by using special photography called *Schlieren* photography, the air flow around the model can be seen.

Wind tunnels are also used for testing the effects of wind on tall structures. Skyscrapers and bridges are being built so high these days that the extra loads on the structures caused by the wind blowing against them cannot be ignored.

# Helicopters

The helicopter is an incredibly versatile machine. It can fly straight up or down, hover like a fly, and go forwards, backwards, or sideways. It therefore needs no long, prepared runway like a normal aeroplane does. It can operate from a flat roof in the city or from a clearing in the jungle with equal ease. However, helicopters are nowhere near as fast as normal fixed-wing aeroplanes and have a top speed of only about 200 miles an hour.

The helicopter has often been called the 'flying windmill' because of the long, rotating blades, or *rotor,* on top of its body. This main rotor serves as both propeller and wings to provide the thrust to drive the helicopter along and the lift to raise it off the ground. The rotor blades have the same aerofoil shapes as aeroplane wings and propellers to generate the required lift and thrust.

When the main rotor rotates, the body of the helicopter tends to spin by reaction in the opposite direction. To counteract this effect, a small rotor is mounted sideways on the tail. The thrust produced by this tail rotor in normal flight just balances the turning tendency of the helicopter body, which therefore remains steady. Both the main and the tail rotors are driven by the helicopter engine through the same gearbox. Another way of maintaining stability in the air is to have twin main rotors, which each rotate in opposite directions. Then the 'spinning' tendency of one cancels that of the other.

Movement of the helicopter in any direction – forwards, backwards, sideways, up or down – is achieved simply by varying the pitch of the rotor blades. The *pitch* is the angle at which the blades slice through or 'bite' the air. To fly straight upwards all the blades are set at the same sharp pitch. In this position they achieve maximum lift. To descend, the pitch of the blades is reduced so that the lift is insufficient to support the helicopter's weight.

The pilot adjusts the pitch for up or down movement by moving the *collective-pitch* stick. This is also linked to the engine throttle so that more power is applied as the pitch increases. This is to overcome the greater air resistance with the sharper pitch. Conversely, the engine is throttled back when the pitch is reduced for descent because there is less air resistance.

To fly forwards, the pitch of the rotor blades is varied as they rotate. For forward motion, the pitch is made least at the front and gradually increased as the blade moves backwards. This has the effect of sweeping

Top: Helicopters are invaluable in air-sea rescue operations, being able to hover above the person in the water and winch them to safety. Below: Mountain-rescue work is another area where the helicopter's 'land anywhere' capability has saved many lives.

the air backwards, and therefore pushing the helicopter forwards. The same goes for any horizontal movement – the pitch is made least in the desired direction of travel. The cyclic-pitch stick controls the pitch of the rotor for horizontal flight.

The turning movements of the helicopter are controlled by means of the tail rotor. Increasing the pitch of the rotor makes the tail swing one way; reducing it makes the tail swing the other. Foot pedals control this operation.

The blades are hinged near the centre of the rotor. This gives the helicopter stability. The blades flap down in the direction of motion and up in the opposite direction. Early helicopters met with disaster because they were built with rigid rotors.

The forerunner of the helicopter was the autogyro, invented by Juan de la Cierva in Spain in 1923. This machine had a normal front propeller, but achieved its lift by means of a freely rotating rotor.

# Balloons and Airships

A long time before man began to fly in heavier-than-air aeroplanes, he had been flying, first in balloons and then in airships. These craft are lighter than air and that is why they can fly. Being lighter than air, they rise upwards through it just as a cork, which is lighter than water, will rise through water.

The first balloons (1780s) were hot-air balloons. As we all know, hot air rises. It rises because it is lighter than the cold air surrounding it. Fill a bag with hot air, and it will rise. And that was the basis of the hot-air balloons. The air was kept warm by a brazier of hot coals slung beneath the open base of the balloon. This, of course, was rather dangerous.

The hydrogen balloon was a great improvement on the hot-air balloon. Hydrogen is the lightest of all gases and is fourteen times lighter than air. It gives a much greater upward 'lift' than hot air. The dangerous disadvantage is that hydrogen burns and explodes extremely readily. Many of today's balloons are filled with helium, which is slightly heavier than hydrogen but is non-flammable. Ballooning was a popular sport right up until the 1930s, and even today there are keen hydrogen and hot-air balloonists.

However, the greatest present-day uses of balloons are in weather-forecasting and scientific research. These 'sounding' balloons carry instruments up into the atmosphere to measure such things as temperature, humidity and pressure at different heights. They also carry a radio transmitter to relay the instrument readings back to scientists on the ground. That is why they are called *radio-sondes*.

The great drawback about balloons is that they can go only where the wind blows them. In 1900, Ferdinand von Zeppelin launched the first really successful *dirigible* (steerable) balloon, or *airship*. It was a 420-foot long, cigar-shaped craft, with a rigid skeleton supporting a gas-tight skin. It was powered by the recently developed petrol engine. Soon Zeppelin's airships were carrying passengers and mail. Britain, America, and Russia built similar craft. But a series of disasters throughout the world in the 1930s showed how dangerous airships were, and they were soon withdrawn from service.

Ballooning is achieving something of the popularity it enjoyed in the '30s. International meetings are held regularly. The above picture shows a hydrogen-balloon meeting. Note the bags of ballast in the left foreground Ballast is taken up in the basket of the balloon and poured out when greater height is required.

Below: Hot-air ballooning also has its enthusiasts, some of whom crossed the Sahara in 1972. Today's hot-air balloonists make use of butane gas burners to heat the air inside the balloon.

# Milestones in Aviation

**1783** Two French brothers Joseph and Étienne Montgolfier invented the hot-air balloon. They released their first balloon in June. In August a fellow Frenchman J. A. C. Charles released the first hydrogen balloon. In October a French nobleman, Pilaître de Rozier made the first human ascent in a captive Montgolfier balloon.

**1804** The English scientist Sir George Cayley built a model glider which is regarded as the first true aeroplane. Earlier he had discovered and described the principles on which heavier-than-air-flight is based.

**1852** A Frenchman, Henri Giffard, made the first controlled, powered flight in an airship over Paris. His airship had a steam engine.

**1853** Sir George Cayley built a glider, the first heavier-than-air machine to carry a man.

**1890s** A German, Otto Lilienthal, made more than two thousand successful glider flights.

**1900** Count Ferdinand von Zeppelin built the first successful dirigible airship, 420 feet long.

**1903** The Wright brothers made the first controlled aeroplane flight on December 17.

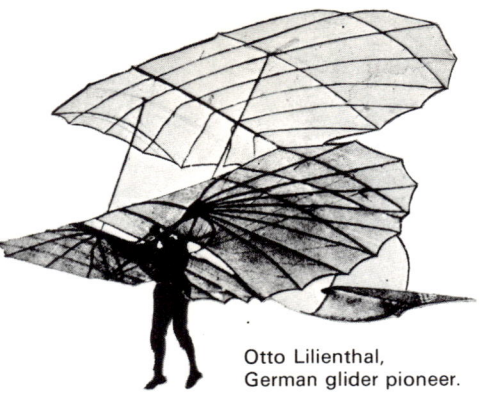
Otto Lilienthal, German glider pioneer.

**1909** The French aviator, Louis Blériot, made the first flight across the English Channel on July 25, in a monoplane which he designed and built himself. His flight took 37 minutes. Blériot was one of the first to build a successful monoplane. He is also credited with the invention of the aileron.

**1913** The Russian engineer Igor Sikorsky built and flew the first four-engined aeroplane on May 13.

**1919** Lieutenant Commander A. C. Read of the United States Navy piloted a flying boat, *N-C-4*, from New York to Plymouth to make the first transatlantic crossing. They set out on May 8 and made stops in Newfoundland, the Azores, and Portugal to complete the trip in 52 hours 31 minutes.

**1919** Captain John Alcock and Lieutenant Arthur Whitten Brown made the first non-stop transatlantic flight on June 14 to 15, in a twin-engined Vickers *Vimy* biplane. They flew 1,936 miles from Lester's Field, near St. John's Newfoundland, to Clifden, Ireland, in 16 hours 12 minutes. Alcock was the pilot and Brown the navigator. They won a £10,000 prize which the *Daily Mail* had offered for the direct crossing in 1913. King George V knighted them both.

**1919** The British airship *R-34* made the first two-way transatlantic flight from Scotland to New York and back, with 32 people on board, in July.

**1919** Aircraft Transport and Travel Ltd. began the first regular international airline service between London and Paris, using converted biplane bombers.

**1923** Juan de la Cierva's Autogyro made its first flight in Madrid on January 9. It was the first practical craft with a rotary wing and foreshadowed the helicopter.

**1926** Lieutenant Commander Richard Evelyn Byrd of the United States Navy and Floyd Bennet made the first successful flight over the North Pole in a Fokker monoplane on May 8 to 9.

**1927** The American aviator Charles A. Lindbergh made the first solo, non-stop transatlantic flight in his monoplane *Spirit of St. Louis* on May 20 to 21. He flew 3,600 miles from Roosevelt Field, New York, to Le Bourget Airport, near Paris, in 33 hours 30 minutes.

---

## The Wright Brothers

Aviation as we know it today, began with the Wright Brothers' first powered flights near Kitty Hawk in North Carolina, USA. It was the younger brother, Orville, born in Dayton, Ohio, in 1871, who made the first flight. It was for only 120 feet and lasted only 12 seconds, but it demonstrated that powered, heavier-than-air flight was possible. The brothers made three further flights that day. Wilbur, born in Indiana in 1867, made the longest one of a fraction under one minute, covering a distance of 852 feet.

Their first plane, the *Flyer*, had a wing span of just over 40 feet. Its two wings were 6 feet apart. The lightweight petrol engine which they built to power the machine was situated just to the right of centre on the lower wing. The pilot had to lie, face down, just to the left of it to maintain the balance. The engine drove two propellers which faced the rear and 'pushed' the plane along.

Control over the plane was achieved by means of a tail rudder of two movable, up-right vanes, and a kind of 'elevator' with two horizontal surfaces which projected in front. The machine looked, in fact, as if it were flying tail-first. Lateral, or side-to-side, stability was provided for by warping the wings. The pilot, by pulling wires, could vary the angles at which his right and left wings met the airstream.

In 1904 they built and flew a second, improved machine at Dayton. Their best flight lasted about 5 minutes. It was their third machine of 1905, however, which was the first practical aeroplane. It could turn and circle with ease. In this machine the brothers made flights lasting more than half an hour. By 1908, the Wrights' improved machines could stay aloft for more than 2 hours.

On September 17 of the same year, at Fort Meyer, Virginia, their first and only disaster occurred. Orville was flying with a passenger when a propeller blade split. One of the wire struts was ripped away, and the rudder collapsed. In the ensuing crash Orville was not too badly hurt, but his passenger, Thomas E. Selfridge, was killed. Selfridge was, in fact, the first man to be killed in an aeroplane disaster.

The Wright brothers grew up in Dayton, Ohio. Both keen mechanics from an early age, they were in 1896 buying, selling and manufacturing bicycles. They became interested in aeronautics that year on reading of the death of the glider pioneer Otto Lilienthal. The early gliders they designed, based on available aerodynamic data, were not successful. They therefore built a wind-tunnel to carry out their own design studies and found that most of the information published about air-flow over wings was inaccurate. Using the new data, they built a really airworthy glider. With the experience they gained, they began designing the machine which made the very first flight at Kitty Hawk in 1903.

Wilbur died of typhoid fever in 1912. Orville died in 1948. He lived long enough to see what a revolution in transport he and his brother had started. By 1948, a new, more powerful kind of engine — the jet — was beginning to replace the petrol engine in aircraft. The sound barrier had been broken. The Wrights' original biplane is now in the National Air Museum in Washington. From 1928 until Orville's death, the plane was on display at the Science Museum in London.

**1928** The Australian pilot Sir Charles Kingsford-Smith and his crew made the first transpacific flight in a Fokker monoplane called the *Southern Cross*.

**1929** The 775-foot-long airship, *Graf Zeppelin*, made the first flight round the world between August 8 and 29.

**1930** A British engineer, Frank Whittle, patented a design for a jet aircraft engine. Amy Johnson became the first woman to fly solo from England to Australia between May 5 and 24 in a second-hand *Gypsy Moth*.

**1932** Amelia Earhart made the first solo transatlantic flight by a woman on May 20 to 21. She flew in a Lockheed *Vega* monoplane from Harbour Grace, Newfoundland, to Ireland in 15 hours 18 minutes.

**1937** The 812-foot-long Zeppelin *Hindenburg* caught fire and exploded while landing in a storm at Lakehurst, New Jersey, on May 6. The first fully controllable helicopter was tested in Germany in July.

**1939** The first jet-propelled aeroplane, the Heinkel *He 178* made a successful flight in Germany in August.

**1939** Igor Sikorsky made the first successful flight in a single-rotor helicopter in September.

**1944** The Germans began the bombardment of London with *V-1* flying bombs, or *buzz bombs*, in June. The V-1s were powered by a form of jet engine known as the pulse jet. In September the Germans began launching their *V-2* rocket bombs against London.

**1947** An American pilot, Captain Charles E. Yeager made the first faster-than-sound flight, in a rocket-powered aircraft the *Bell X-1*.

**1952** The De Havilland *Comet 1* became the world's first jet airliner on British Overseas Corporation's route between London and Johannesburg, South Africa, when it entered regular service on May 2.

**1953** The first turboprop airliner to enter regular service was the Vickers *Viscount*, which began flying on British European Airways route between London and Nicosia on April 12.

**1956** The British Test pilot Peter Twiss set up the first world speed record of more than 1,000 miles an hour when he averaged 1,132 mph on two runs on March 10 in a Fairey *Delta 2*.

**1958** The first transatlantic jet airline service was inaugurated on October 4. BOAC's De Havilland *Comet 4* began operations between London and New York.

**1967** The North American Aviation rocket craft *X-15A-2*, piloted by W. J. Knight, flew at a speed of 4,532 mph on October 3.

**1968** Russia's Tu 144 supersonic airliner, designed to carry 120 passengers at about 1,550 mph made its maiden flight on December 31.

**1969** The Boeing 747 'Jumbo Jet' made its maiden flight on February 9. This giant, 195-foot-long craft can seat more than 400. *Concorde 001*, the Anglo-French supersonic airliner, made its maiden flight on March 2.

**1973** Russia's TU 144 supersonic airliner exploded in mid-air at the Paris Air Show in June, causing several deaths.

Above: Alcock and Brown leaving Lester's Field, St. John's, Newfoundland, at the start of their epic non-stop voyage across the Atlantic in a Vickers Vimy converted twin-engined bomber on June 14, 1919.

Right: The Autogyro, precursor of the helicopter, invented by Juan de la Cierva in 1923.

Right: Canopus, one of a famous line of flying boats, which were in widespread use up to World War 2.

Before crashing near Paris in 1930, the British airship R.101 had been the biggest in the world. The disaster brought Britain's interest in airships virtually to an end.

# Airports

Every year well over 150 million passengers fly on the world's airlines. This figure does not include special charter flights, either, of which there are many. The large international airports such as Kennedy Airport, New York, and London (Heathrow) Airport, handle a phenomenal amount of traffic. London Airport is one of the busiest in Europe, handling in excess of 16 million passengers a year. Fifty-nine airlines from 49 countries operate services into the airport. Aircraft fly from the airport to 93 countries. This complex pattern is repeated for all the other major international airports, too.

You can well imagine, therefore, just how busy a large airport must be and what a problem it is to co-ordinate the flow of planes, passengers, and freight through it. The problems of passenger-handling have become even more acute since the introduction of the massive 'jumbo jets', which can carry almost 500 passengers. As soon as an aircraft has landed on the runway, it moves along a 'side road', or *taxiway* to a parking area, or apron. Immediately, trucks to carry baggage and buses to carry passengers hurry

Above: The Boeing 747 'jumbo' jet, capable of seat-more than 400 passengers. The advent of such planes has accelerated the need for superior passenger-handling facilities.

Below: The runway complex at London Airport (Heathrow), one of the world's busiest. The triangular pattern of runways permits multidirectional take-off and landing.

towards it. Then fuel tankers move in to refill the airliner's fuel tanks, which may hold 20,000 gallons or more for a transatlantic crossing.

A quick turn-around is always aimed for because the airliner is not earning any money while it is on the ground. And there may be other planes waiting to land. The buses carry the passengers to the terminal building, where they collect their baggage and go through Customs. Nearby are the huge hangers, in which the airliners are regularly inspected and overhauled by skilled maintenance staff.

The most important people at the airport are the air-traffic controllers in the control tower, who supervise the landing and take-off of each aircraft. Pilots contact the control tower as they approach the airport. If the runways are full, the controller tells the pilots to circle one above the other, at intervals of 1,000 feet, until the runway is free. This is called *stacking*. As the runway becomes clear, the controller calls in the aircraft at the bottom of the 'stack'. And he instructs the aircraft above to descend to the next level.

The ground controller guides the aircraft down to a safe landing by using radio, radar, and other instruments. Radar screens in the control tower indicate the direction and distance of the aircraft. The controller can therefore tell the pilot what adjustments to make to his height, direction, and speed to bring him safely down along the correct glide path.

Many airports and airliners are now equipped with what is called the *Instrument Landing System*. This helps the pilots to line up their aircraft onto the correct glide path. Certain electronic devices in the aircraft 'lock' onto radio beams coming from the runway along the correct glide path.

In both the radar and the Instrument Landing approach the pilot does the actual landing. In thick fog and bad visibility, when the pilot cannot see the runway on the final approach, both of these systems are too risky. So-called *blind-landing systems* have therefore been developed which bring down an aircraft automatically without the pilot touching the controls at all. Eventually this kind of system will be a standard feature of all major airliners and airports. And pilots will then be able to fly in all weather conditions.

When the aircraft has touched down safely, the controller tells the pilot which taxi-way to use to take him towards the airport terminal. In a similar way the controller guides the aircraft from the apron, along the taxiways, to the main runway, and gives the pilot the 'go-ahead' for take-off. And at any one time there may be scores of aircraft moving along them, taxiing into position for take-off, taking off, landing, and taxiing after landing. Control of aircraft on the ground is therefore as vital as control in the air.

The runways at most major international airports are more than 10,000 feet long (about 2 miles). The large transatlantic airliners need to travel almost that far before they are going fast enough to take off. On landing, they need quite a distance to reduce speed after touch-down. And there must also be a margin of safety included.

Most airports have at least two runways, each one facing in a different direction so that it is possible for an aircraft to take off and land against the wind at all times.

Inside the control tower at London Airport, showing the radar apparatus the controllers use to locate and identify aircraft coming into land. They give instructions to the incoming aircraft by radio. Note also the closed-circuit television screen at top left of the picture which gives arrival and departure details.

**Radar**

Radar is a device for detecting objects by bouncing radio waves off them. The name stands for *radio direction and range*. A pulse of radio waves is beamed from an aerial into the sky. When it hits an aeroplane, it is reflected back into the aerial as a kind of echo. The 'echo' causes an illuminated spot, or *blip* to appear on a screen similar to a television screen. The trough-shaped radar aerial rotates so that it can scan all of the sky.

The ground controller in the control tower of an airport generally has two radar screens. On one, called the *plan position indicator* (PPI), he can tell from the position of the blip the distance and direction of an approaching plane. Another, called the *range height indicator* (RHI), indicates both the distance and the height of the aircraft.

Large planes generally have an on-board radar, which is housed in a dome called a *radome* on the nose. This radar is known as 'weather' radar because the pilot uses it to avoid storms which may lie in his path. Storm clouds show up on the radar screen.

Of course, radar is not used solely for aircraft. Practically all ships carry it, too, to aid navigation. It is invaluable in storm and fog, when otherwise the ship's navigator would be sailing 'blind'. Police use radar to catch speeding motorists. Space scientists use it to track satellites and probes travelling deep into space. Many guided missiles have radar homing devices.

# Modern Farming

Farming is still the world's greatest industry. On average, over half of the population spend their lives ploughing the soil, planting seeds, harvesting crops, and raising animals which provide us with meat, milk, eggs, wool, leather, and many other products. Traditional farming, in which most of the hard work on the farm is done by manual labour and animals, is still practised in many countries. But traditional methods cannot produce enough food to cope with the great 'explosion' in population.

Only with modern methods of farming, using machines to replace men and animals, artificial fertilizers to enrich the soil, and chemicals to keep weeds and diseases under control, can man hope to avoid a catastrophic world food shortage. Farming methods are well advanced in most parts of Europe and North America, and farmers elsewhere are being helped by such organizations as the United Nations Food and Agricultural Organization (FAO) to modernize as quickly as possible.

## Farm Machinery

Machinery has brought about the greatest revolution in farming. By far the most important machine is the *tractor,* which is used for pulling many kinds of implements – ploughs to break up the soil, drills to sow the seeds, trailers to carry manure, and so on. In this respect it does the same kind of work as teams of horses, or oxen once did. But a tractor does more than just pull. It also provides power to drive other machinery, such as mowing machines, balers, and potato harvesters.

The standard type of tractor has two massive wheels at the rear and two small ones at the front, but those designed for travelling over heavy ground or for pulling heavy loads have caterpillar or crawler tracks. Tractors supply power to drive other machinery either through a pulley wheel driven by the engine or by means of what is called power take-off (p.t.o.). In p.t.o. drive, the moving part of the machine the tractor is pulling, such as the knife blade of a mower, is turned directly by a shaft driven by the engine.

Among the many other farm machines, *combine harvesters* are probably the most important. They are so called because they combine the actions of reaping and threshing. Cereals are the main, but not the only crops that are harvested with the combine.

Terraced rice paddy fields in south-east Asia, where traditional high-labour farming is still practised. Even there, though, the farmers have benefited from modern methods, using improved strains of rice to give a better yield.

The advent of selective weedkillers, insecticides and fungicides has probably caused the greatest improvement in crop production. Aerial spraying enables vast areas to be treated quickly and thoroughly.

Legumes, such as peas and beans, can also be harvested with it.

## Crops

Growing plants take their nourishment from the soil. They need certain essential elements, particularly calcium, nitrogen, phosphorus, and potassium, if they are to grow well. It stands to reason that after every crop there will be less of these elements remaining in the soil for a subsequent crop. To ensure a good crop every time, farmers apply *fertilizers* to the soil before planting to replace the essential elements the previous crop removed.

The farmer's traditional fertilizer is farmyard *manure,* a rotted mixture of dung and straw. It improves the texture of the soil as well. But there is not enough manure to supply the needs of a modern farmer, and artificial fertilizers are widely used instead. There are a variety of artificial fertilizers, each supplying certain of the essential elements. *Sulphate of ammonia* supplies nitrogen, *superphosphate of lime* supplies phosphorus, *bone meal* supplies both nitrogen and phosphorus, *sulphate of potash* supplies potassium, and there are many, many more. Superphosphate is probably one of the most widely used of all the fertilizers. Vast quantities of sulphuric acid are used in its manufacture from bone ash or furnace slag. *Lime* is also added to the soil on a large scale. It is valuable not only as a fertilizer but also as a soil conditioner.

Of course, stimulating crops into better growth makes the weeds grow better, too!

Fortunately the chemist has come to the aid of farmers and developed *selective weed-killers*, properly called *herbicides*, for use with their crops. Herbicides have a selective action in that they kill the weeds but leave the crop unharmed. They are used particularly with the cereal crops. Chemists have also developed a wide range of *fungicides* to kill fungus diseases, such as potato blight or rust on wheat, which could devastate the crops. Farmers wage continual war against insects, too. Maggots, beetles, and weevils can do untold damage if they are not checked. Again, modern science has produced powerful *insecticides* to protect crops against insect attack.

## Livestock

The farmer is equally plagued with pests and diseases that threaten to destroy his livestock. And they make the widest use of modern methods to eliminate the danger as far as possible. They dip their sheep in disinfectant and insecticides regularly to kill sheep ticks and other insect parasites that could damage the fleece and harm the animal itself. They are able to innoculate their livestock against some diseases, but there are still diseases, like foot-and-mouth disease in cattle, for example, that cause havoc even in the most advanced farming communities.

Over the years the farmer can improve his livestock by what is called *selective breeding*. He selects and breeds from animals that have the characteristics he would require in their offspring. He may *pure-breed* his livestock from animals within the same breed, or he may *cross-breed* with animals from different breeds. Most dairy cattle are pure-bred; most beef cattle are cross-bred. Cross-bred animals are usually stronger and grow more quickly than pure-bred animals.

Speed of growth is essential if livestock farmers are to make a reasonable profit. And these days, they practice what is called *intensive farming*. They raise large numbers of animals indoors under carefully controlled conditions. Poultry and pigs, especially, are reared by intensive methods. The livestock are housed in large buildings kept at a steady temperature, where there is good ventilation and lighting. They may be kept in groups or in cages, as in the case of hens in an egg-laying *battery*. Feeding and waste-disposal is made to be as automatic as possible to save labor. By carefully controlling the environment of the animals and the quality of their food, farmers can produce exactly the kind of animal they want to get the best market price.

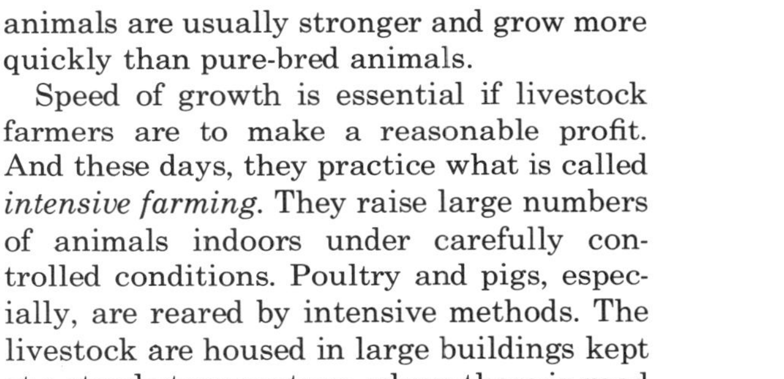

Above: A wealth of machines are employed on modern farms, none of them more useful than the combine harvester, which both reaps and threshes the grain. Combines were developed in the United States as long ago as the 1830s, but did not go into general use there until about 1918. The combine did not go into widespread use in Britain until after World War 2.

Farmers improve their livestock by cross-breeding and keep their animals healthy by treating them with insecticides, vaccines and antibiotics as necessary. The majority of sheep raised these days are cross breeds, for example. Cross breeds show many of the qualities of the pure breeds they were bred from, but are usually much hardier.

# Staple Foods

## Bread

Bread has been part of man's staple diet for thousands of years, and is still. It contains a good proportion of the carbohydrates, proteins, minerals and vitamins that human beings need to be healthy.

Most bread is made out of flour produced by grinding, or milling wheat or rye. These two cereals are suitable for bread-making because they contain a protein called *gluten* which gives bread its characteristic spongy texture. Rye contains less gluten than wheat, resulting in a heavier bread. Rye bread is also very dark.

White flour is made by removing the darker-coloured outer parts – the *bran* and the *germ* – from the wheat kernel (grain), leaving the pure white inner part, called the *endosperm*. Flour for brown bread is produced by milling the whole kernel. It has a greater food value than pure white flour.

Much bread these days is made in large factory-type bakeries using continuous, automatic processes. The flour and the other ingredients go in one end, and golden-brown loaves come out of the other. Machines do what the baker does, only on a much larger scale.

The first stage of bread-making is activating the yeast. Yeast is a kind of plant which feeds on sugar to produce alcohol and carbon dioxide by a process called *fermentation*. In bread-making the yeast is added to a liquid mixture which encourages fermentation. When 'working' well, the yeast is mixed thoroughly with the other ingredients of the bread, flour, more water, salt, and so on, to form a *dough*. The dough is then thoroughly kneaded, which distributes the yeast evenly and gives the dough an elastic texture. Then it is allowed to ferment.

The carbon dioxide produced by the yeast becomes trapped inside the dough and causes it to rise. After a further brief kneading, the partly risen dough is put into individual baking pans and allowed to ferment, until it has risen to the required height. This is called *proving*.

The pans then go into the oven for baking. The oven is very hot, over 230°C (445°F). This kills the yeast in the bread, preventing the bread rising too high, and burns the crust on the loaf.

## Milk

Milk from the cow is one of the world's most important and most nourishing foods. It is as good for humans as it is for calves.

Above: Blobs of butter remaining in the churn after the buttermilk has drained away.

Below: Automation has reached the bakery. The modern baker directs the operation from a control panel.

Its vitamins, proteins, fats, carbohydrates, and minerals ensure a healthy diet. Dairy cows, on average, each produce about a thousand gallons of milk a year.

Milk from the farm is taken to the dairy by tanker or in churns as *raw* milk. There it is first filtered, or clarified to remove dust, bits of hair, and so on. Then it may be

Above: Freshly churned, smooth-textured butter ready for removal from the churn and packaging.
Below: Preparing the curd in cheesemaking. A bacteria culture has been added to pasteurized milk in the vat, which is stirred to ensure even distribution.

Above: A Swiss cheese-maker putting a newly made cheese into a curing cellar, where it will be left for several months to ripen, or develop its characteristic flavor and texture. In general a short curing period produces a cheese with a mild flavor and a smooth texture. An extended curing period will produce a cheese with a sharp 'bite' and a crumbly texture.

*homogenized*, a treatment whereby the milk is forced under pressure through tiny holes. This disperses the fat throughout the milk and prevents it forming a layer of cream on the top.

The milk then passes to the *pasteurizing* unit. Pasteurization is a process in which the milk is heated at 71°C (161°F) for 15 seconds and then quickly cooled. (There is also a slower, lower-temperature method.) This treatment destroys bacteria which could cause disease, although they do not destroy the bacteria which cause milk to go sour. More prolonged heat treatment would be required for such sterilization. After pasteurization the milk is ready to be bottled, which is done in high-speed automatic machines. Some of it is concentrated into evaporated and condensed milk. A lot is made into butter and cheese.

## Butter

Butter is made in a *creamery*, an appropriate name because cream is what butter is made from. In large commercial creameries the cream is separated from the milk in a centrifugal separator, a machine which whirls the milk round at high speed. The separated cream is then pasteurized in the same way as milk and kept cool until it is required.

Churning is the next stage of butter-making. The mechanical churns are large stainless steel cylinders, with flat 'baffle' plates inside. They churn the cream as they revolve. After about three-quarters of an hour blobs of butter begin to separate, leaving a liquid known as *buttermilk*. The butter is washed and salted (if necessary) and then churned some more to give it an even texture. Then it is ready for packaging.

## Cheese

The milk for cheese-making must first be pasteurized. The pasteurized milk is led to a vat, and a 'culture' of lactic acid bacteria (the ones that turn milk sour) is added to it. After a while a substance called rennet is added to the vat and causes the milk very soon to curdle. The solid part is called the *curd,* and the liquid part the *whey*. After the whey has been drained off, the curd is pressed into moulds to make large, round 'cheeses', as they are usually called. Then the cheese is put into curing rooms or cellars where it is allowed to mature. Here is where the characteristic flavour and texture of cheeses develop. An ordinary Cheddar cheese may take anything from a few days to a few years to mature.

The great selection of cheeses we have today are made by varying the basic cheese-making process in different ways. Different moulds or bacteria are added to give a characteristic flavour or appearance, such as the blue veins of Stilton or Roquefort.

# Food Preservation

Leave a piece of fresh fish lying around for a few days in the warmth of summer and try to forget about it. You can't! Within a day or so it starts to decay and gives off the indescribable smell only rotten fish can give off. Pick pears from a tree and forget to eat them and soon your pears will be 'sleepy', or over-ripe and mushy, and useless for any purpose. These are just two common examples of food spoilage that may occur.

Fortunately today there are a great many ways of preserving foods that enable us to enjoy all kinds of meats, fruits, and vegetables out of season and from other countries.

## Food Spoilage

Just how and why do foods spoil? Why does fish go rotten, butter turn rancid, milk sour, and pears 'sleepy'? Basically there are two means by which food spoilage occurs. One is by the growth of micro-organisms on the outside, and the other is by chemical changes on the inside. Micro-organisms such as yeasts, moulds, and bacteria are always present in the air around us. They settle on food and, if the conditions are right, begin to break it down. Some yeasts tend to make food ferment. They change sugar into alcohol, which then turns into acid, making the food taste sour. Moulds tend to grow on moist foods. You have probably seen the greenish-colored mould on bread and on cheese. It starts as tiny white spores and grows into a greenish mass, on the surface at first and then gradually right through. The food develops a musty taste.

The sour and musty foods are not necessarily harmful. But some kinds of bacterial spoilage can be dangerous. Certain bacteria produce toxins which cause food poisoning. The worst thing about them is that they may be present even where there are no outward signs of decay. Botulism is the most deadly form of food poisoning of this kind. It is almost always fatal. But the toxins are destroyed by cooking, as are most toxins.

Chemical spoilage may occur as a result of several processes. Contact with the air oxidises foods such as butter which contain fats and oils, and makes them rancid. Substances called enzymes, which are present in all living things, are responsible for food spoilage, too. Their action, if limited, is beneficial. For example, they act on meat and make it more tender.

All methods of food preservation aim to prevent the growth of any micro-organisms, and prevent or arrest the action of other agents causing decay. *Canning, bottling* and *freezing* are the most common methods of food preservation today. *Drying* and *curing* are methods that have been practised for centuries in one form or another, and they are still very important. Two up-to-date methods use *radiation* and *antibiotics* (especially for poultry) to sterilize and preserve foodstuffs, but they are not used on anywhere near the same scale as the other methods. *Pasteurization* is another method of temporary preservation used for milk.

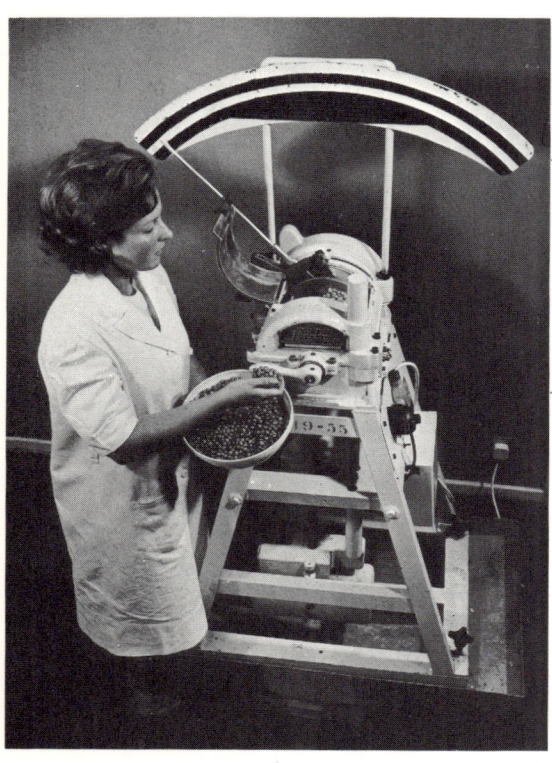

Peas are among the most common vegetables preserved, and their flavor is not appreciably altered during preservation. Dehydration is a common method of preservation. The picture below right shows part of the dehydration plant for Surprise peas.

Below left: In some canning plants an unlikely sounding instrument called a tenderometer is used to gauge the tenderness of peas prior to canning. What it actually does is measure shearing force.

## Canning, Bottling, and Freezing

*Canning and bottling* aim to sterilize the food – that is, heat it to kill any micro-organisms it may contain – and then to keep air away from it so that no further organisms can attack it. Methods of canning and bottling are very similar, except that in canning the food is packed in sealed tin cans and in bottling, in bottles.

As in all methods of preservation, the foods must be carefully prepared. Excess fat, gristle, bones, bruised skins, cores, stones, and stalks must be carefully removed. And the food must be at exactly the right stage of ripeness. Some foods, such as green vegetables, are *blanched,* or scalded with steam or boiling water, to prevent enzyme action while they are being processed.

The tin cans themselves are made of tinplate, sheet mild steel coated with tin. They are made astonishingly quickly on automatic machines, and are usually sprayed with a lacquer on the inside. After the preparation, the whole canning process may be done automatically. Machines fill the can with exactly the right amount of food, and sauce, syrup or gravy as necessary. Then the cans are steamed or otherwise heated briefly to drive out much of the air, and sealed. The sealed cans pass into large heating units to be sterilized. These units may take the form of huge pressure cookers.

*Freezing* prevents the growth of micro-organisms and slows down enzyme action. The freezing process must be quick if loss of texture and flavour are to be avoided. During slow freezing large ice crystals may form inside the food and damage the cells. And enzyme action may continue for long enough to spoil the flavour. Quick-freezing methods employ temperatures down to $-40°C$. The food to be frozen must be packaged carefully to conserve the moisture in it. Otherwise it will dry out, with consequent loss in texture.

## Drying and Curing

*Drying* preserves food because it removes the moisture which micro-organisms need in order to grow. Dried fruit such as sultanas, raisins, and currants are examples of dried foods we are all familiar with. They are actually different kinds of dried grapes. The grapes are laid out for days in the hot sun until they lose most of their water and shrivel up. This method of sun-drying has been practised for thousands of years. Nowadays many foods are dried in specially heated kilns. An even more up-to-date method is called *accelerated freeze drying.* The food is first deep-frozen and then placed in a vacuum. Under these conditions, the ice formed inside the food sublimes; that is, it changes into water vapour without melting first. The food is heated slightly at the same time to accelerate the process. This method preserves the original texture of the food, unlike normal drying methods.

*Curing* slows down or prevents the growth of micro-organisms in the food. Salt, vinegar, sugar, and wood smoke are the commonest curing agents. Since they are all strong-flavoured, they alter the taste of the food. *Salting and smoking* are widely used for meat and fish – bacon is often both salted and smoked to give it its characteristic flavour. *Sugar* needs to be added in large quantities to preserve jams and bottled fruit. *Vinegar* is used for pickling all kinds of things from walnuts and onions to gherkins and herrings.

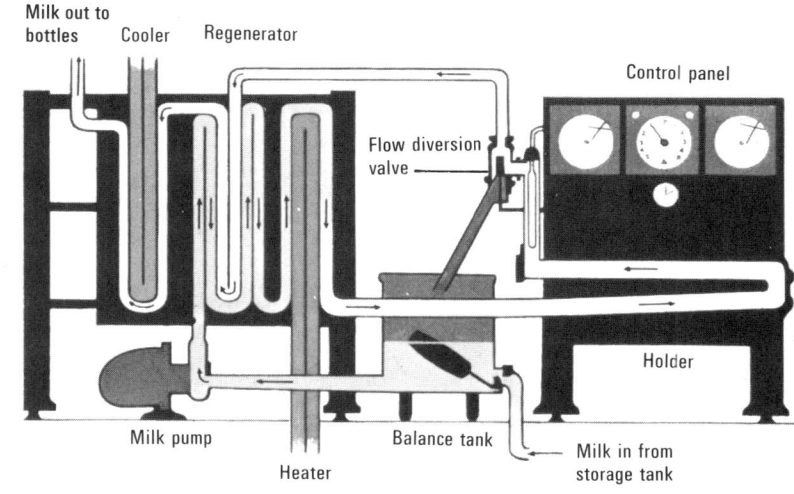

A simplified diagram of a high-temperature, short-time pasteurization unit. The heater, regenerator and cooler consist of a series of thin stainless steel plates. The milk flows over one side, and the heating or cooling fluid over the other. The heater heats the milk to about 71°C and it is held at this temperature for 15 seconds. The hot milk then flows through the regenerator, where it gives up heat to incoming milk, and thence through the cooler.

Below: Sun drying is a traditional method of preservation used for fruit, meat and, as here, fish. Drying removes moisture from the food, and without moisture the organisms that cause decay cannot grow.

# Inventions

**Inventors and Inventions**

Necessity, they say, is the mother of invention, and this is often true. For example, the need in the southern American states to produce more cotton to satisfy the demands of the spinners in England in the late 1700s led Eli Whitney to invent the cotton gin. Some men have a naturally inventive mind. They are so imbued with curiosity that they are unwearying in their search for new things to tax their brilliant minds. Not all the inventors listed here were like this. But one was, Thomas Edison, without doubt the most brilliant of all inventors. A brief look at his life reveals a little of the inventive mind at work.

| Date | Invention | Inventor |
|---|---|---|
| 1450 | Movable Type | Johannes Gutenberg (Ger.) |
| 1550? | Spinning Wheel | China? |
| 1590? | Compound Microscope | Zacharias Janssen (Dutch) |
| 1593 | Thermometer | Galileo Galilei (Ital.) |
| 1608/9 | Refracting Telescope | Hans Lippershey (Dutch) |
| | | Galileo Galilei (Ital.) |
| 1670 | Reflecting Telescope | Isaac Newton (Brit.) |
| 1681 | Pressure Cooker | Denis Papin (Fr.) |
| 1698 | Steam Pump | Thomas Savery (Brit.) |
| 1712 | Beam Engine | Thomas Newcomen (Brit.) |
| 1733 | Flying Shuttle | John Kay (Brit.) |
| 1760 | Lightning Conductor | Benjamin Franklin (US) |
| 1767 | Spinning Jenny | James Hargreaves (Brit.) |
| 1768 | Spinning Frame | Richard Arkwright (Brit.) |
| 1779 | Spinning Mule | Samuel Crompton (Brit.) |

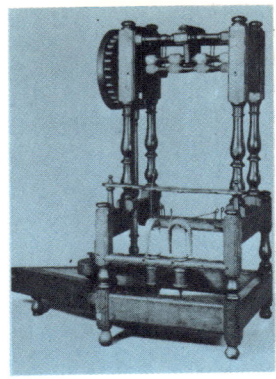

Richard Arkwright's spinning frame (1768), invented just a few months after James Hargreaves had patented his spinning jenny.

Alexander Graham Bell in 1876, the year in which he patented the telephone.

Below: Eli Whitney's cotton gin could separate cotton seeds from the fibers some 50 times faster than a man could.

| Date | Invention | Inventor |
|---|---|---|
| 1780s | Double-acting Steam Engine | James Watt (Scot.) |
| 1783 | Hot-air Balloon | Montgolfier Brothers (Fr.) Joseph Michel Jacques Étienne |
| 1787? | Power Loom | Edmund Cartwright (Brit.) |
| 1793 | Cotton Gin | Eli Whitney (US) |
| 1800? | Electric Battery | Alessandro Volta (Ital.) |
| 1800 | Lathe | Henry Maudsley (Brit.) |
| 1804 | Glider | George Cayley (Brit.) |
| 1804 | Steam Locomotive | Richard Trevithick (Brit.) |
| 1807 | Steamboat | Robert Fulton (US) |
| 1812 | Mechanical Printing Press | Friedrich König (Ger.) |
| 1815 | Safety Lamp | Humphry Davy (Brit.) |
| 1816 | Stethoscope | René T. H. Laënec (Fr.) |
| 1818 | Tunnelling Shield | Marc Isambard Brunel (Brit.) |
| 1830s | Photography | Joseph Nicephore Niépce (Fr.) Louis J. M. Daguerre (Fr.) William Fox Talbot (Brit.) |
| 1836 | Revolver | Samuel Colt (US) |
| 1836 | Screw Propeller | John Ericcson (Swed.) |
| 1837 | Telegraph | William Cooke (Brit.) Charles Wheatstone (Brit.) Samuel Morse (US) |
| 1839 | Steam Hammer | James Nasmyth (Scot.) |
| 1845 | Sewing Machine | Elias Howe (US) |
| 1852 | Safety Lift | Elisha Graves Otis (US) |
| 1856 | Bessemer Steel-Making Process | Henry Bessemer (Brit.) |
| 1865 | Yale Lock | Linus Yale (US) |
| 1867 | Dynamite | Alfred Nobel (Swed.) |
| 1867 | Typewriter | Christopher L. Sholes (US) |
| 1876 | Four-Stroke Internal Combustion Engine | Nikolaus August Otto (Ger.) |
| 1876 | Telephone | Alexander Graham Bell (US) |
| 1877 | Phonograph | Thomas Alva Edison (US) |
| 1878 | Cathode-Ray Tube | William Crookes (Brit.) |
| 1878/9 | Electric Lamp | Joseph Swan (Brit.) Thomas Alva Edison (US) |
| 1880s | Machine Gun | Hiram Stevens Maxim (US) |
| 1884 | Steam Turbine | Charles Algernon Parsons (Brit.) |
| 1885 | Gas Engine | Karl Benz (Ger.) Gottlieb Daimler (Ger.) |
| 1888 | Pneumatic Tyre | John Boyd Dunlop (Brit.) |
| 1891 | Roll Film | George Eastman (US) |
| 1893 | Diesel Engine | Rudolf Diesel (Ger.) |
| 1895 | Radio | Guglielmo Marconi (Ital.) |
| 1903 | Powered Aircraft | Wilbur Wright (US) Orville Wright (US) |
| 1904 | Radio Valve (Tube) | John Ambrose Fleming (Brit.) |
| 1926 | Liquid-Propelled Rocket | Robert Hutchings Goddard (US) |
| 1926/8 | Television | John Logie Baird (Scot.) Vladimir Zworykin (US) |
| 1930 | Cyclotron | Ernest Orlando Lawrence (US) |
| 1930 | Jet Engine | Frank Whittle (Brit.) |
| 1935 | Radar | Robert Watson-Watt (Brit.) |
| 1936/9 | Helicopter | Heinrich Focke (Ger.) Igor Sikorsky (US) |
| 1944 | Digital Computer | Howard Aiken (US) |
| 1947 | Polaroid Camera | Edwin H. Land (US) |
| 1950s | Rotary Gas Engine | Felix Wankel (Ger.) |
| 1955 | Hovercraft | Christopher Cockerell (Brit.) |
| 1959 | Fuel Cell | Francis Bacon (Brit.) |

Above: Frank Whittle, who patented a design for a jet engine in 1930. At right is a modern jet engine able to develop more than 30 times the thrust of Whittle's first practical engine (850 pounds).

Left: Another notable pioneer of aerospace, Robert Hutchings Goddard, who developed the liquid-propellant rocket in 1926.

## Edison

If necessity be the mother of invention, Thomas Alva Edison (1847-1931) could well be called its father! In the space of 60 years Edison patented more than 1,100 inventions – the most memorable being the phonograph (1877) and the incandescent lamp, or electric-light bulb (1879). He was an intensely practical person, a shrewd businessman, and a tireless worker. He would work for days with scarcely any sleep at all. Well did he define genius as 'One per cent inspiration and ninety-nine per cent perspiration'.

Edison received much of his early teaching from his mother, who encouraged his natural inclination to experiment and to try out for himself the discoveries of others. When he was 12, he began working as a newsboy and sweetseller on a train. He began printing a weekly newspaper in his 'laboratory' in the luggage van of the train. Edison later trained as a telegraphist and had a succession of jobs in various parts of the Mid-West. He was continually on the move but never stopped experimenting with telegraphy equipment.

In 1868, Edison invented an electric vote-recording machine, but had no success with it. But not long afterwards he invented an improved stock-ticker machine which was used by brokers to telegraph gold prices. With the proceeds from selling this invention he opened up a workshop in Newark, New Jersey. There he developed an improved metal typewriter In 1876 Edison moved to a new laboratory at Menlo Park, New Jersey. There he made his great inventions, the first of which was a carbon transmitter for the telephone. In 1877 Edison made what he called his favourite invention – the *phonograph*. He came upon it as a result of his work on telegraph repeaters. In his 'improvement in the automatic telegraph', a rotating disc of paper was embossed with the dots and dashes of an incoming message, and could be played back over another telegraph wire. While working on this device, Edison noticed that the embossed marks gave off a noise when struck. This led him, within 6 months, to his invention of the phonograph (see page 104).

Two years later Edison worked out a successful system for the production of an electric light. His difficulty for years had been to produce a filament for the light bulb which would stand up to hours of burning without disintegrating. Eventually he found the answer in carbonized sewing thread. Later he found better materials for the filament. By 1882 he was supplying a whole district of New York with electricity, having built the power station, incorporating generators of his own design.

Thomas Alva Edison, the greatest of inventors, pictured with his favourite invention, the phonograph.

While experimenting with his light bulb one day, Edison inserted a metal plate in the bulb. To his astonishment he found that current would flow from the filament to this plate if the plate were given a positive charge. This effect is now termed the Edison effect and is fundamental to electronics.

Edison's later inventions were less spectacular but important nonetheless. They included one of the earliest motion-picture cameras, called the kinetoscope and later a combined camera and phonograph which was a forerunner of the 'talkies'. Batteries, cement-mixers, dictophones and duplicators – these were a few of the other machines which Edison experimented with and improved before he died.

# Weights and Measures

For years in English-speaking countries two different systems of measurement have co-existed, the *English* and the *metric*. The scientist has used the metric, measuring in meters and centimeters, weighing in grams, and so on. The layman and engineer, though, have favored the English system, measuring in feet and inches and weighing in pounds. Now the movement is towards complete metrication, with eventual abandonment of the English system. Despite the obvious advantages of the metric system, based as it is on units of 10, the demise of the English system must be regretted. The gill and pint, league and fathom, peck and bushel will soon have had their day, like the cubit and digit before them. And not even the chemist will have any scruples.

## Metric System

### Weight
| | |
|---|---|
| 10 milligrams (mg) | = 1 centigram (cg) |
| 10 centigrams | = 1 decigram (dg) |
| 10 decigrams | = 1 gram (g) |
| 10 grams | = 1 decagram (Dg) |
| 10 decagrams | = 1 hectogram (hg) |
| 10 hectograms | = 1 kilogram (kg) |
| 1000 kilograms | = 1 ton (metric ton) |

### Linear Measure
| | |
|---|---|
| 10 millimeters (mm) | = 1 centimeter (cm) |
| 10 centimeters | = 1 decimeter (dm) |
| 10 decimeters | = 1 meter (m) |
| 10 meters | = 1 decameter (Dm) |
| 10 decameters | = 1 hectometer (hm) |
| 10 hectometers | = 1 kilometer (km) |

### Capacity Measure
| | |
|---|---|
| 10 milliliters (ml) | = 1 centiliter (cl) |
| 10 centiliters | = 1 deciliter (dl) |
| 10 deciliters | = 1 liter (l) |
| 10 liters | = 1 decaliter (Dl) |
| 10 decaliters | = 1 hectoliter (hl) |
| 10 hectoliters | = 1 kiloliter (kl) |

(1 liter = 1000 cubic centimeters (cc)
1 kiloliter = 1 cubic meter)

## English, or Imperial System

### Avoirdupois Weight
| | |
|---|---|
| 1 dram (dr) | = 27·34375 grains (gr) |
| 16 drams | = 1 ounce (oz) |
| 16 ounces | = 1 pound (lb) |
| 14 pounds | = 1 stone |
| 28 pounds | = 1 quarter |
| 4 quarters | = 1 hundredweight (cwt) = 112 pounds |
| 20 hundredweights | = 1 long ton = 2240 pounds |
| 100 pounds | = 1 short hundredweight (USA) |
| 2000 pounds | = 1 short ton (USA) |

### Troy Weight
| | |
|---|---|
| 1 pennyweight (dwt) | = 24 grains |
| 480 grains | = 1 ounce |
| 12 ozs | = 1 pound |

### Apothecaries' Weight
| | |
|---|---|
| 20 grains | = 1 scruple |
| 3 scruples | = 1 drachm |
| 8 drachms | = 1 ounce |

### Linear Measure
| | |
|---|---|
| 1 nail | = $\frac{1}{16}$ yard (yd) |
| 1 link | = 7·92 inches (in.) |
| 12 inches | = 1 foot (ft) |
| 3 feet | = 1 yard |
| 22 yards | = 1 chain |
| 10 chains | = 1 furlong |
| 8 furlongs | = 1 mile |
| 3 miles | = 1 league |
| 1 mile | = 1,760 yards |
| 1 mile | = 5,280 feet |

### Square Measure
| | |
|---|---|
| 144 square inches | = 1 square foot |
| 9 square feet | = 1 square yard |
| 30¼ square yards | = 1 square rod |
| 160 square rods | = 1 acre |
| | = 4,840 square yards |
| 640 acres | = 1 square mile |

### Liquid Capacity Measure
| | |
|---|---|
| 4 gills | = 1 pint (pt) |
| 2 pints | = 1 quart (qt) |
| 4 quarts | = 1 gallon (gal) |

(1 imperial gallon is the volume of water weighing 10 pounds.)

| | |
|---|---|
| 1 US gallon | = 0·83 imperial gallon |
| 1 imperial gallon | = 1·2 US gallon |

### Dry Capacity Measure
| | |
|---|---|
| 2 pints | = 1 quart |
| 4 quarts | = 1 gallon |
| 2 gallons | = 1 peck |
| 4 pecks | = 1 bushel |
| 8 bushels | = 1 quarter |
| 36 bushels | = 1 chaldron |

### Cubic Measure
| | |
|---|---|
| 1,728 cubic inches | = 1 cubic foot |
| 27 cubic feet | = 1 cubic yard |

## English-Metric Equivalents
| | |
|---|---|
| 1 ounce | = 28·35 grams |
| 1 pound | = 0·45 kilogram |
| 1 hundredweight (112 lb) | = 50·8 kilograms |
| 1 ton (20 cwt) | = 1·02 tons |
| 1 inch | = 2·54 centimeters |
| 1 foot | = 30·48 centimeters |
| 1 yard | = 0·91 meter |
| 1 mile | = 1·61 kilometers |
| 1 square inch | = 6·45 centimeters |
| 1 square foot | = 9·29 square decimeters |
| 1 square yard | = 0·84 square meter |
| 1 acre | = 0·41 hectare |
| 1 square mile | = 2·59 hectares |
| 1 pint | = 0·57 liter |
| 1 gallon | = 4·55 liters |
| 1 cubic inch | = 16·39 cubic centimeters (cc) |
| 1 cubic foot | = 0·28 cubic meter |
| 1 cubic yard | = 0·76 cubic meter |

## Metric-English Equivalents
| | |
|---|---|
| 1 gram | = 15·43 grains |
| 1 kilogram | = 2·21 pounds |
| 1 ton | = 2,204·6 pounds |
| 1 ton | = 0·98 ton |
| 1 centimeter | = 0·39 inch |
| 1 meter | = 39·37 inches |
| 1 meter | = 3·28 feet |
| 1 meter | = 1·09 yards |
| 1 kilometer | = 0·62 mile |
| 1 square centimeter | = 0·16 square inch |
| 1 square meter | = 10·76 square yards |
| 1 hectare | = 2·47 acres |
| 1 liter | = 1·76 pints |
| 1 liter | = 0·22 gallon |
| 1 cubic centimeter (cc) | = 0·06 cubic inch |
| 1 cubic meter | = 1·31 cubic yards |

The units of measurement used for many years in science were termed CGS units, for they used the centimeter, gram and second as basic units. Volume, for example, was expressed in cubic centimeters, velocity in centimeters per second, density in grams per cubic centimeter, and so on.

Later the CGS system gave way to what was termed the MKSA system, whose basic units were the meter, kilogram, second and ampere. In 1960 an extended MKSA system became known as the Système Internationale d'Unités (International System of Units), abbreviated to SI. The following table shows basic and supplementary units in the system together with units derived from them.

| Quantity | Unit |
|---|---|
| Length | meter (m) |
| Mass | kilogram (kg) |
| Time | second (s) |
| Electric current | ampere (A) |
| Temperature | degree Kelvin (°K) |
| Luminous intensity | candela (cd) |
| Plane angle | radian (rad) |
| Solid angle | steradian (sr) |

## The International System (SI) of Units

| Quantity | Unit |
|---|---|
| Area | square meter |
| Volume | cubic meter |
| Frequency | hertz (Hz) |
| Density | kilogram per cubic meter |
| Velocity | meter per second |
| Angular velocity | radian per second |
| Acceleration | meter per second per second |
| Force | newton (N) |
| Pressure | newton per square meter |
| Work, energy, quantity of heat | joule (J) |
| Power | watt (W) |
| Quantity of electricity | coulomb (C) |
| Electric tension, potential difference, electromotive force | volt (V) |
| Electric resistance | ohm ($\Omega$) |
| Electric capacitance | farad (F) |
| Magnetic flux | weber (Wb) |
| Inductance | henry (H) |
| Magnetic flux density | tesla (T) |
| Magnetomotive force | ampere |
| Luminous flux | lumen (lm) |
| Illumination | lux (lx) |

The following table gives the symbols used in the International System to indicate multiples and sub-multiples of units.

| Name | Symbol | Equivalent |
|---|---|---|
| tera | T | $10^{12}$ |
| giga | G | $10^{9}$ |
| mega | M | $10^{6}$ |
| kilo | k | $10^{3}$ |
| hecto | h | $10^{2}$ |
| deca | da | 10 |
| deci | d | $10^{-1}$ |
| centi | c | $10^{-2}$ |
| milli | m | $10^{-3}$ |
| micro | $\mu$ | $10^{-6}$ |
| nano | n | $10^{-9}$ |
| pico | p | $10^{-12}$ |
| femto | f | $10^{-15}$ |
| atto | a | $10^{-18}$ |

## Melting and Boiling Points of the Elements

| Element | Melting point (°C) | Boiling point (°C) |
|---|---|---|
| Actinium | 1230 | 3100 |
| Aluminum | 660·1 | 2400 |
| Americium | >850 | * |
| Antimony | 630·5 | 1440 |
| Argon | −189·4 | −185·9 |
| Arsenic | sublimes at 610 | |
| Astatine | 250 | 350 |
| Barium | 710 | 1600 |
| Berkelium | * | * |
| Beryllium | 1280 | 2500 |
| Bismuth | 271·3 | 1500 |
| Boron | 2030 | 3700 |
| Bromine | −7·3 | 58·2 |
| Cadmium | 321 | 767 |
| Caesium | 28·6 | 690 |
| Calcium | 850 | 1450 |
| Californium | * | * |
| Carbon | 3500 | 3900 |
| Cerium | 804 | 2900 |
| Chlorine | −101 | −34·1 |
| Chromium | 1900 | 2600 |
| Cobalt | 1492 | 2900 |
| Copper | 1083 | 2580 |
| Curium | * | * |
| Dysprosium | 1500 | 2300 |
| Einsteinium | * | * |
| Erbium | 1525 | 2600 |
| Europium | 830 | 1450 |
| Fermium | * | * |
| Fluorine | −219·6 | −188·1 |
| Francium | 30 | 650 |
| Gadolinium | 1320 | 2700 |
| Gallium | 29·8 | 2250 |
| Germanium | 958 | 2850 |
| Gold | 1063 | 2660 |
| Hafnium | 2000 | 5300 |
| Helium | −269·7 | −269·0 |
| Holmium | 1500 | 2300 |
| Hydrogen | −259·2 | −252·8 |
| Indium | 156·6 | 2000 |
| Iodine | 113·6 | 183 |
| Iridium | 2443 | 4550 |
| Iron | 1539 | 2800 |
| Krypton | −157·3 | −153·4 |
| Lanthanum | 920 | 3400 |
| Lawrencium | * | * |
| Lead | 327·3 | 1750 |
| Lithium | 180 | 1330 |
| Lutetium | 1700 | 3300 |
| Magnesium | 650 | 1100 |
| Manganese | 1250 | 2100 |
| Mendelevium | * | * |
| Mercury | −38·9 | 356·6 |
| Molybdenum | 2620 | 4600 |
| Neodymium | 1024 | 3100 |
| Neon | −248·6 | −246·1 |
| Neptunium | * | * |
| Nickel | 1453 | 2800 |
| Niobium | 2420 | 5100 |
| Nitrogen | −210·0 | −195·8 |
| Nobelium | * | * |
| Osmium | 3000 | 4600 |
| Oxygen | −218·8 | −183 |
| Palladium | 1552 | 3200 |
| Phosphorus | 44·2 | 280 |
| Platinum | 1769 | 3800 |
| Plutonium | 640 | * |
| Polonium | 254 | 960 |
| Potassium | 63·2 | 760 |
| Praseodymium | 935 | 3000 |
| Promethium | 1000 | 2700 |
| Protactinium | 1200 | 4000 |
| Radium | 700 | 1140 |
| Radon | −71 | −62 |
| Rhenium | 3180 | 5600 |
| Rhodium | 1960 | 3700 |
| Rubidium | 38·8 | 710 |
| Ruthenium | 2300 | 4100 |
| Samarium | 1050 | 1600 |
| Scandium | 1400 | 2500 |
| Selenium | 217 | 685 |
| Silicon | 1410 | 2500 |
| Silver | 960·8 | 2180 |
| Sodium | 97·8 | 883 |
| Strontium | 77 | 1450 |
| Sulphur | 119 | 444·6 |
| Tantalum | 3000 | 5500 |
| Technetium | 2100 | 4600 |
| Tellurium | 450 | 997 |
| Terbium | 1356 | 2500 |
| Thallium | 304 | 1460 |
| Thorium | 1700 | 4200 |
| Thulium | 1600 | 2100 |
| Tin | 231·9 | 2600 |
| Titanium | 1680 | 3300 |
| Tungsten | 3380 | 5500 |
| Uranium | 1133 | 3800 |
| Vanadium | 1920 | 3400 |
| Xenon | −111·9 | −108·1 |
| Ytterbium | 824 | 1500 |
| Yttrium | 1500 | 3000 |
| Zinc | 419·5 | 907 |
| Zirconium | 1850 | 4400 |

(* denotes unobtainable, as with certain unstable man-made elements)

## Coefficients of Expansion

The coefficient of linear expansion is the proportional increase in length of a body per degree centigrade rise in temperature. The coefficient is usually denoted by $a$. In most cases the coefficient increases with rise in temperature, and values are often quoted over a certain temperature range. The coefficient of volume, or cubical expansion, is roughly three times the coefficient of linear expansion. Below are the coefficients of linear expansion of some common substances. Note the greater expansion of the metals, except for nickel steel (invar), which expands scarcely at all.

| Substance | $a \times 10^6$ |
|---|---|
| Aluminum | 23 |
| Brass | 19 |
| Brick | 5 |
| Bronze | 18 |
| Chromium | 7 |
| Concrete, cement | 13 |
| Copper | 16·7 |
| Diamond | 1·3 |
| Glass (ordinary) | 8·5 |
| Gold | 14 |
| Lead | 29 |
| Nickel | 12·8 |
| Nickel-steel (36% Ni) | 0–1 |
| Platinum | 8·9 |
| Portland stone | 3 |
| Pyrex glass | 3 |
| Rubber | 220 |
| Silver | 19 |
| Silica (fused) | 0·4 |
| Stainless steel | 16·4 |
| Steel (mild) | 11 |
| Tin | 9 |
| Tungsten | 4·5 |
| Wood | 3–5 |

# Nobel Prizes

Each year Nobel prizes are awarded to people in the fields of physics, chemistry, physiology and medicine, and literature for their outstanding contributions to those fields. There is also a Nobel peace prize awarded to an individual or an organization for the most effective work in advancing the cause of world peace. Below appear the Nobel prize winners in physics and chemistry from 1901, the year in which Nobel prizes were first awarded.

The Nobel prizes are financed from a fund set up by the inventor of dynamite, the Swede, Alfred Bernhard Nobel. As a result of his invention Nobel became one of the world's richest men. But he suffered with remorse for having developed such a substance, which could be used for widespread death and destruction. And this prompted him to establish the Nobel Foundation, the income from which provides money for the prizes. Depending on the income in a particular year, a Nobel prize may be worth £20,000 or more.

The prize winners are chosen by various organizations in the Swedish capital Stockholm. The Royal Academy of Science awards the prizes for physics and chemistry. The Caroline Institute awards the prize for medicine. The Swedish Academy of Literature awards the literature prize. And a committee of the Norwegian parliament (the Storting), awards the peace prize. The prizes are presented in Stockholm and Oslo each year on December 10, the anniversary of Nobel's death (in 1896).

Alfred Nobel was born in Stockholm in 1833, and studied in St Petersberg (now Leningrad) and in the United States. When Nobel was a young man, explosives such as nitroglycerin were very dangerous to handle. Nobel began experimenting to try to make a safe nitroglycerin explosive. He did not really solve the problem until 1867. The breakthrough came when he discovered that a kind of chalky earth called kieselguhr could absorb about three times its weight of nitroglycerine. The resulting mixture (dynamite) was very much less sensitive to shock than nitroglycerin itself and could be handled in comparative safety. It required a detonator to set it off.

Nobel improved his original dynamite by using woodpulp and sawdust instead of kieselguhr to absorb the nitroglycerin. In 1875 he combined nitrocotton with nitroglycerin to produce what he called blasting gelatine. Later, by adding woodpulp and sodium nitrate to the gelatine, he produced the kind of dynamite we call gelignite

## PHYSICS PRIZE

| Year | Winner(s) |
|---|---|
| 1901 | W. K. Roentgen (Ger) |
| 1902 | H. A. Lorentz and P. Zeeman (Dutch) |
| 1903 | A. H. Becquerel and P. and M. Curie (French) |
| 1904 | Lord Rayleigh (Brit) |
| 1905 | P. Lenard (Ger) |
| 1906 | Sir J. J. Thomson (Brit) |
| 1907 | A. A. Michelson (Amer) |
| 1908 | G. Lippmann (French) |
| 1909 | G. Marconi (Ital) K. F. Braun (Ger) |
| 1910 | J. D. van der Waals (Dutch) |
| 1911 | W. Wien (Ger) |
| 1912 | N. Dalen (Swed) |
| 1913 | H. K. Onnes (Dutch) |
| 1914 | M. T. F. von Laue (Ger) |
| 1915 | Sir W. H. Bragg and Sir W. L. Bragg (Brit) |
| 1916 | No award |
| 1917 | C. Barkla (Brit) |
| 1918 | M. Planck (Ger) |
| 1919 | J. Stark (Ger) |
| 1920 | C. E. Guillaume (French) |
| 1921 | A. Einstein (Ger) |
| 1922 | N. Bohr (Danish) |
| 1923 | R. A. Millikan (Amer) |
| 1924 | K. M. G. Siegbahn (Swed) |
| 1925 | J. Franck and G. Hertz (Ger) |
| 1926 | J. B. Perrin (French) |
| 1927 | A. H. Compton (Amer) and . . . |
| 1927 | C. T. R. Wilson (Brit) |
| 1928 | O. W. Richardson (Brit) |
| 1929 | L. V. de Broglie (French) |
| 1930 | Sir C. V. Raman (Indian) |
| 1931 | No award |
| 1932 | W. Heisenberg (Ger) |
| 1933 | P. Dirac (Brit) E. Schrödinger (Austrian) |
| 1934 | No award |
| 1935 | Sir J. Chadwick (Brit) |
| 1936 | C. D. Anderson (Amer) V. F. Hess (Austrian) |
| 1937 | C. Davisson (Amer) G. Thomson (Brit) |
| 1938 | E. Fermi (Ital) |
| 1939 | E. O. Lawrence (Amer) |
| 1940 | No award |
| 1941 | No award |
| 1942 | No award |
| 1943 | O. Stern (Amer) |
| 1944 | I. I. Rabi (Amer) |
| 1945 | W. Pauli (Austrian) |
| 1946 | P. W. Bridgman (Amer) |
| 1947 | Sir E. V. Appleton (Brit) |
| 1948 | P. M. S. Blackett (Brit) |
| 1949 | H. Yukawa (Jap) |
| 1950 | C. F. Powell (Brit) |
| 1951 | Sir J. D. Cockcroft (Brit) E. T. S. Walton (Irish) |
| 1952 | F. Bloch and E. M. Purcell (Amer) |
| 1953 | F. Zernike (Dutch) |
| 1954 | M. Born and W. Bothe (Ger) |
| 1955 | W. E. Lamb and P. Kusch (Amer) |
| 1956 | J. Bardeen, W. H. Brattain, and W. Shockley (Amer) |
| 1957 | T. D. Lee and C. N. Yang (Amer) |
| 1958 | P. A. Cherenkov, I. M. Frank and I. Y. Tamm (Russ) |
| 1959 | E. Segrè and O. Chamberlain (Amer) |
| 1960 | D. A. Glaser (Amer) |
| 1961 | R. Hofstadter (Amer) R. L. Mössbauer (Ger) |
| 1962 | L. D. Landau (Russ) |
| 1963 | M. Goeppert-Mayer (Amer) J. H. D. Jensen (Ger) E. P. Wigner (Amer) |
| 1964 | N. Basov and A. Prokhorov (Russ) C. H. Townes (Amer) |
| 1965 | S. Tomonaga (Jap) J. S. Schwinger and R. P. Feynman (Amer) |
| 1966 | A. Kastler (French) |
| 1967 | H. A. Bethe (Amer) |
| 1968 | L. W. Alvarez (Amer) |
| 1969 | M. Gell-Mann (Amer) |
| 1970 | H. O. G. Alfven (Swed) L. E. F. Néel (French) |
| 1971 | D. Gabor (Brit) |
| 1972 | J. Bardeen, L. N. Cooper and J. R. Schrieffer (Amer) |

## CHEMISTRY PRIZE

| Year | Winner(s) |
|---|---|
| 1901 | J. H. van't Hoff (Dutch) |
| 1902 | E. Fischer (Ger) |
| 1903 | S. A. Arrhenius (Swed) |
| 1904 | Sir W. Ramsay (Brit) |
| 1905 | A. von Baeyer (Ger) |
| 1906 | H. Moissan (French) |
| 1907 | E. Buchner (Ger) |
| 1908 | E. Rutherford (Brit) |
| 1909 | W. Ostwald (Ger) |
| 1910 | O. Wallach (Ger) |
| 1911 | M. Curie (French) |
| 1912 | F. A. V. Grignard and P. Sabatier (French) |
| 1913 | A. Werner (Swiss) |
| 1914 | T. W. Richards (Amer) |
| 1915 | R. Willstätter (Ger) |
| 1916 | No award |
| 1917 | No award |
| 1918 | F. Haber (Ger) |
| 1919 | No award |
| 1920 | W. Nernst (Ger) |
| 1921 | F. Soddy (Brit) |
| 1922 | F. W. Aston (Brit) |
| 1923 | P. Pregl (Austrian) |
| 1924 | No award |
| 1925 | R. Zsigmondy (Ger) |
| 1926 | T. Svedberg (Swed) |
| 1927 | H. O. Wieland (Ger) |
| 1928 | A. Windaus (Ger) |
| 1929 | Sir A. Harden (Brit) H. A. S. von Euler-Chelpin (Ger) |
| 1930 | H. Fischer (Ger) |
| 1931 | C. Bosch and F. Bergius (Ger) |
| 1932 | I. Langmuir (Amer) |
| 1933 | No award |
| 1934 | H. C. Urey (Amer) |
| 1935 | F. and I. Joliot-Curie (French) |
| 1936 | P. J. W. Debye (Dutch) |
| 1937 | Sir W. N. Haworth (Brit) P. Karrer (Swiss) |
| 1938 | R. Kuhn (Ger) |
| 1939 | A. Butenandt (Ger) L. Ružička (Swiss) |
| 1940 | No award |
| 1941 | No award |
| 1942 | No award |
| 1943 | G. von Hevesy (Hung) |
| 1944 | O. Hahn (Ger) |
| 1945 | A. Virtanen (Fin) |
| 1946 | J. B. Sumner, W. M. Stanley and J. H. Northrop (Amer) |
| 1947 | Sir R. Robinson (Brit) |
| 1948 | A. Tiselius (Swed) |
| 1949 | W. F. Giauque (Amer) |
| 1950 | O. Diels and K. Alder (Ger) |
| 1951 | E. M. McMillan (Amer) and . . . |
| 1951 | G. T. Seaborg (Amer) |
| 1952 | A. J. P. Martin and R. Synge (Brit) |
| 1953 | H. Staudinger (Ger) |
| 1954 | L. C. Pauling (Amer) |
| 1955 | V. du Vigneaud (Amer) |
| 1956 | Sir C. Hinshelwood (Brit) N. N. Semenov (Russ) |
| 1957 | Sir A. Todd (Brit) |
| 1958 | F. Sanger (Brit) |
| 1959 | J. Heyrovsky (Czech) |
| 1960 | W. F. Libby (Amer) |
| 1961 | M. Calvin (Amer) |
| 1962 | J. C. Kendrew and M. F. Perutz (Brit) |
| 1963 | G. Natta (Ital) K. Ziegler (Ger) |
| 1964 | D. C. Hodgkin (Brit) |
| 1965 | R. B. Woodward (Amer) |
| 1966 | R. S. Mulliken (Amer) |
| 1967 | M. Eigen (Ger) R. G. W. Norrish and G. Porter (Brit) |
| 1968 | L. Onsager (Amer) |
| 1969 | D. H. R. Barton (Brit) O. Hassel (Norwegian) |
| 1970 | L. F. Leloir (Argent) |
| 1971 | G. Herzberg (Ca) |
| 1972 | C. Anfinsen, S. Moore and W. H. Stein (Amer) |

# Index

## A

Abacus 112
A-bomb, see Atomic bomb
Abrasive 126, 147
Absolute zero 71
Absorption spectrum 47
AC, see Alternating current
Accelerated freeze drying 241
Accelerator, particle 31
Accumulator 81, 209
Acetate rayon 36, 192
Acetylene 141
Acetyl salicylic acid 176
Acid 167, 168, 170, 176
'Acrilan' 36, 193
Acrylic fibre 191
Acrylic resin 175
Acrylics 36
ACTH 177
Actinide series 28
Actinium 27, 28, 245
Adobe 92
Adrenalin 177
Adrenocorticotropic hormone, see ACTH
Advanced passenger train 52
Aerial ropeway 61
Aerodynamics 219
Aero engines, see Aircraft propulsion
Aerofoil 218, 220, 222, 230
Aeroplane 218–229
Aérospatiale 228
Aérotrain 53
AFD, see Accelerated freeze drying
Afterburner 222
Agave 191
Aiken, H. 242
Aileron 219, 221
Aircraft 218–233
Aircraft construction 226
Aircraft instruments 224
Aircraft propulsion 222
Aircraft Transport and Travel Ltd 232
Air-cushion vehicle 123
Airport 234
Airscrew, see Propeller (aircraft)
Airship 231
Air-speed indicator 224
Alcock, Sir J. 232
Alcohol 168
Alder, K. 246
Aldrin, E. 19
Alfven, H. O. G. 246
Algol 113
Alkali 28, 168

Alkali metal 143, 144, 145
Alkaline earth metal 143, 144, 145
Alkyd resin 175
Alkylation 79
Allergy 177
Alloy 134
Alloy steel 134
Alnico 142
Alpaca 191
Alpha-particle 28, 29, 30, 31, 32
Alternating current 54, 83, 84
Alternator 84, 209
Altimeter 225
Alumina 93, 126, 131, 143
Aluminum 27, 71, 126, 130, 143, 218, 245
Aluminum bronze 134
Alvarez, L. W. 246
Alweg monorail 53
AM, see Amplitude modulation
Amalgam 144, 170
Amatol 173
Americium 27, 31, 245
Ammonal 173
Ammonia 74, 167, 169, 184
Ammonia-soda process 170
Ammonium chloride 80, 171
Ammonium sulphate, see Sulphate of ammonia
Ampère, A. M. 81, 83
Ampere (unit) 81
Amphibole asbestos 126
Amplitude 109, 114
Amplitude modulation 109
Anaesthetic 177
Analgesic 177
Analog(ue) computer 112
Anchor escapement 201
Anders, W. 22
Anderson, C. D. 246
Andromeda 187
Aneroid barometer 117
Anfinsen, C. 246
Angledozer 151
Angle of attack 218
Angora goat 191
Angstrom unit 43
Aniline 198
Anion 167
Annealing glass 98
Anode 80, 106, 133
Antenna 13
Anthracite 74
Antibiotic 177, 237, 240
Antifreeze 210
Antihistamine 177

Antimony 27, 134, 143, 245
Antinode 115
Apennines (Moon) 24
Apennine tunnel 163
Aperture 49
Apogee 13
Apollo (god) 178
Apollo (spacecraft) 18, 21, 24, 179
Appendicectomy 177
Appleton, Sir C. V. 246
APT, see Advanced passenger train
Aqueduct 160, 165
Arc furnace 133
Arc welding 140
Arch bridge 154
Arch dam 156
Archimedes' principle 116
Argon 27, 243
Aries, see Ram, The
Arizona meteor crater 185
Arkwright, Sir R. 200
Armature 84
Armstrong, N. 19
'Arnel' 36, 192
Arrhenius, S. A. 246
Arsenic 27, 143, 245
Artemis 178
Artificial element 26
Artificial horizon 224
Artificial satellite, see Satellite (artificial)
Asbestos 126
Asphalt 158
Aspirin 176
Assembly-line 204, 215
Astatine 27, 245
Asteroid belt 182, 184
Asthma 177
Aston, F. W. 246
Astronaut 18, 21, 179, 180
Astronomy 178
Atlas (rocket) 9
Atmosphere 178
Atmospheric pressure 117
Atmospheric railway 56
Atom 26, 166
Atomic bomb 32, 33, 34
Atomic clock 143, 203
Atomic energy 32–35
Atomic number 26, 28
Atomic theory, 166
Atomic weight 26
Atom-smasher 31
Attitude gyro 224
Atto 245
Audio-frequency 108
Auriga, see Charioteer, The
Autobahn 158

Autoclave 193
Autogyro 232
Automatic transmission 207
Automation 214
Automobile 204–213
Automobile battery 67, 81, 209
Autoroute 158
Autostrada 158
Autumnal equinox 179
Aviation 218–233
Avogadro, A. 166
Avogadro's hypothesis, 167
Axial-flow turbine 73
Axminster carpet 197
Azurite 126

## B

Babbage, C. 112
Bacon, F. 242
Bacon, R. 173
Bacteria 176, 239
Bactrian camel 191
Baekeland, L. H. 36
Baeyer, A. von 246
Bailey bridge 155
Baird, J. L. 242
'Bakelite' 36, 40
Balance wheel 201
Ballastite 173
Balloon 230
Bandot, E. 100
Band saw 61, 62
Banking (aircraft) 221
Bank of Manhattan 163
Barbiturate 177
Bardeen, J. 246
Barium 27, 47, 126, 143, 245
Barkla, C. 246
Barometer 117
Barringer crater, see Arizona meteor crater
Barton, D. H. R. 246
Barytes 126, 143
Bascule bridge 155
Basic-oxygen process 132
Basov, N. 246
Battery 80
Battery (hens) 237
Bauxite 126, 128, 143
Bayonne bridge 163
Bean, A. 22
Becquerel, A. H. 29, 32, 246
Beech 64
Belka 20
Bell, A. G. 102, 242
Bell X-1 233
Bennet, F. 232
Benz, K. 204, 242
Bergius, F. 246
Berkelium 27, 245
Beryl 143
Beryllium 27, 28, 31, 143, 245
Bessemer, Sir H. 242
Bessemer process 132, 242
Best Friend of Charleston 56
Beta-rays 29
Betatron 31
Bethe, H. A. 246
Big-bang theory 189
Big-Ben 201, 202
Big-end 206
Binary numbers 113
Binary star 188
Biological shield 86
Birch 64
Biscuit firing 91
Bismuth 27, 143, 245
Bisque firing 91
Bit 113

Bituminous coal 74
Blackett, P. M. S. 246
Blackpool Tower 163
Black powder 173
Blanching 241
Blast furnace 130
Blasting gelatin 246
Bleaching 198
Bleaching powder 165
Blende 127
Blending 169
Blériot, L. 232
Blind-landing system 235
Bloch, F. 246
Blow moulding 40
Blue gas 75
Blue john 126
Blue shift 186
Boeing-707 226
Boeing-747 234
Bog-iron ore 127
Bohr, N. 32, 246
Boiler 72, 120
Boll 190
Bone ash 89
Bone china 89
Bone meal 236
Booster 9, 24
Borax 142
Boric oxide 94
Boring 147
Borman, F. 22
Born, M. 246
Boron 27, 28, 245
Borosilicate glass 94
Bosch, C. 246
Bothe, W. 246
Bottling 240
Botulism 240
Boulton, M. 73
Box camera 48
Boyle's law 117
Bragg, Sir W. H. 246
Bragg, Sir W. L. 246
Brakes 212
Branca, G. 73
Brass 134
Brattain, W. H. 246
Braun, K. F. 246
Braun, W. von 20
Brazing 142
Bread 238
Breakwater 161
Breeder reactor 35
Brick 92
Bridge 154
Bridgman, P. W. 246
Bright-line spectrum 47
Britannia tubular bridge 56
Brittleness 138
Brocade 197
Bromide (drug) 177
Bromine 26, 27, 245
Bronze 135
Brown, Sir A. W. 232
Brunel, I. K. 56
Brunel, Sir M. I. 242
BTU 71
Buchner, E. 246
Bull, The 187
Bull chain 61
Bulldozer 150
Bull's eye glass 97
Buoyancy 116, 122
Butadiene 38, 67
Butane 76, 231
Butenandt, A. 246
Butter 238
Buttermilk 239
Buttress dam 157
Butt weld 141
Buzz bomb 33
Byrd, Lt-Cdr R. E. 232

## C

CAB see Cellulose acetobutyrate
Cablegram 100
Cable-laying ship 101
Cable railway 53
Cadmium 27, 34, 143, 245
Caesium 27, 143, 203, 245
Caisson 153
Calcite 126, 166
Calcium 27, 28, 47, 126, 143, 236, 245
Calendering
 cloth 193, 199
 paper 65
 plastics 40
 rubber 67
Calfdozer 150
Californium 27, 245
Caloric 70
Calorie 71
Calvin, M. 246
Cam-and-peg steering 212
Camera
 ciné 51
 photographic 48
 television 109
Camphor 36
Camshaft 206
Canal 161
Canis Major, see Great Dog, The
Canis Minor, see Little Dog, The
Canning 240
Canopus 233
Cantilever bridge 155
Cap frame 196
Carbide 93, 139
Carbohydrates 238
Carbolic acid 36
Carbon 26, 27, 31, 34, 38, 80, 126, 130, 245
Carbon boil 132
Carbon dioxide 167
Carboniferous Period 74
Carbonization 195
Carbon monoxide 74, 130
Carbon steel 134
Carborundum 147
Carburettor 208
Cardiac pacemaker 29
Carding 195
Car ferry 119
Cargo ship 119
Carpet 197
Carrier wave 108
Cartwright, E. 200, 242
Cascade tunnel 163
Casein 36, 192
Cashmere goat 191
Cassiopeia 187
Cassiterite 126, 129
Casting
 glass 94
 metal 136, 138
 plastics 40
Cast iron 134, 136
Catalyst 167, 168
Catalyst paint 175
Catalytic cracking 79
Caterpillar tractor 150
Cathode 80, 106, 133
Cathode-ray tube 107, 109, 110
Cation 167
Caustic soda 28, 170
Cavendish Laboratory 32
Cayley, Sir G. 232, 242
Cayley Plains 22
CCTV, see Closed-circuit television

Celestial navigation 124
Celluloid 36, 40
Cellulose acetate 36, 40, 48
Cellulose aceto-butyrate 36
Cellulose nitrate 36
Celsius scale 70
Cement 164
Cement paint 175
Centi 245
Centigrade scale 70
Centimetre-gramme-second, see CGS system
Central Pacific Railroad 56
Centre of curvature 44
Cepheid 188
Cepheus 187
Ceramics 88, 92
Cereal 236
Cerium 27, 245
Cetus, see Whale, The
CGS system 245
Chadwick, Sir J. 32, 246
Chaffee, R. 21
Chain, engineer's 149
Chain reaction 33
Chain saw 60
Chain surveying 149
Chalcopyrite 126
Chalk 126, 129, 179
Chamberlain, O. 246
Change of state 71
Charioteer, The 187
Charles, J. A. C. 232
Charles's law 117
Chartres cathedral 99
Cheese 239
Chemical change 167
Chemical combination 166
Chemical element 26
Chemical industry 168
Chemical operations 168
Chemical pulp 64
Cherenkov, P. A. 244
Chesapeake Bay bridge 163
Chevreul, M. E. 172
China 88
China-clay 126
China stone 89
Chincom 22
Chlorination 165
Chlorine 26, 27, 127, 245
Chloroform 177
Chloromycetin 177
Choke valve 206
Cholera 165
Christmas tree, 77
Chrome iron ore 143
Chromite 143
Chromium 27, 134, 143, 245
Chromosphere 181
Chronometer 124, 203
Chrysler building 163
Chrysotile asbestos 126
Cierva, J. de la 228, 232
Ciné camera 51
Ciné projector 51
Cinnabar 144
Circular accelerator 31
Civil engineering 148
Clay 88
Clock 201–203
Closed-circuit television 109, 235
Clutch 205, 211
Coagulation 66
Coal 74
Coal gas 74
Coal tar 75
Cobalt 27, 30, 135, 182, 245
Cobol 113
Cochineal 198
Cockcroft, Sir J. D. 32, 246
Cockerell, Sir C. 242

Codeine 177
Coefficient of expansion 70, (table) 245
Cofferdam 153
Coir 191
Coke 74, 130
'Coke-bottle' design 226
Collimator 47
Collins, M. 22
Color 46
Color film 50
Colt, S. 242
Columba, see Dove, The
Columbite 144
Columbium, see Niobium
Combine harvester 236
Combing 195
Comet 185
*Comet* (aircraft) 228, 233
Command module 18, 24
Communications satellite 13
Commutator 84
Compactation 153
Compass 83, 124, 181, 224
Complementary colour 46, 51
Compound 166
Compound microscope 242
Compression-ignition 216
Compression moulding 40
Compression ratio 216
Compressor 223
Compton, A. H. 246
Computer 58, 112, 147, 242
Concave lens 45
Concave mirror 44
*Concorde* 218, 219, 224, 225, 226, 227, 228, 229, 230
Concrete 164, 165
Condensed milk 238
Condenser (optical) 45
Condenser (steam engine and steam turbine) 72, 73
Conduction 71
Conifer 60
Connecting rod 206
Conrad, C. 22
Constant composition, law of 166
Constellation 187
Contact-breaker 209
Contact process 170
Containers 58
Continuous casting 136
Continuous spectrum 47
Control rod 34
Convection 71
Convex lens 45
Convex mirror 44
Coon Butte crater, see Arizona meteor crater
Cooper, L. N. 246
Co-ordinate valency 167
Copernicus, N. 178
Copernicus (crater) 17, 181
Copolymer 38
Copper 27, 28, 126, 127, 132, 134, 245
Copper pyrites 126
Cordite 173
Cornish stone 89
Corona 181
Cortisone 177
Corundum 126
Cosmonaut 18
*Cosmos* 21
Cotton 190, 192
Cotton gin 190, 215, 242
Cotton linters 36, 40
Count-down 24
'Courtelle' 36, 193
Covalency 167
Coventry Cathedral 99
Cowpox 177

Cracking 79, 168
Crane 152
Crankcase 207
Crankshaft 205, 207
Crater 180, 185
Crawler tractor 150
Creamery 239
Critical angle 44
Critical mass 34
Critical path analysis 148
Crompton, S. 200, 242
Crookes, Sir W. 242
Crops 236
Cross-breeding 237
Cross-Channel packet 119
Crude oil 76
Crude rubber 66
Crust (Earth's) 179
Cryolite 126, 131, 179
Crystal 166
Cuprite 126
Cupronickel 135
Cuprum 167
Curd 239
Curie, M. 29, 32, 246
Curing (food) 241
Curium 27, 245
Curtain walling 159
Cut-and-cover 57, 59, 160
Cutting, glass 97, 98
Cutting oil 146
Cyclic process 170
Cyclotron 31, 242
Cygnus, see Swan, The
Cylinder block 206
Cylinder head 206

## D

'Dacron' 37, 193
Daguerre, L. J. M. 242
Daimler, G. 204, 242
Dalen, N. 246
Dalton, J. 166
Dam 117, 156
Damask 197
Data processing 112
Dative valency 167
Davisson, C. 246
Davy, Sir H. 242
Day 179, 182
DC, see Direct current
Dead reckoning 124
Death's head watch 202
De Broglie, L. V. 246
Debye, P. J. W. 246
Deca 245
'Decca' navigator 124
Deci 245
Decibel 115
Decorating pottery 90
Dee 31
De Havilland *Comet* 233
Dehydration, see Drying (food)
De Lesseps, F. 162
Depolarizer 81
Depth of field 49, 51
Destructive distillation 74, 169
Detector (radio) 108
Detergent 172
Detonator 173, 246
Deuterium 34
Developing 48
Diabetes 177
*Diamant* rocket 21
Diamond 126
Diana 178
Die 136, 137, 139
Die-casting 136
Dielectric 80

Diels, O. 246
Diesel, R. 242
Diesel-electric locomotive 54
Diesel engine 216, 242
Diesel-hydraulic locomotive 54
Diesel locomotive 52, 54
Diesel-mechanical locomotive 54
Diesel oil 76, 79
Differential 211
Digital computer 112
Dihedral angle 221
Diode 106
Dipping (sheep) 237
Diptheria 177
Dirac, P. 246
Direct current 54, 84, 106
Disc brake 212
Discharge tube 82
Disintegration, radioactive 29
Dispersion (colour) 42
Dissolving 169
Distillation 79, 169
Distributor 209
Diving, deep-sea 117
Dobrovolsky, F. 22
Dock 119, 125
Dolomite 93, 144
Domagk, G. 177
Dope 36
Doppler effect 115
Dough 238
Dounreay reactor 35
Dove 187
Draco, see Dragon, The
Drag 12, 120, 123, 218, 220
Dragline 74, 128
Dragon, The 187
Drawing (metal) 139
Dredging 119, 129, 161
Drier (paint) 175
Drill (fabric) 197
Drilling 147
Drilling rig 76, 77, 78
Drip-dry fabric 193
Drogue 25
Drop forging 137
Drug 176
Drum brake 212
Dry battery 80
Dry dock 125
Drying
  chemicals 169
  food 241
Dry spinning 193
Ductility 138
Duke, C. 22
Dunlop, J. B. 242
Duralumin 134
Dyeing 198
Dynamite 173, 242, 246
Dynamo 84, 209
'Dynel' 193
Dysentry 165
Dysprosium 27, 245

## E

Earhart, A. 233
Earth 179
Earthen dam 156
Earthenware 88, 90
Earth-moving 148–150
Eastman, G. 242
East Rand gold mine 129
Ebonite 67
*Echo* satellite 20
Eclipse 181
Eclipsing binary 187
Edison, T. A. 104, 242, 243
Edison effect 243

Efficiency (engine) 72, 216
Eiffel Tower 162, 163
Eigen, M. 246
Einstein, A. 32, 246
Einsteinium 27, 245
Einstein's equation 32, 33
Electric car 217
Electric clock 202
Electric current 80
Electric discharge 82
Electric-eye 48, 51
Electric furnace 133
Electricity 80–84
Electricity transmission 84, 85, 87
Electric locomotive 54, 55
Electric motor 84
Electrodes 80, 106, 140, 141
Electrolysis 81, 82, 131, 171
Electrolyte 81
Electromagnet 83, 84, 102, 105, 108
Electromagnetic wave 29, 42
Electromagnetism 83, 84
Electron 26–32, 80, 106, 166
Electron-beam welding 142
Electron gun 107, 109, 111
Electronics 106
Electron microscope 107
Electron shell 28
Electron structure 28
Electroplating 82
Electrostatics 81
Electrovalency 167
Element 26–33
Elements, melting and boiling points of 245
Elements, table of 27
Elevator (aircraft) 219, 221
Elliptical galaxy 186
Elm 64
Embankment dam 156
Embroidery 197
$E=mc^2$ 32, 33
Emerald 143
Emery 126
Emission spectrum 47
Empire State Building 162, 163
Emulsion (photographic) 48–51
Emulsion paint 175
Endosperm 238
Engineering bricks 92
Engineering drawing 148
Engraving (glass) 97, 98
Enlarger 50
Enzyme 240
Epoxy resin 36, 164
Equation, chemical 167
Eratosthenes (crater) 181
Erbium 27, 245
Ericcson, J. 242
Eridanus 187
Escapement 201
Escape tower 24
Escape velocity 17
Escape wheel 201
Esparto grass 64
Ester 168
Esterification 168
Etching 98
Ethane 76
Ether 177
Ethylene 37
Euler-Chelpin, H. A. S. von 246
Europa bridge 155
Europium 27, 245
Euston tower 163
Evaporation 169
Evening star 182
Excavator 74, 128, 151

Exhaust manifold 209
Expanded polystyrene 36
Expanding universe 187
*Explorer 1* satellite 20
Explosive 128, 173
Exposure 49
Exposure meter 49
Extender 176
Extracting (metals) 130–132
Extraction (timber) 61
Extrusion
  metals 138
  plastics 40
  rubber 67
Eyepiece 45

## F

Face shovel 151
Fahlberg, C. 177
Fahrenheit scale 70, 71
Fairey *Delta 2* 233
FAO, see Food and Agricultural Organization
Faraday, M. 83
Farming 236
Farm machinery 236, 237
Fats 238
Fatty acid 172
Feldspar 126
Felling 60
Felt 197
Femto 245
Fermentation 238
Fermi, E. 246
Fermium 27, 245
Ferromanganese 144
Ferrous oxide 105
Ferry 118, 119, 123
Fertilizer 236
Feynman, R. P. 246
Ffestiniog railway 52
Fibre glass 36, 37, 41, 71, 97, 99
Fibres 190–193
Fillet weld 141
Filtering 169
Filter press 169
Film 48–51
Film setting 68
Firebrick 92
Fireclay 92
Fire-tube boiler 72, 73, 120
Fischer, E. 246
Fischer, H. 246
Fission 33–34, 121
Fixing (photography) 50
Flaps 220
Flax 190
Fleming, Sir A. 176
Fleming, Sir J. A. 242
Float chamber 208
Float glass 97, 98
Floating dock 125
Flowchart 113
Fluid flywheel 207
Flume 62
Fluorescence 126
Fluorescent paint 175
Fluoridation 165
Fluorine 27, 245
Fluorite 126, 127
Fluorspar 126, 127
Flux 94, 140
*Flyer* aircraft 232
Flyer frame 196
Flying boat 233
Flying shuttle 200, 242
Flywheel 206, 207, 211
FM, see Frequency modulation
f-number 49

Foam rubber 67
Focke, H. 242
Fokker 232, 233
Food 238, 241
Food and Agricultural Organization 236
Food preservation 240
Fool's gold 126, 127
Foot-and-mouth disease 237
Ford, H. 204, 215
Forest fire 63
Forestry 60
Forging 137
Formaldehyde 36
'Formica' 38, 41
Forth rail bridge 163
Forth road bridge 163
Fort Peck dam 156
Fortran 113
'Fortrel' 37, 193
Foundations 152,
Fourdrinier machine 65
Four-stroke cycle 205, 216
Fractionating 79, 170
Fra Mauro (crater) 22
Frame saw 62
*France* (liner) 118, 123
Franck, J. 246
Francis turbine 87
Francium 27, 245
Frank, I. M. 246
Franklin, B. 81, 242
Frasch process, see Sulphur mining
Fraunhofer, J. von 47
Fraunhofer lines 47
Free fall 16
Free-turbine turboprop 223
Freezing 240, 241
Freighter 119
Frequency 43, 114
Frequency modulation 109
Fuel cell 14, 217, 242
Fullers' earth 172
Fulton, R. 242
Fundamental (sound) 115
Fundamental particles, see Electron, Neutron, Positron, Proton
Fungicidal paint 175
Fungicide 236
Funicular 53
Furnace 92, 130–133
Fused quartz 96
Fusee 201
Fusion (atomic) 33, 34, 181
Fusion, latent heat of 71
Fusion welding 140

## G

Gabardine 197
Gabor, D. 246
Gadolinium 27, 245
Gagarin, Y. 9, 18, 21
Galaxy 181, 186–189
Galena 127, 132, 144
Galileo 178, 202, 242
Gallium 27, 28, 144, 245
Galvanization 145
Gamma-ray photography 29
Gamma-rays 29, 30, 43
Gangue 130
Gas 74, 75
Gaseous volumes, law of 166
Gas laws 117
Gasoline, see Petrol
Gasometer 75
Gas turbine 52, 55, 223
Gas-turbine car 217
Gas welding 141
Gay Lussac, J. 166

Gearbox 205, 211
Gelignite 246
Gell-Mann, M. 246
Gelmo, P. 177
Gemini, see Twins, The
*Gemini* spacecraft 9
General anaesthetic 177
Generator, electric 83–85, 209, 222
Generator, radioactive 14, 30
Geology 179
George Washington bridge 163
Gerhardt, C. 176
Germ (wheat) 238
Germanium 27, 107, 144, 245
G-forces 18
Giants' Causeway railway 57
Giauque, W. F. 246
Giffard, H. 232
Giga 245
Gilbert, W. 82
Glaser, D. A. 246
Glass 94–99
Glass blowing 94, 96
Glass wool 98, 99
Glazing (pottery) 91
Glenn, J. 21
Glider 232, 242
Globular cluster 187
Glost firing 91
Gluten 238
Glycerine 172
Goddard, R. H. 20, 242, 243
Goeppert-Mayer, M. 246
Gold 27, 28, 129, 130, 144, 245
Golden Gate bridge 162, 163
Gordon, R. 22
Governor 73, 242
Grab 151
'Grace' 103
*Graf Zeppelin* 233
Grain (wood) 63
Gramophone 104
Grand Coulee dam 157
Grande Dixence dam 148
Grandfather clock 201
Granite 126, 179
Graphite 127
Graving dock 124
Gravitation, law of 20, 178
Gravity (Moon's) 180
Gravity-arch dam 157
Gravity dam 157
Gravity meter 77
Gravure printing 69
Great Bear, The 187
Great Dog, The 187
Great Western Railway 56
Greenockite 143
Gresley, Sir N. 57
Grid (electricity) 87
Grid (electrode) 106
Grignard, F. A. V. 246
Grinding (chemicals) 171
Grinding (metal) 147
Grisson, V. 21
Groundwood pulp 64
Group (periodic table) 28
Grout 153
Gudgeon pin 206.
Guided missile 11
Guillaume, C. E. 246
Gulf of California 14, 15
Gunpowder 173
Gutenberg, J. 68
Gutta-percha 101
Gypsum 127
*Gypsy Moth* 233
Gyrocompass 125, 224
Gyroscope 124, 224

## H

Haber synthesis 169
Hackling 195
Hadley Rill 22
Haematite 126
Hafnium 27, 144, 245
Hahn, O. 246
Haise, F. 22
Hairspring 201
Hale telescope 178
Half-life 30
Half-tone 69
Halite 127
Halley, E. 185
Halley's comet 185
Halogen 28
Hammaguir base 21
Harbour 161
Harden, A. 246
Hard-paste porcelain 88
Hardware (computer) 112
Hare, The 187
Hargreaves, J. 200, 242
Harmonic 115
Hassel, O. 246
Haworth, W. N. 246
Hay fever 177
H-bomb, see Hydrogen bomb
Heat 42, 70
Heat barrier 225
Heat engine 70
Heat exchanger 34, 86
Heat-resistant paint 175
Heat shield 19, 93
Heat transfer 71
Heat treatment 135
Heavy chemicals 169–171
Heavy hydrogen 34
Heavy spar 126
Heavy water 34
Hecto 245
Heinkel *He-78* 233
Heisenberg, W. 246
Helicopter 220, 229, 232, 242
Helium 26–29, 33, 42, 47, 186, 231, 245
Hemlock 64
Hemp 191
Henry, J. 83
Herbicide 237
Heroin 177
Hertz, G. 246
Herzberg, G. 246
Hess, W. F. 246
Hevesy, G. von 246
Heyrovsky, J. 246
High-density polythene 37
Highway 158
*Hindenburg* airship 233
Hinshelwood, Sir C. 246
Hiroshima 34
Histamine 177
Hodgkin, D. C. 246
Hofstadter, R. 246
Holmium 27, 245
Homogenized milk 239
Honing 147
Horizontal stabilizer, see Tailplane
Hormone 177
*Hornet Moth* 218
Horse's Head nebula 188
Hot-air balloon 116, 231, 232
Hour glass 202
Hovercraft 52, 123, 242
Howe, E. 242
Howrah bridge 163
Hydra, see Water serpent, The

Hydraulic brakes 117, 212
Hydraulic-fill dam 157
Hydraulic press 116, 117, 137
Hydrocarbon 76, 79, 168
Hydroelectric power 85, 86, 156, 157
Hydrofoil 123, 220
Hydroforming 79
Hydrogen 26–28, 42, 47, 74, 81, 181, 186, 231, 245
Hydrogen, liquid 10
Hydrogenation 168
Hydrogen balloon 231, 232
Hydrogen bomb 33, 34, 181
Hydrogen chloride 166, 167
Hydrogen sulphide 74
'Hydrolastic' suspension 213
Hydrometer 117
Hydrophilic 172
Hydrophobic 172
Hydrophone 114
Hydroplane 122
Hydropulper 64, 65
Hydrostatics 116
Hypo 50

I

Iceland spar 126
Igneous rock 179
Ignition system 208, 209
Ignition coil 209
Ilmenite 127, 145
ILS, see Instrument Landing System
Image orthicon (television camera tube) 109
Impact extrusion 137
Impulse turbine 73, 86
Incandescent-filament lamp 82
Independent suspension 213
Indigo 198
Indium 27, 144, 245
Induction 84
Induction coil 84
Induction furnace 132
Industrial Revolution 194
Inert gas 28
Inert-gas welding 141
Inertial guidance 124
Infra-red rays 43, 70, 71, 96
Ingot 135, 137
Injection moulding 40
Injector (diesel engine) 216
Inlet manifold 209
Inner planets 182
Insecticide 237
Instrument Landing System 235
Insulator, electric 80, 85, 193
Insulator, heat 71
Insulin 177
Intaglio printing 69
Integrated circuit 107
*Intelsat* satellite 12, 13
Intensive farming 237
International system (SI) of units 245
Invar 71, 135
Iodine 27, 245
Ionosphere 109
Iridium 27, 144, 245
Iris diaphragm 49
Iron, 26, 27, 130, 134, 179, 185, 245
Iron ore 126–128, 130, 132
Ironstone 127
Irwin, J. 22
Isotope 26
Isotope, radioactive, 29–32

J

Jackblock method 159
Jacquard loom 196
*Jaguar* trainer 222
Janssen, Z. 242
Jenner, E. 177
Jensen, J. H. D. 246
Jet engine 11, 123, 223
Jet propulsion, principle of 11
Jetty 161
Jig dyeing 198
Jigger 90
John Hancock Center 163
Johnson, A. 233
Joining metal 140, 141
Joliot-Curie, F. 246
Joliot-Curie, I. 246
Jolley 90
Joule, J. P. 70
Jumbo drilling rig 160
Jumbo jet 218, 233, 234
Jupiter 17, 182–184
Jute 64, 191

K

Kagoshima Space Centre 22
Kaldo process 133
Kaolin 89, 126
Kaplan turbine 87
Karrer, P. 246
Kastler, A. 246
Kay, J. 200, 242
Kendrew, J. C. 246
Kennedy Airport 234
Kepler, J. 20, 178
Kerosene 10, 76, 79
Kidney ore 127
Kieselguhr 246
Kiln 89–91
Kilo 245
Kinetic energy 71
Kinetiscope 242
Kingsford-Smith, Sir C. 233
Kitty Hawk 218, 232
Klystron 107
Knight, W. J. 233
Knitting 197
Komarov, V. 21
König, Friedrich 242
Kraft pulp 64
Krypton 27, 28, 245
Kuhn, R. 246
Kusch, P. 246

L

Lac 40
Lace 197
Lacquer 36
Lactic acid 239
Laënec, R. T. H. 242
Laika 20
Lamb, W. E. 246
Laminated glass 96
Laminating 41
Land, E. H. 242
Landau, L. D. 246
Langmuir, I. 246
Lanolin 176
Lanthanum 27, 245
Lap 194
Lapping 147
Lap weld 141
Latent heat 71
Latex 66
Lathe 147, 242
Laue, M. T. F. von 246
Laughing gas 177
Lawrence, E. O. 242, 246

Lawrencium 27, 245
LD process 132
Leaching 131
Lead 27, 30, 94, 127, 130, 132, 134, 143, 144, 245
Lead-acid battery 81, 209
Lead glance 127
Lead glass 94
Lead paint 175
Leaf springs 213
Leap year 179
*Le Capitol* locomotive 58
Leclanché cell 80
Lee, T. D. 246
Legume 236
Lehr 98
Leloir, L. F. 246
Lenhard, P. 246
Lenoir, J. E. 204
Lens 44, 48
Leo, see Lion, The
Leonov, A. 21
Lepus, see Hare, The
Letterpress printing 68
Leukaemia 32
Level (instrument) 149
Libby, W. F. 246
Life-support system 18
Lift (aerodynamic) 218
Lift, safety 242
Lift-slab 159
Light, properties of 44
Light, velocity of 33, 43
Light bulb 82, 242, 243
Light meter 49
Lightning 81
Lightning conductor 81, 242
*Lightning* fighter 218, 225
Lightship 124
Light-year 186
Lignite 74
Lignocaine 177
Lilienthal, O. 232
Lime 94, 236
Limestone 94, 126, 129, 130
Limonite 127
Lindbergh, C. A. 232
Linear accelerator 31
Linear induction motor 55
Linen 190, 192
Liner 118
Linotype 68
Linseed oil 175
Lint 190
Linters 36, 40
Lion, The 187
Lippershey, J. 242
Lippmann, G. 246
Liquefied petroleum gas, see LPG
Liquid hydrogen 10
Liquid-liquid extraction 170
Liquid oxygen 10, 173
Liquid propellant 10, 11
Lithium 27, 28, 144, 245
Lithopone 126
Litho printing 68
Little Bear, The 187
Little Dog, The 187
Liverpool and Manchester railway 56
Liverpool docks 125
Livestock 237
Llama 191
Local anaesthetic 177
Lock (canal) 161
Lockheed *Vega* 233
*Locomotion No. 1* locomotive 56
Locomotive 52
Lodestone 82, 121
Log 124
Log boom 61

London Airport (Heathrow) 234, 235
Long-case clock 201
Long-chain molecule 38
Longitudinal wave 42
Long wave 43
Loom 195, 196
Loran (navigation) 124
Lorentz, H. A. 246
Lotschberg tunnel 163
Loudspeaker 104, 105, 108
Lovell, J. 22
Lower Zambesi bridge 163
LPG 76
*Luna* probes 21
Lunar module 18, 19, 23–25
Lunar orbital rendezvous 24
*Lunar Orbiter* probes 17, 21, 180
Lunar rover 22
*Lunik* probes 20
*Lunokhod* moonwalker 22
Lutetium 27, 245
Lye 171
Lyra, see Lyre, The
Lyre, The 187

M

Machine gun 242
Machine-tool 136, 146, 147
Machining 146, 147
Mach number 225
Mackinac Straits bridge 163
Madder 198
M and B 177
McMillan, E. M. 246
Magnesite 93, 144
Magnesium 27, 131, 134, 218, 245
Magnesium chloride 131
Magnetic field 83, 179
Magnetic north 124
Magnetic pole 82
Magnetic tape 104, 105, 110, 113
Magnetism 82–84
Magnetite 126, 127
Magneto 223
Magnetometer 77
Magnetron 107
Magnifying glass 45
Mahogany 62
Mains electricity 80, 84
Malachite 127
*Mallard* locomotive 57
Malleability 138
Mandrel 138, 139
Manganese 27, 132, 144, 245
Manganese dioxide 80
Manila hemp 64
Man-made fibre 192
Mantle (Earth's) 179
Manure 236
Maracaibo Lake bridge 163
Marble 128, 129, 179
Marconi, G. 108, 242, 246
Maria 22, 180
Marine paint 175
*Mariner* probes 16, 17, 21, 22, 182, 184
Mars 16, 182–184
Marshalling yard 58
Martin, A. J. P. 246
Mass production 214
Mattingly, T. 22
Maudsley, H. 242
Maxim, H. S. 242
Medium wave 43
Mega 245
Melamine-formaldehyde 36

251

Melting points of elements 245
Melt spinning 193
Memory (computer) 112
Mercerizing 199
Mercury (planet) 17, 182, 184
*Mercury-Atlas* rocket 9
Mercury barometer 117
Mercury fulminate 173
Merino 191, 192
Meteor 19, 184, 185
Meteorite 180, 185
Methane 74, 76
Methyl salicylate 176
Metric system 242
Metropolitan line (Underground) 57
MF, *see* Melamine-formaldehyde
Mica 127
Michelson, A. A. 246
Micro 245
Micro-organism 177, 240
Microphone 102, 108
Microscope 45
Microwave 43, 107
Mild steel 134
Milk 238
Milky Way 181, 186, 187
Milli 245
Millikan, R. A. 246
Milling 147, 228
Milometer 112
Mineral dressing 130
Minerals 126
Miniature railway 52
Mining 128
Miraflores lock 162
Mirage 44
Mirror 44, 45
Missile 11
Mitchell, E. 22
Mixing 169
MKSA system 245
*Model-T* Ford 204
Moderator 34
Modulation 108, 109
Module (spacecraft) 18
Mohair 191
Moissan, H. 246
Molecule 166
Molybdenum 27, 135, 144, 245
Monazite 144
Monochromatic light 42
Monomer 38, 169
Monorail 53
Mono recording 104
Monotype 68
Mont Blanc tunnel 160, 162, 163
Mont Cenis tunnel 163
Montgolfier brothers 232, 242
Month 179, 180
Moon 12, 16, 17, 24, 25, 178, 180
Moore, S. 246
Mordant 198
Morning star 182
Morphine 177
Morse, S. F. B. 242
Morse code 100
Mossbauer, R. L. 246
Motion picture 51
Motorway 158
Mould (on food) 240
Moulding
  plastics 40
  rubber 67
Mount Palomar Observatory 178

Mount Washington cog railway 53
Movies 51
Muck-shifting 150
Mullikin, R. S. 246
Multiple proportions, law of 166
Multistage rocket 9, 10

**N**

Nano 245
Narcotic 177
Nasmyth, J. 242
National Physical Laboratory 203
National Westminster Tower block 163
Native metals 130
Natta, G. 246
Natural gas 76–79
*Nautilus* (submarine) 122
Naval architect 120
Navigation 124, 235
Nebula 187, 188
Néel, L. E. F. 246
Negative 50
Neodymium 27, 245
Neon 27, 245
Neptune 182, 184
Neptunium 27, 31, 245
Nernst, W. 246
Network planning 215
Neutron 26–34
Newcomen, T. 73, 242
New Orleans bridge 163
Newsprint 64
Newton, Sir I. 20, 178, 242
Nickel 27, 82, 134, 144, 179, 185, 245
Niépce, J. N. 242
*Nimbus* satellite 13
Niobium 27, 144, 245
Nitration 168
Nitrile rubber 67
Nitrocellulose 36, 173, 175
Nitrogen 26–28, 31, 236, 245
Nitrogen iodide 173
Nitroglycerin 173, 246
Nitrous oxide 177
Nobel, A. B. 242, 246
Nobelium 27, 245
Nobel prize winners 246
Node 115
Noise 114
Norrish, R. G. W. 246
Northrop, J. H. 246
North star, *see* Pole star
Nova 189
Nuclear reactor 34, 35, 85, 86, 121
Nucleus (atom) 26–31
Nucleus (comet) 185
Numerical control 146
Nursery (forest) 60
Nylon 36, 38, 40, 193

**O**

Objective 45
Oersted, H. C. 83
Offset-litho 68
Offshore drilling rig 76
Ohm, G. S. 81
Ohm (unit) 81
OHV, *see* Overhead valve engine
Oil, *see* Petroleum
Oil, lubricating 210
Oil of wintergreen 176

Oil prospecting 77
Oil refinery 8, 78, 79
Olds, R. E. 215
Oleum (concentrated sulphuric acid) 172
Onnes, H. K. 246
Onsager, L. 246
Opencast mining 74, 128
Open cluster 186
Open-hearth furnace 93, 132
Opium 177
Orbit 9, 12
Orbital velocity 12
Ore 130
Organic acid 168
Orion 187
Orion Nebula 188
'Orlon' 36, 193
Oscilloscope 107, 115
Osmium 27, 144, 245
Ostankino TV tower 163
Ostwald, W. 246
*Osumi* satellite 22
Otis, E. G. 242
Otto, N. A. 204, 242
Otto cycle 205
Outer planets 182
Overburden 128
Overdrive 211
Overhead-valve engine 206
Overtone 115
Oxidant 10
Oxidation 168
Oxyacetylene torch 120, 141
Oxygen 26–28, 31, 40, 245
Oxygen, liquid 10
Oxyhydrogen torch 141

**P**

Pacemaker 29
*Pacific* class (locomotive) 57
Paint 46, 169, 171, 174, 175
Paint testing 175
Palladium 27, 144, 245
Panama canal 161, 162
Panning 129
Pantograph 55
Paper 64, 65
Papin, D. 242
Paraffin 10, 76, 77
Paraffin wax 38, 79
Parallax 186
Para-typhoid 165
Parison mould 97
Parsons, Sir C. A. 73, 242
Particle accelerator 31
Pasteurization 239–241
Patina 127
Patseyev, V. 22
Pauli, W. 246
Pauling, L. C. 246
Pavement 158
Peacock ore 126
Peat 74
Peenemünde 20
Pegasus 187
Pelton wheel 85, 86
Pendulum 201, 202
Penicillin 176, 177
Penicillium mould 176
Penstock 86
Pentlandite 144
Pentode 106
Perigee 13
Period (in periodic table) 28
Period (of satellite) 13
Periodic table 26–28
Perrin, J. B. 246
Perseus 187
'Perspex' 36, 40
Perutz, M. F. 246

Petrol 76, 79
Petrol engine 205–210
Petroleum 76–79
Petuntze 89
Pewter 134
PF, *see* Phenol-formaldehyde
Pharmaceutical 176, 177
Phases (Moon) 180
Phenol 36, 172
Phenol-formaldehyde 36, 38
Phenolic resin 175
Phonograph 104, 242, 243
Phosphor III
Phosphorescence 32
Phosphorus 27, 130, 132, 236, 245
Photocell 48, 51, 105
Photoelectric effect 109
Photoengraving 69
Photography 48–51, 242
Photogravure printing 69
Photosphere 181
Phthalic anhydride 175
Physical operations 169
Pickling (food) 241
Picric acid 173
Pig iron 130
Pigment 46, 174
Pile 151, 153
Pile driving 151, 153
Pine 64
*Pioneer* probes 20, 22
Pipeline 79
Piston 204–206
Piston rings 206
Pitch (sound) 115
Pitchblende 29, 32, 127
Pitching 221
Placer 126, 129
Planck, M. 32, 246
Plane mirror 44
Planet 178, 182
Planetoid 184
Planing 147
Planographic printing 69
Plan position indicator 234
Plaster of Paris 127
Plasticity 138
Plasticizer 67
Plastisol 40
Plate glass 97
Platinized asbestos 170
Platinum 27, 28, 71, 130, 139, 144, 245
Pleides 187
Plough, The 187
Plumbum 167
Pluto 182, 184
Plutonium 14, 27, 31, 33, 34, 245
Plywood 63
PMMA, *see* 'Perspex'
Pneumatic caisson 153
Pneumatic tyre 242
Pneumonia 177
Poincaré, H. 32
Polaris *see* Pole star
Pole star 182, 187
Poliomyelitis 177
Polonium 27, 32, 245
Polyamide 36, 193
Polyester 36, 37, 41, 193
Polyethylene, *see* 'Polythene'
Polymer 31, 169, 193
Polymerization 38, 40, 79, 169, 193
Polymethyl methacrylate, *see* 'Perspex'
Polypropylene 37
Polystyrene 37, 38, 40, 168, 175
Polytetrafluorethylene, *see* PTFE

'Polythene' 37, 38, 40
Polyurethane 37
Polyurethane resin 175
Polyvinyl chloride, see PVC
Pontchartrain Causeway 163
Pontoon bridge
Poplar 64
Porcelain 88–89, 91
Porter, G. 246
Portland cement 164
Positron 29
Post Office Tower 163
Post-tensioning concrete 164
Potassium 27, 28, 47, 144, 236, 245
Potassium chloride 131
Potassium sulphate, see Sulphate of potash
Potato blight 236
Pottery 88–91
Powder metallurgy 139
Power loom 200
Power station 85–87
Power take-off 236
PPI, see Plan position indicator
Praseodymium 27, 245
Precast concrete 164
Pregl, P. 246
Pressed bricks 92
Pressing (metal) 137
Pressure cooker 242
Pressure testing 228
Pressurization 228
Pretensioning concrete 164
Primary cell 81
Primary colour 46, 51, 110, 111
*Princess Margaret* hovercraft 123
Principal focus 45
*Principia* 20
Printing 68
Printing, photographic 49
Printing, textile 199
Prism 42, 43, 46, 47
Prismatic binoculars 45
Prismatic compass 45
Procaine 177
Producer gas 171
Program 112
Projector 51
Prokhorov, A. 246
Promethium 27, 245
Prominence 181
Propane 76, 141
Propellant 10
Propeller
  aircraft 220
  ship's 120
Propeller shaft 211
Protactinium 27, 30, 33, 245
Proteins 238
Proton 26–32
Proxima Centauri 186
Psilomelane 144
PTFE 37, 38
PTO, see Power take-off
Pulsar 178, 188
Pulsating star 187
Pulse jet 233
Punched card 113
Punched tape 113
Purcell, E. M. 246
Pure breeding 237
Push rod 206
PVC 36–38, 40, 41, 175
Pylon 85, 87, 92, 93
'Pyrex' 71, 84
Pyrites 126, 127, 170
Pyrites, copper, see Chalcopyrite
Pyrolusite 144

## Q

Quality (sound) 115
Quantum theory 32
Quarrying 128, 129
Quartz 96, 127
Quartz clock 203
Quasar 178, 188
Quasi stellar radio sources, see Quasar
Quebec bridge 163
*Queen Elizabeth* 118
*Queen Elizabeth 2* 13, 123
Quicksilver 144

## R

*R.34* 432
*R.101* 233
Rabi, I. I. 246
Rack-and-pinion steering 212
Rack railways 52, 53
Radar 43, 124, 234, 235, 242
Radar tracking 13
Radial-flow turbine 73
Radiation, alpha-, beta- and gamma- 29
Radiation, heat 71
Radiator (automobile) 210
Radiator (heat) 71
Radical 167
Radioactive generator 14, 30
Radioactivity 28–32
Radio altimeter 224
Radio astronomy 189
Radio broadcasting 108, 109, 242
Radio direction finder 124
Radio-frequency 108
Radio-isotope 29–31
Radiosonde 230
Radio telescope 178, 188, 189
Radiotherapy 30
Radio wave 43
Radium 27, 29, 30, 32, 144, 245
Radome 235
Radon 27, 245
Rag paper 64
Railways 52–55
Ram, The 187
Raman, Sir C. V. 246
Ramie 64
Ramsay, Sir W. 246
Rance tidal power station 87
Range-finder 49
Range height indicator 235
*Ranger* probe 21
Rare earth 28, 144
Rate-of-climb indicator 224
Rayleigh, Lord 246
Rayon 36, 60, 191, 192
Reaction 10, 20
Reaction turbine 73, 86
Read, Lt Cdr A. C. 232
Reaming 147
Recirculating-ball steering 212
Record player, see Gramophone
Rectification 106
Rectifier 54
Red lead 175
Red Planet 184
Red shift 186
Red Spot 184
Reduction 168
Re-entry 19
Refinery 8, 78, 79
Refining metals 132, 133
Refining oil 79

Reflecting telescope 178, 242
Reflection 44
Reforming 79
Refraction 44
Refractive index 44
Refractory 93
Regenerated fibre 192
Regulator 201, 202
Reinforced concrete 164
Relativity theory 32
Relief printing 69
Rennet 239
Repeater (cable) 101
Reservoir 148, 165
Resistance (electrical) 80, 81
Resistance welding 140
Resonance 115
Retrobraking 25
Retrograde motion 182
Retting 191
Revolver 242
Rhenium 27, 144, 245
RHI, see Range height indicator
Rhodes, C. 56
Rhodium 27, 144, 245
Richardson, O. W. 246
'Rigmel' 199
Rill 22, 180
Ring-spinning 195
'Ripple' 30
Rippling 190
Riveting 120, 140, 228
Road 158
Robinson, Sir R. 246
Rock crystal 127
Rock drill 129
Rockefeller Center 163
Rocker arm 206
Rocket 9–11
Rocket fuel 10
*Rocket* locomotive 56
Rock-fill dam 157
Rocks 179
Rock salt, see Halite
Roentgen, W. K. 246
Rokko tunnel 163
Roll film 48, 242
Rolling (aircraft) 221
Rolling (metal) 137
Roosa, S. 22
Roquefort 239
Rosin 40
Rotor, helicopter 220, 230
Rotor, turbine 72, 73
Rotor, Wankel engine 216
Rotor arm 209
Rotary drilling 77
Roving 194
Rozier, P. de 232
Rubber 66, 67
Rubidium 27, 144, 245
Ruby 126
Ruby copper 126
Rudder 221
Ruthenium 27, 145, 245
Rutherford, Lord 29, 31, 32, 246
Ružička L. 246
Rye 238

## S

Sabatier, P. 246
Saccharin 177
Safège monorail 53
Safety glass 96
Safety lamp 242
Saint Gotthard tunnel 163
Saint Lawrence Seaway 162
Sal ammoniac, see Ammonium chloride

Salazar bridge 163
Salicylic acid 176
Salisbury cathedral 163
Salt 127
Salting (food) 241
*Salyut* space station 22
Samarium 27, 245
Sand glass 202
Sandstone 179
'Sanforize' 199
San Francisco Bay Area Rapid Transit system 59
Sanger, F. 246
Saponification 172
Sapphire 126
Satellite
  artificial 9, 12–15, 102
  natural 182
Satin 197
Saturn (planet) 182, 183, 184
*Saturn V* rocket 9, 10, 24
Savery, T. 73
Sawmill 61
Scandium 27, 28, 245
Scanning 109, 110
Scarlet fever 177
Scheelite 145
Schlieren photography 229
Schrieffer, J. R. 246
Schrödinger, E. 246
Schwinger, J. S. 246
Scotch boiler 121
Scott, D. 22
Scouring 195
Scraper 151, 152
Screw propeller 242
Scutching 191
Seaborg, G. T. 246
Seam welding 141
Sea of Showers 22, 25
Sears building 159
Seasoning 63
Seasons 179
Secondary cell 81
Sedative 177
Sedimentary rock 179
Segrè, E. 246
Seikan tunnel 163
Seismometer 22, 77
Selective breeding 237
Selective weedkillers 237
Selenium 27, 245
Selfridge, T. E. 232
Semenov, N. N. 246
Semi-conductor 107
Service module 18, 24, 25
Servo-brakes 213
Seven Sisters 187
Severn bridge 154, 163
Sewage disposal 165
Sewing machine 242
Sextant 124
Shadow mask 111
Shaping 147
Sheep 191, 237
Sheepsfoot roller 152
Sheep-shearing 192
Sheet pile 151, 153
Shellac 40
Shepard, A. 21, 22
Shipbuilding 120, 121
Ships 118–121
Shock absorber 213
Shockley, W. 246
Shock waves 225
Sholes, C. L. 242
Shooting star 19
Short wave 43
Shuang Chen Tsu 22
Shutter 48
Shuttering 159
Siderite 127

253

Side-valve engine 206
Siegbahn, F. M. G. 246
Signal box 58
Sikorsky, I. 232, 233, 242
Silica 93, 127
Silica glass 94
Silicon 14, 27, 37, 40, 130, 245
Silicon bronze 134
Silicon carbide 147
Silicone rubber 67
Silicones 36, 37, 175
Silk 191, 192, 196, 198
Silk-screen printing 199
Silkworm 191, 192
Silver 27, 28, 48, 50, 127, 130
Silver solder 142
Simple cell 80
Simplon tunnels 163
Sintering 131, 139
Sisal 191
SI units 245
Size 65
Skimmer 151
Skin milling 227, 228
*Skylab* project 22
Skyscraper 159, 229
Slag 130
Slip ring 83, 84
Sliver 195
Slotting 147
Sludge 165
Slush moulding 40
Smallpox 177
Smelting 131
Smoking (food) 241
Snowdon Mountain Railway 53
Snowy Mountains scheme 162
Soaking pit 135, 137
Soap 172
Soapstone 127
Soda 170
Soda ash 94
Soda-lime glass 94
Soddy, F. 32, 246
Sodium 27, 47, 145, 171, 186, 245
Sodium bicarbonate 171
Sodium carbonate, *see* Soda
Sodium chloride 127
Sodium hydroxide 28
Sodium hypochlorite 165
Sodium nitrate 246
Sodium palmitate 172
Sodium pentothal 177
Sodium phenate 176
Sodium thiosulphate 50
Soft-paste porcelain 88
Software (computer) 112, 113
Soil mechanics 148, 152, 153
Solar cell 13, 14, 16, 217
Solar system 178, 181
Solder 134
Soldering 142
Solenoid 83, 203
Solid propellant 11
Solstice 179
Sonic bang 225
Sonic speed 225
Sound 42, 114, 115
Sound barrier 225
Sounding rocket 11
Sound recording 104
Sound track 105
*Southern Cross* 233
*Soyuz* spacecraft 9, 21, 22
Space probes 16
Spacesuit 18, 19
Sparking plug 209

Specific heat 71
Spectroscope 47, 186
Spectrum 42, 43, 46, 47, 186
Specular ore 127
Speed of light 33, 43
Speed of sound 114
Speedometer 112
Sphalerite 127
Spillway 157
Spin-dyeing 198
Spinneret 192, 193
Spinning 190, 194, 196
Spinning jenny 200
Spinning mule 195, 200
Spinning wheel 194, 242
Spiral galaxy 186
*Spirit of St. Louis* 232
Spot welding 141
Spraying (paint) 174
Spray steelmaking 133
Spruce 64
Spun silk 196
*Sputnik* satellite 9, 20
*SR-N4* hovercraft 123
Stabilizer 119
Stacking (aircraft) 234
Stage, microscope 45
Stage, rocket 9
Stage, turbine 73
Stained glass 99
Stainless steel 134
Stalling 219
Stanley, W. M. 246
Staple foods 238, 239
Star 47, 178, 181, 186–189
Stark, J. 246
Static electricity 81
Stationary orbit 13
Staudinger, H. 246
STD 103
Steady-state theory 187
Steamboat 242
Steam engine 72, 73
Steam hammer 137, 242
Steam locomotive 53, 54
Steam pump 73, 242
Steam turbine 72, 73, 85, 86, 121
Steel 132
Steel-arch bridge 154, 163
Steering system 212
Stein, W. H. 246
Stellite 143
Step-down transformer 84
Step rocket 9
Step-up transformer 84
Stephenson, G. 56
Stephenson, R. 56
Stereo 104
Sterilization 239, 240
Stern, O. 246
Stethoscope 242
Stibnite 143
Stilton 239
Stockton and Darlington Railway 56
Stoneware 88, 91
Stop (camera) 49
Stop bath 50
Stoving paint 175
Stratosphere 228
Streamlining 220
Strelka 20
Strip mining 128
Strontium 27, 245
Styrene 37, 66, 169
Styrene-butadiene rubber 67
Subgrade 158
Submarine 122
Subscriber trunk dialling, *see* STD
Subtractive colour process 46, 51

SU carburettor 206
Sucramine 177
Suez Canal 162
Sulky 61, 63
Sulpha drugs 177
Sulphanilamide 177
Sulphate of ammonia 236
Sulphate of potash 236
Sulphate pulp 64
Sulphite pulp 64
Sulphonamide, *see* Sulpha
Sulphur 27, 67, 127
Sulphur dioxide 170, 171
Sulphuric acid 80, 171
Sulphur mining 129
Sulphur trioxide 170, 171
Summer solstice 179
Sumner, J. B. 246
Sump 206
Sun 42, 47, 178, 185, 186
Sundial 202
Sunspot 181
Supernova 189
Superphosphate 236
Supersonic speed 114, 218, 219
Surface-piercing hydrofoil 123
Surfactant 172
Surveying 77, 148, 149
*Surveyor* probe 16, 21, 22
Suspended monorail 53
Suspension bridge 154, 155
Suspension system (automobile) 213
Svedberg, T. 246
Swan, Sir J. W. 242
Swan, The 187
Swigert, J. 122
Swing bridge 155
Sydney Harbour bridge 163
Sympathetic vibration 115
Synchrocyclotron 31
Synchromesh 21
Synge, R. 246
Synthetic fibre 192, 193
Synthetic gems 93
Synthetic resin 175
Synthetic rubber 67
Syphilis 177

**T**

Tailfin 221
Tailplane 219
Talbot, W. H. F. 242
Talc 127
Tamm, I. Y. 246
Tanker 79, 119, 161
Tantalite 144
Tantalum 27, 144, 145, 245
Tape recorder 105
Tappet 206
Tapping (rubber) 66
Tarmac 158
Taurus, *see* Bull, The
Tay bridge 163
Technetium 27, 245
Teeming ingots 135
'Teflon' 37
Telegram 100
Telegraphy 100, 101, 242
Telemetry 14
Telephone 102, 103
Telephone exchange 103
Teleprinter 100
Telescope 45, 47, 178, 242
Television 109–111, 235, 242
Telex 101
Tellurium 27, 145, 245
*Telstar* satellite 21
Temperature 70, 71

Tera 245
Terbium 27, 245
Tereshkova, V. 21
Terramycin 177
'Terylene' 37
Test rig 227
Tetanus 177
Tetrode 106
Textiles 190–197
Thallium 27, 145, 245
Thatcher Ferry bridge 163
Theodolite 149
Thermal cracking 79
Thermionic emission 106
Thermodynamics, laws of 70
Thermoforming 40
Thermometer 70, 71, 112, 242
Thermoplastic 40
Thermoset, *see* Thermosetting plastic
Thermosetting plastic 40
Thermosoftening plastic, *see* Thermoplastic
Thermostat 70, 71, 210
Thinner (paint) 174
Thomson, G. 246
Thomson, Sir J. J. 32, 246
Thorite 145
Thorium 27, 30, 33, 145, 245
Throwing (pottery) 89
Throwing (silk) 196
Thrust 218, 220
Thulium 27, 245
Tidal power 87
Tide 181
Timber 60–63
Tin 27, 97, 126, 134, 141, 143, 145, 245
'Tin Lizzie' 204
Tin oxide 91
Tinstone 126
Tinting (paint) 174
*Tiros* satellite 20
Tiselius, A. 246
*Titan* rocket 9
Titanium 27, 28, 127, 137, 145, 245
TNT (trinitrotoluene) 33, 34, 173
Todd, Sir A. 246
Tokyo TV tower 163
Tomonager, S. 246
Tonsilectomy 177
Torque converter 54
Toughened glass 96
Tower Bridge 155
Tower subway 57
Town gas 75
Townes, C. H. 246
Toxin 240
Tracking station 13
Track rod 212
Tractor 151, 236
Train ferry 118, 119
Tramp 119
Tranquillizers 177
Transfer orbit 17
Transformer 84, 87
Transistor 106, 107
Transit 182
Transition series 28
*Transit* satellite 13
Transparency 50
Transmission of electricity 87
Transmission system (automobile) 205, 210, 211
Transmutation 29, 30
Trans-Siberian Railway 56
Transverse wave 42
Trepanner 75
Trevithick, R. 56, 73, 242

Triacetate 192
Triangle, The 187
Triangulation 149
Triangulum, see Triangle, The
'Tricel' 36, 192
Triode 106
Truss bridge 154
*Tu 144* airliner 233
Tufted carpet 197
Tug 118, 119
Tungsten 27, 93, 135, 139, 143, 145, 245
Tungsten carbide 139
Tunnel 160
Tunnelling shield 59, 160, 242
Turbine
   gas 52, 55, 217, 223
   steam 72, 73, 85, 86
   water 85
   wind 85
Turbofan 222, 223
Turbogenerator 34, 85
Turbojet 218, 223
Turboprop 223
Turbotrain 52
Turn-and-bank indicator 224
Turning (machining) 146
Turpentine 175
Tuyeres 130
Twill 196
Twin-lens reflex camera 48
Twins, The 187
Twiss, P. 233
Two-stroke cycle 205, 216
Type 68, 69, 242
Type metal 134
Typewriter 242
Typhoid 165, 177

## U

UF, see Urea-formaldehyde
UHF 43, 110
Ultra high frequency, see UHF
Ultrasonic 115
Ultra-violet rays 43, 96
Undercarriage 220
Underground mining 129
Underground railway 57
Union Pacific Railroad 55, 56
Universal joint 211
Universe 178, 181, 187
Upthrust 116
Uranium 26–34, 127, 145
Uranus 182, 184
Urea-formaldehyde 37, 38
Urey, H. C. 246
Ursa Major, see Great Bear, The
Ursa Minor, see Little Bear, The

## V

*V-1* flying bomb 233
*V-2* rocket bomb 20, 233
Vaccine 177, 237
Valency 167, 168
Valve (engine) 206
Valve (radio) 106, 242
Vanadium 27, 135, 143, 245
Vanadium pentoxide 170
Van Allen belts 14, 20
Van der Graaf accelerator 31
Van der Waals, J. D. 246

Van't Hoff, J. H. 246
Vaporization, latent heat of 71
Variable star 186, 188
Varnish 174
Vehicle (paint) 174
Velocity of light 33, 43
Velvet 197
Veneer 63
Venus 17, 182
Vernal equinox 179
Verrazano-Narrows bridge 155, 163
Vertical stabilizer, see Tail-fin
Very high frequency, see VHF
Vfr 20
VHF 43, 110
Vickers *VC-10* 226
Vickers *Vimy* 232
Vicker's *Viscount* 233
Victoria line (Underground) 59
Vicuna 191
Video-tape 110
Vigneau, V. du 246
Vinegar 241
Vinyl 38
Virtanen, A. 246
Virtual image 44, 45
Viscose 191
Vitamins 238
Vitreous enamel 175
Vitrification 88
Volkov, V. 22
Volk's electric railway 57
Volt 80
Volta, A. 80, 242
Voltage regulator 209
Voltaic cell 81
*Voshkod* spacecraft 21
*Vostok* spacecraft 9, 18, 20
Vulcanization 67

## W

Wallach, O. 246
Wall Tower 163
Walton, E. T. S. 32, 246
Wankel, F. 242
Wankel engine 216, 217, 242
Warp 197
'Wasp-waisted' design 226
Watch 202
Water, electrolysis of 166
Water frame 200
Water gas 75, 171
Watermark 65
Water pressure 117
Water Serpent, The 187
Water-tube boiler 72, 120
Water turbine 87, 156
Watson-Watt, Sir R. 242
Watt, J. 73, 242
Wavelength 43
Weather satellite 12, 13
Weaving 190, 196, 197
Weft 197
Weighting 198
Weightlessness 19
Weights and measures 242
Welding
   metal 120, 140–142
   plastics 41
Werner, A. 246
Weston cell 143
Wet spinning 192
Whale, The 187

Wheatstone, Sir C. 242
Whey 239
White, E. 21
White lead 175
White light 42, 43
White spirit 175
Whitney, E. 215, 242
Whittle, Sir F. 233, 243
Whooping cough 177
Widnes-Runcorn bridge 163
Wieland, H. O. 246
Wien, W. 246
Wigner, E. P. 246
Wilson, C. T. R. 246
Wilton carpet 197
Windaus, A. 246
Wind tunnel 155, 226, 229, 232
Winter equinox 179
Wireless 108, 242
Wood, see Timber
Wood-pulp 40, 60, 64, 65, 192, 246
Wood's metal 143
Woodward, R. B. 246
Woof, see Weft
Wool 191
Woollen yarn 195
Worden, A. 22
World Trade Center 162, 163
Worm-and-nut steering 212
Worsted yarn 195
Wright brothers 218, 232, 242
Wrought iron 134
Wuppertal line (monorail) 53

## X

*X-1* aircraft 233
*X-15A-2* aircraft 233
Xenon 27, 245
X-rays 29, 32, 43

## Y

Yale, L. 242
Yale lock 242
Yang, C. N. 246
Yawing 221
Yeager, Capt C. E. 233
Year 179, 182
Yeast 238, 241
Young, J. 22
Ytterbium 27, 245
Yttrium 27, 245

## Z

Zeeman, P. 246
Zeigler, K. 246
Zein 192
Zeppelin, F. von 232
Zeppelin airship 231, 232
Zernike, F. 246
Zinc 27, 28, 80, 130, 132, 134, 145, 245
Zinc blende 127, 132, 145
Zinc chloride 80
Zinc oxide 91
Zinc paint 175
Zirconia 145
Zirconium 27, 28, 145, 245
*Zond* probes 16, 22
Zsigmondy, R. 246
Zworykin, V. 242

# Acknowledgements

The Publishers express their gratitude to all the people and organizations who have provided photographs and assisted in the preparation of illustrations for this encyclopedia, particularly those mentioned below:

Aerofilms, Australian News and Information Bureau, Boeing Aircraft Corporation, Bowater Paper Company, British Aircraft Corporation, British Broadcasting Corporation, British Leyland Motor Corporation, British Overseas Airways Corporation, British Paints, British Petroleum Company, British Rail, British Steel Corporation, British Travel and Holiday Association, Burmeister and Wain Marine Diesel Company, Canadian Government Travel Board, Caterpillar Tractor Company, Central Electricity Generating Board, Central Office of Information, CIBA, Cotton Board, Courtaulds, Electrical and Musical Industries, Esso Petroleum Company, *Flight International*, W. K. Folkes, Ford Motor Company, Granada Television, Greek Tourist Office, Grundig, Dick Hampton, Imperial Chemical Industries, Institute of Geological Sciences, International Business Machines, Robin Kerrod, William Kerrod, Kodak, London Transport Board, Sir Alfred McAlpine & Son, Mansell Collection, Marconi Wireless and Telegraph Company, Mercedes-Benz, Midland Silicones, Ministry of Defence, Mount Wilson and Palomar Observatories, National Aeronautics and Space Agency, National Coal Board, National Physical Laboratory, Novosti Press Agency, Pilkington Brothers, Post Office, Reed Paper Group, Riker Laboratories, Royal Astronomical Society, Science Museum, Shell International Petroleum Company, Steel Company of Wales, Swiss Railways, Societé Nationale des Chemins de Fers, Tarmac, Trinity House, Unilever, Union Pacific Railroad, United Kingdom Atomic Energy Authority, United States Information Services, Vickers, Volkswagen Motors, Josiah Wedgwood and Sons, Westminster Dredging Company, Carl Zeiss